Solid State Science

Past, Present and Predicted

Solid State Science

Past, Present and Predicted

Edited by

D L Weaire

Department of Pure and Applied Physics,
Trinity College, Dublin

and

C G Windsor

AERE Harwell, Oxfordshire

Adam Hilger, Bristol

British Library Cataloguing in Publication Data

Solid state science: past, present and predicted.
1. Solid state physics——History
I. Weaire, D.L. II. Windsor, C.
530.4′1′09 QC176

ISBN 0-85274-584-2
ISBN 0-85274-586-9 (pbk)

Published under the Adam Hilger imprint by IOP Publishing Ltd
Techno House, Redcliffe Way, Bristol BS1 6NX, England

Printed in Great Britain by J W Arrowsmith Ltd, Bristol

Contents

Contributors

Professor Ernst Braun Österreichische Akademie der Wissenschaften, Institut für Sozio-Ökonomische Entwicklungsforschung, Fleischmarkt 20, 1. Stiege, 3. Stock, A-1010 Vienna, Austria.

Professor R W Cahn Department of Materials Science and Metallurgy, University of Cambridge, Pembroke Street, Cambridge CB2 3Q2, UK.

Dr T D Clark Physics Building, University of Sussex, Falmer, Brighton BN1 9QH, UK.

Professor Brian Henderson University of Strathclyde, Department of Physics, John Anderson Building, 107 Rottenrow, Glasgow G4 0NG, UK.

Dr M J Kelly GEC Research Laboratories, Hirst Research Centre, East Lane, Wembley, Middlesex HA9 7PP, UK.

Dr S M McMurry Department of Applied Mathematics, Trinity College, Dublin 2, Ireland.

Dr D Shoenberg Cavendish Laboratory, Madingley Road, Cambridge CB3 0HE, UK.

Dr Cyril Stanley Smith 31 Madison Street, Cambridge, Massachusetts 02138, USA.

Professor D L Weaire Department of Pure and Applied Physics, Trinity College, Dublin 2, Ireland.

Dr C G Windsor Building 521.2—MP and M Division, AERE Harwell, Oxon OX11 0RA, UK.

Professor E P Wohlfarth Department of Mathematics, Imperial College, Prince Consort Road, London SW7 2BZ, UK.

Professor M M Woolfson Department of Physics, University of York, Heslington, York YO1 5DD, UK.

Dr J Yeomans Department of Theoretical Physics, Oxford University, 1 Keble Road, Oxford OX1 3NP, UK.

Pencil sketches were drawn by **Hugh Masterson.**

Preface

HISTORY AND PREHISTORY

The articles in this book sketch the history, present achievements and potential of the science of the solid state. The authors are drawn from many corners of the subject and have been chosen to provide a balanced overview, accessible in part or in whole to a wide readership.

All too often diverse styles and purposes have obscured the unity of solid state science. We see a need to affirm its identity, unity and importance in science today. The last twenty years have seen it replace nuclear physics and radiocommunication as the branch of science most affecting our daily lives. Whilst it has become a cliché to speak of the 'Age of Silicon', how many people understand how our present devices came to be invented, and in what directions they are evolving at the present time? Our aim, therefore, is to trace the roots of solid state science, follow the development of its main branches and indicate some probable future developments.

Most histories of early physical science concentrate on the more idealised models and fundamental laws, neglecting the great mass of work which has been concerned with the investigation of real materials. This is the prehistory of solid state science, about which Cyril Stanley Smith writes in Chapter 2, and has written for many years with eloquence and insight. He has also offered inspiration to those who follow: 'In the history of the pure sciences, the richest opportunities lie in the area of solid state physics. Not only is this the largest sub-field of physics today, but it is changing the concept of what physics can be. The boundary conditions no longer need be those of idealised atomism and simple symmetry, but include imperfections and real structures.'

While the study of the early predecessors of solid state science offers fascinating insights, the subject did not acquire a separate identity or a coherent basis until the 1930s. It needed the methods of quantum mechanics which were developed in the 1920s to turn the arts of using materials in metallurgy, engineering, crystallography, chemistry, magnetism, optics and electronics into sciences. In every case the same quantum mechanical rules for the formation of solids and their resulting properties offer insight, understanding and predictive power. All were revealed as branches of the single new field of solid state science; as illustrated in figure 1. In relation to

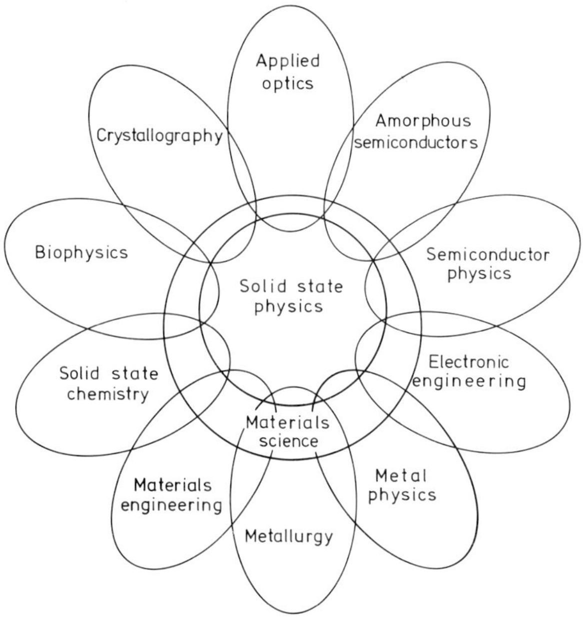

Figure 1 Some of the major overlapping research fields within Solid State Science. Solid State Physics is at the core of nearly all fields, while Materials Science embraces the wider application of the basic studies.

many of the problems which it hoped to solve, solid state physics was therefore a late starter and it was for long preoccupied with explanation rather than prediction. A good example is that of the photographic process, which is based on some very subtle physical and chemical effects in silver halide crystals. Invented in the nineteenth century, it was tentatively explained only in the late 1930s, by Mott and Gurney. N F Mott, in his recent autobiography, notes how belated his theory was from an industrial point of view, however intellectually satisfying it may be. A delightful extension of their theory came much later when direct photography from a positive became available. The migration of *electrons* in the classic process makes a *negative*, but holes migrate in the opposite direction and can be persuaded to make a *positive*. Today, the tables are turned; solid-state-related industries neglect basic research at their peril.

Even that rich period of the development is poorly understood and little appreciated, despite the fact that some of its most heroic figures, such as Mott, are still with us. The current International Project on the History of Solid State Physics will do much to remedy this and its archives will provide the raw material for a generation of science historians. In due course, a coherent picture will emerge. This book attempts only a modest contribution towards that end.

DIVERSITY AND DIVERGENCE

Why has this process, by which the identity of such an important field becomes established, proved so laborious? The reason must lie in the bewildering diversity of the solid state both as regards physical phenomena and applications. No central single quest—'split the atom', 'plasma fusion', 'find new particles'—has focused the solid state scientist's activities. His explorations have been like those of the Victorian plant collectors, searching for exotic and unusual specimens, and adding them to his collection. The subject has lacked a Charles Darwin who could classify all the information around a single grand theme. Instead we divide it up as we see fit for the purposes at hand; for the subjects of conferences, for publishing, or for the convenience of university administration. Figure 2 shows a classification by *structure*. It fails completely to recognise the diversity which nature gives us within even an identical structure, such as that of a single crystal.

Figure 3 attempts a classification by *type of material*. The four sectors certainly point as four signposts to different directions of solid state science. Again they are useful in practice—as in the journals of The Institute of Physics. However, as a classification, they are too coarse. A single crystal insulator may be studied in order to probe its structure, mechanical properties, magnetism, conductivity, or optical properties. This suggests an interleaving classification based on the *object of study*, as attempted in figure 4. It is this type of classification which is used in abstracting journals such as *Physics Abstracts*. This is a more contentious classification since it is continually evolving, often undergoing fission into separate subfields. Such a tendency is sometimes deplored, but is probably quite healthy. Returning to

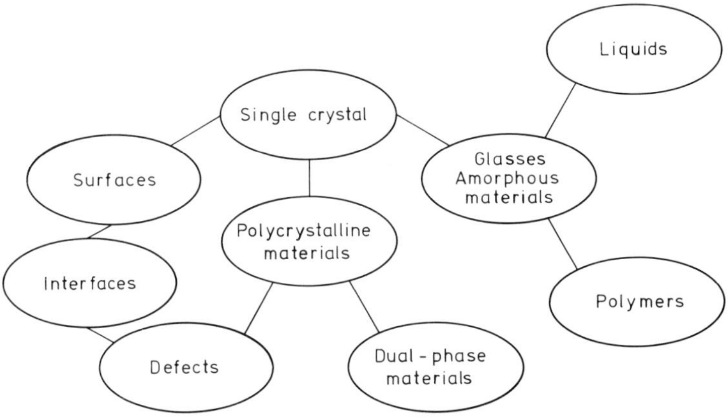

Figure 2 A classification by structure. There are relatively few distinct types of structure encountered in solid state science. A single crystal sample can be viewed using a host of different techniques and so contribute to our understanding of many branches of solid state science.

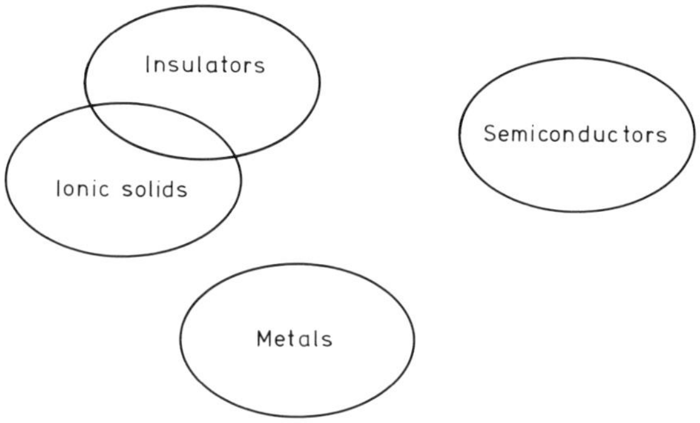

Figure 3 A classification by type of material. The basic electronic properties of materials provide a basis for a useful classification according to the nature of electron states in the material. Again any one type of material, such as a metal, can be examined using a host of techniques for a variety of objects of study.

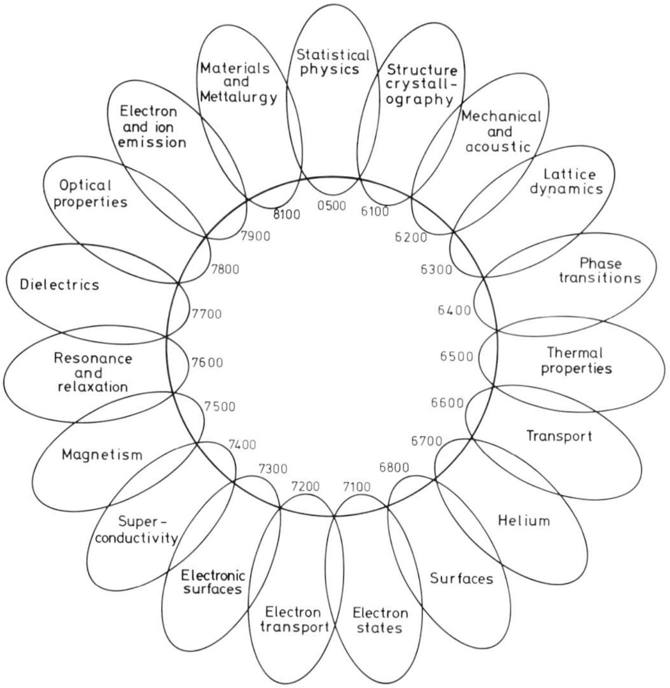

Figure 4 A classification by object of study. While it is clearly useful, this classification is open-ended and is boundaries are largely a matter of convention and opinion. The choice here is the official one devised by *Physics Abstracts* and used by many journals to help the working scientist find papers close to his chosen interests.

our botanical analogy, gardeners will find a ready parallel in the growth of perennial plants. As each clump expands its centre dies, partly of old age, but essentially by being denied the resources for growth by the vigorous outer ring of young growth. The prudent gardener divides the young growth into separate pieces, each of which has room to grow. This process of definition and division is often defeated by further growth, causing individual subfields to coalesce. Many key words of our subject (magneto-optics, thermo-electricity) carry this message. The same process brings a subfield into closer contact with neighbouring fields and hybrid subjects are formed. This presents problems of classification. Does the transport of ions in solids belong to physics or chemistry? Do semiconductor devices belong to engineering or physics?

Perhaps the least satisfactory classification (but one which classifies the scientists themselves rather well) is that determined by *technique*. This topic is pursued in Chapter 12. Some techniques, like the de Haas–van Alphen methods, are closely identified with an object of study. Others, like neutron scattering, are very broad and one hardly knows whether they belong to physics, chemistry or engineering.

In the search for new effects or in the isolation of familiar ones the solid state physicist has gone to great extremes. High pressures, low temperatures, high magnetic fields; all define frontiers of his explorations. Above all, high purity has been the recipe for success. Today's great temples of solid state science, the fabrication laboratories of the semiconductor industry, attended by their white-robed Vestal Virgins, are among the cleanest places on Earth. On the other hand, materials science recognises the complicated character and composition of most practically useful materials and the necessity for merely phenomenological methods, *faute de mieux*, in describing them. The conventional solid state physicist's tastes have recently included complexity (in certain refined forms, such as that of the simplest glasses) but he has some way to go before he has much to say about a common brick!

The changing course of science is not at all clear as it happens. Even on a yearly time-scale its movement is almost imperceptible, except for the sudden appearance and often rapid growth of new shoots. On the other hand, over a decade, its movement is clear enough, and can be documented by, for example, plotting the numbers of research students supported by the Science and Engineering Research Council in universities as in figure 5. Solid state science is new enough that it does not yet appear as a separate classification, but it would contain the lion's share of 'mainstream' physics after nuclear and space physics have been excluded. After the growth period of the 1960s, the total numbers of mainstream physics and materials studies have remained roughly constant. The steady shift over the period from physics to materials studies reflects an important trend towards making physics relevant to industry. The proportion of nuclear physics has remained unchanged. Chemistry has followed a similar pattern, although biology has grown substantially over the period.

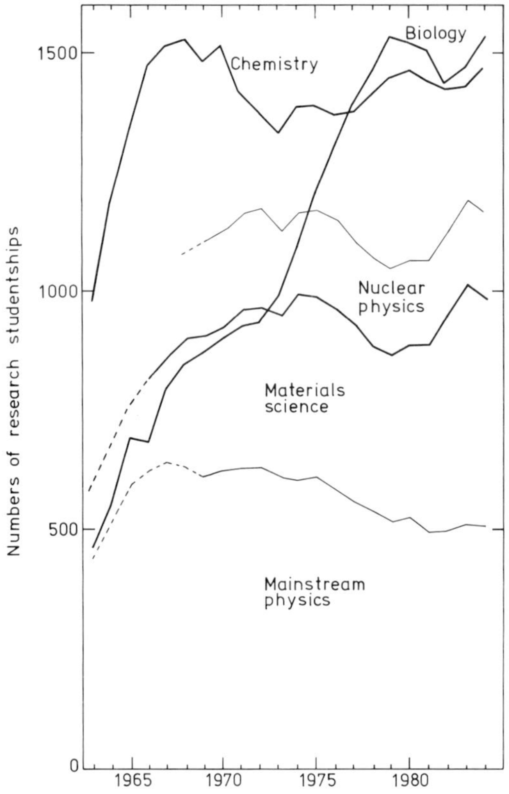

Figure 5 The numbers of Research Studentships awarded by the Science and Engineering Research Council for solid state science and related disciplines. Mainstream physics is largely solid state physics but also includes perhaps a third atomic and molecular physics. Nuclear physics is given separately. Space, radio and astronomy have been excluded. Materials science includes the official categories of materials, metallurgy and materials and materials science and technology. The dashed lines are estimates of the numbers made by interpolation when the classification system used by the Research Council was changed.

How is the content of solid state physics itself changing over the years? Figure 6 shows the numbers of papers indexed in the solid state physics journals *J. Phys. C* (*Solid State Physics*) and *J. Phys. F* (*Metal Physics*), according to the standard *Physics Abstracts* classification scheme over the last dozen years. It is interesting that these show a steady growth over the period, despite the steady numbers of research students—physics is evidently doing its share in raising productivity. The subjects underlying our understanding of metals and semiconductors, for example electron states and electron

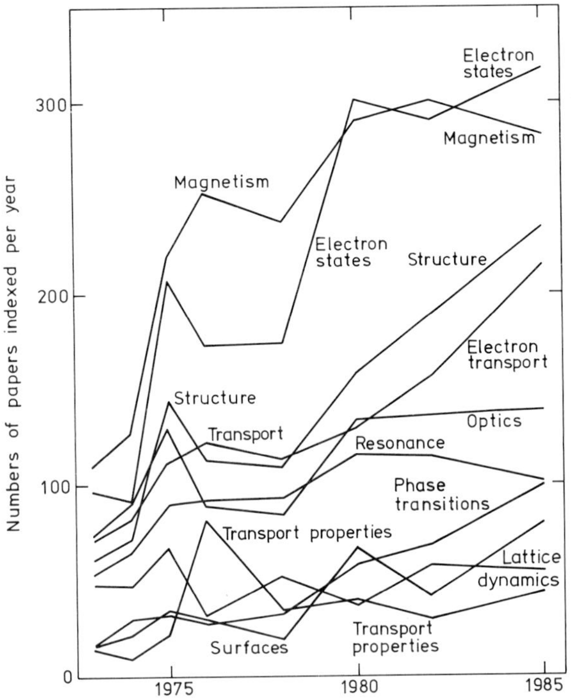

Figure 6 The numbers of papers indexed in the solid state science related journals *J. Phys. C* (*Solid State Physics*) and *J. Phys. F* (*Metal Physics*). The subjects are classified according to the *Physics Abstracts* list illustrated in figure 4. Subjects not rated over 50 during the period are not shown.

transport, continue to grow rapidly. While the calculation of electron energy levels in perfect crystals, the subject of 'band theory', may now be regarded as unexciting, several of its applications, for example the calculations of total energies and hence the prediction of structure, are at a very interesting stage. Partly because simple one-electron theories have worked much better than expected, we now stand on the brink of an era in which reliable calculations can be made on a variety of complex materials for the first time. The consequences of this will be deep and wide, since there are whole areas of the solid state, such as the study of interfaces, where experimental probes are inadequate, and theory may leap-frog ahead of experiment with the help of supercomputers.

Semiconductor device technology continues to stimulate solid state research both intellectually and financially. One might think that by now the emphasis would be very much on the applied side, but the pace of technological advance is such that it is making fresh demands on fundamental physics all the time. The progressive miniaturisation of present devices is currently expected to enter a regime in which the conventional wisdom, acquired from experience at larger scales and dating largely from the 1930s, will break down.

The optical properties of solids have continued as a growth field stimulated by the communications market. Laser light sources have come better with every passing year, giving both smaller and cheaper lasers for communications. Equally valuable are the larger high-powered lasers suitable for non-linear optics. This is a new and sophisticated field, in which theory has often lagged behind experiment. It includes solid state lasers and those special crystals which produce non-linear effects on light transmitted through them. Of these, some of the most exciting are the bistable semiconductor devices recently discovered. Potentially, the field offers the promise of the necessary materials for integrated optics in which light and electrical connections will coexist within the same semiconductor chip.

The study of phase transitions has risen to become a fashionable topic over the decade. Much of the work is academic in nature, with the interest concentrated on critical phenomena, that is, the precise nature of phase transitions. It has proved one of the success stories of the decade, with the new 'renormalisation group' theories providing a real insight into the previously mysterious critical exponents and scaling laws which had occupied experimenters over the decade. Time will tell if this subject has now run its course. It is perfectly desirable and natural that subjects grow under the promise of new understanding that can bring together experimental measurements, and then wither as that understanding is found and applied, eventually reaching that body of accepted knowledge enshrined in the standard texts. Lattice dynamics provides a classic example. It represented a large fraction of the effort in solid state physics at the start of our graph around 1973, yet the 'phonons' of the theory proved easy to measure by neutron inelastic scattering and vibrational spectroscopy became a standard technique of solid state physics. In this rather reduced role it continued to contribute to other fields such as surface studies and phase transitions during the decade.

Magnetism has been one of the strongest fields throughout the period. It would be tempting to assert that this has been promoted by the applications in recording and storage media, but in the UK it would be closer to the truth to say that it had been stimulated by theoretical interest in phase transitions and electronic structure. Its future seems assured while so many of its major problems remain unsolved. For example, the magnetism of even a simple transition metal like iron has not succumbed to any simple explanation, and is indeed the subject of just as much effort, both from theory and experiment, as it was a decade ago.

If all of this has a common theme, it is surely to be found in the aspect of solid state science which was stressed in our first classification, that of structure, emphasised in Chapter 2. It may be argued that the invention of the x-ray diffraction technique constituted the very birth of our subject, yet it still holds great fascination for many. The Nobel Prize for Chemistry in 1985 was given for the refinement of the technique of solving structures by direct methods, by Karl and Hauptman, as described in Chapter 12. This, the eventual fulfilment of an old dream (an impossible dream, according to Bragg) makes an interesting contrast with Von Klitzing's Physics Prize of the same year. The latter also related to the solid state, but his work on the Quantum Hall Effect was a total surprise, a sudden flash of brilliance, reminding us that young men will always find new things . . . .

In addition to the classic method of diffraction, we now possess a versatile armoury of new structural probes for special purposes. Many of these relate to the new field of surface science.

Structure is, in fact, often the key to our understanding, and structural studies are rising to become another of the major areas. This is largely caused by the newer techniques being developed over the period, such as neutron and electron diffraction. New surface techniques, many using synchrotron radiation, have caused a large growth in surface and interface studies over the decade, parallelling the development of bulk property measurement over the period from 1920 to 1970. The development of the field has been dominated by the steady improvement in the basic experimental methods, such as low energy electron diffraction, extended X-ray absorption fine structure and Auger spectroscopy from ion bombardment, all combining exciting possibilities with deplorable acronyms. From the applied side comes the general motivation of the desirability of a better understanding of catalysis. Physics and Chemistry are therefore interwoven throughout much of surface physics. Even the physicist wanting to study pure surfaces has to be expert in the removal of surface adsorbates. The subject will surely further expand as the experimental techniques are refined still further.

THE BOOK

The present volume, which attempts to encompass some of these diverse topics, arose out of discussions of the Solid State Subcommittee of The Institute of Physics. As editors, we were confronted with a difficult task in selecting what was to be covered. No single volume of this size can really represent the whole discipline. For further reading, we would recommend *The Beginnings of Solid State Physics* (London: Royal Society, 1980). Also recommended are Cotterill's *Cambridge Guide to the Material World* (Cambridge: Cambridge University Press, 1985), Braun and MacDonald's *Revolution in Miniature* (Cambridge: Cambridge University Press, 1982). We all await with interest whatever emerges from the International Project on the History of Solid State Physics.

We hope that undergraduates, in particular, will in future be given more of a historical perspective in their studies and realise that the knowledge we possess today is the product of a long process and a wide community of scientists. The intractable and inscrutable solid has not yielded easily to the scientific method but it is at last within our power to shape, control and understand. When we realise how long this has been awaited and how recently it has been achieved, it becomes clear that we have only begun to explore the consequences.

Denis Weaire
Colin Windsor
June 1986

1 Introduction

Ernst Braun

Most of the present volume *Solid State Science–Past, Present and Predicted* has been written by physicists for physicists. Thus it is a double honour for a student of social and historical aspects of science and technology—albeit an ex-physicist—to be asked to write an introductory chapter to this narrative; an honour to have retained the trust of the physics community and an honour to be heard as an historian.

The history of solid state physics has, of late, received little of the attention it so richly deserves. Through the efforts of Professor Margaret Gowing and of the doyen of the British physics community, Sir Nevill Mott, the memoirs of some leading solid state physicists were collected and published in 1980 by the Royal Society in a volume called 'The Beginnings of Solid State Physics'. At a meeting which preceded this publication, it was decided to mount an international effort to write a properly researched history of solid state physics and to locate and preserve as much archival material as possible. The international project indeed got under way, with financial support from several foundations and, in the United States, industrial firms. The results of this effort are intended for publication as a major tome by Oxford University Press and as a source book of archives by the American Institute of Physics. In a sense, this introductory chapter owes its existence to the above project, as it was in association with it that my own knowledge of the history of solid state physics was forged.

Despite the enormous economic significance of solid state physics in its contributions to numerous technical achievements—with modern microelectronics taking the pride of place—the history of this branch of science lacks the heroic qualities of, say, the history of astronomy. Solid state physics did not cause the world to change its shape and structure at one fell swoop, nobody had to answer for their heresies, and nobody was burned at the stake. Yet I would argue that solid state physics has played a greater role as a determinant of human well-being, human happiness and human misery than all cosmological theories put together. In a sense, solid state physics is the humble unsung hero of science and this book is an attempt to turn the

Cinderella into a princess, I hope an enduring one. Copernican astronomy has changed our view of the world and thereby, indirectly, the world itself has changed. Solid state physics has changed the world and thereby, indirectly, our view of the world has changed.

Solid objects have been the subject of enquiry since the dawn of science. The Greek 'electron', or amber, aroused curiosity because it manifested electrostatic phenomena, while the lodestone attracted attention because it exhibited magnetism. There is no continuity, however, between this early fascination and modern science. Interest in the properties of solids in the modern sense began with several separate strands of enquiry. The mechanical properties of solids, such as the modulus of elasticity, Poisson's ratio, the propagation of sounds in solids and such like, were one strand, known in the English literature as 'properties of matter'. The structural properties of metals constituted another strand and, curiously, the engineering science of metallurgy went its own way almost entirely independently of physics. Similarly isolated was the largely observational and, later, mathematical science of crystals. Even some basic properties of semiconductors were experimentally investigated in the nineteenth century, beginning with Michael Faraday, who found in 1833 that silver sulphide was a poor conductor of electricity and that its conductivity increased with temperature, in contradistinction to metals whose conductivity decreases with temperature. Further properties of semiconductors, such as the photovoltaic effect, were also discovered in the nineteenth century.

The many separate strands had a degree of unity in some minds, but no formal recognition of any commonalty. The first truly unifying idea came with the introduction of quantum mechanics. By the efforts of many theoretical physicists it became clear that the electrical and thermal properties of all solids could—in broad outline—be understood by the application of quantum mechanics. This extension of its area of validity was itself a source of strength to quantum theory. It gave a unified explanation to whole fields of physics whose applicability to the theory had not been foreseen by its inventors. Indeed quantum mechanics provided this first unity to solid state physics, which found expression in the first monograph bearing the name of the subject with the newly found identity: *The Modern Theory of Solids*, written between 1936 and 1940 by Frederick Seitz. In this book quantum mechanics is brought to bear on all aspects of solids for which it is suited. Those aspects of solids which are structure-sensitive, and thus not amenable to a very general treatment, are left out.

The understanding of solid state physics had, by that time, made very large strides both within the scope of quantum mechanics and without. By the invention and development of x-ray diffraction, the understanding of crystal structures had vastly increased. By the development of techniques of crystal growth, monocrystals of many substances could be obtained. By combining the two techniques it had been possible, for the first time, to make real progress in understanding the puzzle of what gave solids their strength and to elucidate the mechanisms of plastic deformation. The breakthrough came with the postulate of the dislocation in 1934, though it was to take another twenty and more years before the existence of this hypothetical crystal defect could be proven beyond reasonable doubt. Even without definite proof of the existence of dislocations, the accumulated detailed observational results on the plasticity of crystals began, by the mid-1930s, to make some sense with this hypothesis.

The dislocation came to be regarded as a linear defect and thus one member of a whole family of defects. Though no single umbrella theory of defects could be devised, it nevertheless became clear what types existed and what role they played. Gradually the concept of the point defect—a singularity at a particular point of the crystal lattice—emerged. Such point defects can consist of missing, misplaced or foreign atoms, ions or molecules. The fundamentals of the behaviour of point defects, be it in determining optical properties of alkali halides or electrical properties of semiconductors were, in broad outline, understood by the end of the inter-war period.

Finally, the achievement of very low temperatures, obtained through the liquefaction of hydrogen and helium, led to new insights into the properties of lattice vibrations and cohesive forces in crystals. It became both necessary and possible to classify crystals according to the cohesive forces which held them together and thus made the solids solid. For the different classes of solids, including metals, an understanding of these binding forces emerged and thus some answers to the basic puzzles posed by solids were obtained.

Much painstaking work was required to lift solid state physics out of the morass into which structure-sensitive properties were leading it. For what could be worse for science than the dependence of the result of every measurement upon the detailed—and unknown—composition and structure of each sample? Theories in physics are built upon consistent and repeatable experimental results; inconsistencies of unknown origin are troublesome. On the other hand, it is precisely the resolution of inconsistencies, the finding of causes of apparently erratic behaviour, which leads to deeper understanding. By painstaking experimentation over many years the role of impurities and imperfections was elucidated, despite the fact that the tools of analytical chemistry lagged considerably behind the needs of the solid state physicists.

The outstanding example to illustrate the laborious experimental investigations of solids, aided by rudiments of theory and plenty of experimental skill and patience, is the school of R W Pohl in Göttingen. Between about 1920 and 1940, combined measurements of electrical and optical properties of alkali halide crystals gradually yielded a consistent picture of the role and nature of point defects and other imperfections in these crystals. At the same time, the cohesive energy of ideal ionic crystals was calculated by theoreticians such as Born and Madelung. Ionic crystals thus became the model substance and a reasonably clear understanding of all their properties existed by the late 1930s.

We may distinguish essentially two aspects of solid state physics which had reached a first metastable state of basic understanding, with much detail to be elucidated, by the beginning of World War II.

Quantum mechanics had provided the band theory of the electronic structure of solids. This explained the behaviour of electrons in metals, as well as the distinction between metals and insulators. Semiconductors, for the first time, found a natural niche between metals and insulators. The theory of lattices and of lattice dynamics was sufficiently advanced to understand problems of specific heats, diffusion and the propagation of acoustic waves. By combining these two quantum mechanical theories, a picture of the cohesive forces in different types of solids was obtained and the scattering of electrons by lattice vibrations (phonons) could be understood, at least in principle.

The second branch of development was concerned with structure-sensitive properties, by which we mean properties which depend on deviations from the ideal crystal structure, i.e. upon lattice defects, including impurities. Such properties include the famous F-centres (colour centres), dislocations which determine the plastic deformation and the strength of real solids, photoconductivity and luminescence and, last but not least, the detailed behaviour of semiconductors and their contacts with metals.

Crystallography, low temperature measurements and the art of crystal growth were important activities which contributed to the above general picture of solids. General developments in chemistry, vacuum technique and measuring instruments of all kinds had, of course, all contributed to the above successes.

Grand elegant theories on band structures and lattice dynamics were constructed, but the picture obtained from such theories had to be augmented in many ways by smaller scale theories of less general validity. Some of these were transferred from other branches of physics. The theory of colour centres, for example, leaned heavily on the theory of atomic spectra and the theory of lattice vacancies was a special application of the theory of thermodynamic equilibria. The history of solid state physics provides plenty of examples of both successful and unsuccessful attempts to transfer theoretical—and experimental—procedures from one area of activity to another.

The theory of the rectifying contact between a metal and a semiconductor may serve as an example. Knowledge of band theory, work functions and of thermodynamic equilibrium was applied to this case and led to quite successful calculations of the rectifying properties of such contacts, even if much detail remained to be elucidated in the post-war period. The winning theory assumes that electrons move as a result of electrostatic potential differences as well as a result of chemical potential differences. In other words, electrons diffuse as particles and drift as charge carriers. Their motion is thus determined by the electrochemical potential. Some researchers, including very eminent scientists, attempted to view the current through the barrier layer at a metal–semiconductor contact as a quantum mechanical tunnelling process. The concept of tunnelling was quite new then and had been successfully applied not only to radioactivity but also to cold electron emission from solids, a problem not at all dissimilar to that of the metal–semiconductor contact. Thus rival theories, both equally plausible, were transferred from other areas of physics. Only detailed experimentation over a long period of time eventually could lead to a decision as to which theory was more efficient in explaining the observed facts and could therefore be regarded as 'correct'.

When the dreadful political events of the 1930s began to take their course, the emerging science of solid state physics felt some of the repercussions of these upheavals. For one thing, the lunacy unleashed in Germany in the 1930s led to a considerable transfer of knowledge 'on the hoof', as it were. Because Nazi Germany had no use for even its most eminent scientists unless they were of impeccable origin (from the peculiar Nazi point of view) and carried no 'undesirable' political ballast, a wave of emigration was let loose. From world-famous and eminent professors to humble research students, all those acutely threatened or sickened beyond endurance by the new regime fled to wherever an opening occurred. Partly, at least, by this mechanism, the centre of gravity of research in solid state physics shifted decisively from Germany to the United States and Britain. Perhaps the shift was inevitable, as the centre of economic power had shifted and American industry was more conscious than any other of the scientific base of many of its newer products. Nevertheless, a scientific tradition is more easily destroyed than built up and the old and venerable German tradition is as good and as sad a case in point as can be found anywhere. The injection of new blood vitalised British and American science and accelerated its ascendant motion, while German science suffered a near fatal blow from which it took a generation to recover.

When war broke out, all those activities of solid state physics which were not considered to be of direct utility to the war effort virtually ceased. Physicists were called up either for military service or to help with applied research of direct use to the war machines. The only branch of solid state physics which gained from this concentration of forces was semiconductor physics or, more accurately, the physics necessary to produce better semiconductor devices. Thus, while World War II was arguably good for semiconductor and nuclear physics, it was downright bad for other branches of physics, let alone other branches of knowledge. So, for example, the fundamentals of the mechanical behaviour of solids, the theory of dislocations, made no progress whatsoever during the war. Although many physicists were engaged on problems of manufacture of artillery shells and many others on improving armour plate, the basic theory of the subject was not considered to be of practical utility and was therefore neglected. It is interesting to note that even in semiconductor physics the German effort was very small and was entirely of a utilitarian empirical nature.

The very first solid state electronic device was the so-called 'cat's whisker' detector. In the early days of radio, at the beginning of this century, the crystal detector, consisting of a pointed wire pressed onto a crystal of galena or some other semiconductor, was the only practical device for receiving radio signals. It had replaced the rather weird coherer, which could merely detect the presence of electromagnetic waves and could not be used in applications beyond simple wireless telegraphy. The crystal detector was made obsolete by the advent of valve electronics, as valves provided a more reliable method of detection and had the added invaluable advantage of being able to amplify the received signal. Thus the crystal detector became something of a toy for radio hams, though it never quite disappeared from the scene. Solid state rectifiers, as opposed to detectors, especially cuprous oxide and selenium, became true industrial products in the inter-war years. The cuprous oxide rectifier served as the model substance for theories of the metal–semiconductor contact.

The ancestral device of semiconductor electronics, the cat's whisker detector, was brought back into service when the need for detection of shorter and shorter waves arose. As the transit times of electrons in valves and the capacities of large area solid state rectifiers rendered these devices useless for the detection of centimetre waves, the cat's whisker detector came back into use when radar receivers had to be constructed for military purposes. Although by the late 1930s galena was replaced mostly by polycrystalline silicon, because of its greater ability to withstand the necessary currents, the crystal detector remained a somewhat unsatisfactory empirically produced artefact. Nevertheless, when necessity became the mother of invention and detectors were required for high frequency radar signals, the crystal detector was resurrected. Industrially built military hardware was incompatible with the fiddly techniques required to make cat's whisker detectors operate, delightful as these methods had been to the enthusiastic radio amateur of the early part of the century. Yet pure empiricism did not get beyond somewhat fragile devices of moderate rectifying qualities. Under these circumstances the Americans obtained British know-how and embarked upon a major programme of fundamental research, with the aim of improving understanding of semiconductor materials and the rectifying contact, in order to manufacture better detector devices. This was one of the early instances where pure physics was deliberately and successfully used to improve technology.

Not only did this research effort lead to the production of much improved radar detectors, but it also laid the foundations in knowledge and in research methods toward the revolutionary developments which, after the war, led to the transistor and from it to microelectronics. The fortunes of war are uncertain and so are the fortunes of science in war. What is certain is merely that the urgent requirement for better detectors mobilised financial and human resources to tackle problems of semiconductor physics and this mobilisation yielded results very much more quickly than they might otherwise have been obtained. The mechanism which innovation theorists call 'market pull' became operative and proved singularly efficient.

The field of solid state physics as a whole, though it had gained knowledge merely in the single area of semiconductor devices, including infrared photoconductors for night vision gun-sights and similar applications, emerged greatly strengthened from the war. This applies, of course, only to American and British science. In Germany, science had been virtually destroyed and in other countries had suffered badly. The newly found strength of science was caused mainly by the prowess it had shown during the war. In submarine warfare, in aircraft and radar, in operational research; in all these, science had proved its utility in support of Allied aims and dire needs. As if that were not enough, physics provided its supreme proof of almost boundless might with the destruction of Hiroshima and Nagasaki.

With the end of the war, it was hoped that the proven destructive power of science could be turned to beneficial use and that science would provide the tool to build a better world. No matter whether we now regard this dream as addled or come true, in 1945 it was dreamt by all and for the next nearly thirty miraculous years of uninterrupted growth it certainly looked a pretty sound faith. Solid state physics shared in the general growth of physics in the post-war period, when science was hailed as the motor of the economy and flourished accordingly. At first it was particularly nuclear physics, with its promise of unlimited supplies of cheap energy, which had the reputation of laying scientific foundations for new engineering departures. As the promise of nuclear energy began to pale, solid state physics, which started off as a somewhat mundane and humdrum science of ordinary materials and their ordinary properties, gave birth to prodigious inventions and innovations which have changed the face of society. Whereas the promise of cheap and prolific energy turned sour, the humdrum gave rise to the spectacular and the age of microelectronics was born, arguably out of solid state research.

As one of the expressions of faith in semiconductor physics at the end of the war, a large new effort was launched at Bell Laboratories, one of the most prestigious of American industrial research laboratories. The aim was the fulfilment of an old dream, often invented but never realised before, of designing a solid state amplifier which would replace the clumsy, power-consuming and fragile valve. There were many patents for a solid state amplifier, but no working device. When the first post-war attempts at Bell Laboratories failed yet again, they did not simply give up but investigated the reason for the failure. And while investigating this, they hit almost accidentally upon an alternative way to build a solid state amplifier. Thus the point contact transistor was born and with it the first link in a chain of development which led to the present microelectronics revolution.

This momentous event has been described in numerous publications and the story was briefly retold here only to show two aspects of the development of science and to discuss the tremendous fillip which solid state physics received from this feat.

William Shockley, one of the inventors of the transistor, has called the scientific method which led to the discovery 'creative failure methodology'. When the apparently fool-proof design failed to produce the predicted amplification effect, the investigators set out to modify the theory which had proved a failure. This was not a scientific revolution or a change of paradigm—nothing as dramatic—just an everyday failure of theory which called for modification or addition. No fundamental principle was overthrown, it simply became clear that in this particular instance something was happening that theory had not taken into account.

The amplifier in question was supposed to work on the principle of electrostatic induction of additional charge carriers into a thin slab of semiconductor, thus modulating its conductivity. When this failed to happen, it seemed unlikely that the well tried principle of electrostatic induction should be at fault—the cause of failure had to lie elsewhere. John Bardeen, another inventor of the transistor, came up with the very plausible theory that the charges were indeed induced but failed to contribute to the conductivity of the semiconductor because they became trapped and immobilised in surface states. It meant going back to the drawing board, but even before a new design for a solid state amplifier was completed, the new hypothesis of surface states had to be investigated in detail. Indeed, a new line of research was opened up. The Bell researchers, however, kept their eyes open and never lost sight of their aim to produce an amplifier, however fascinating the theory of surface states might be. Thus, while Walter Brattain, the third of the official trio of inventors of the transistor, was investigating surface effects with the aid of closely spaced electrodes on the surface of germanium, he observed that the potential on one electrode could affect the potential on an adjacent one. Thus the point contact transistor was born. The open mind, rather than a detailed and inflexible plan of research, had led to the new discovery.

When the transistor was announced in the summer of 1948, solid state physics had taken the decisive step into the big league of science. Industrial laboratories all over the world started or augmented semiconductor research and development programmes and a whole new semiconductor industry, with innovation chasing innovation, developed in an amazingly short time.

With only a short lag, the universities began their own research programmes and their graduate and undergraduate teaching activities. Soon solid state physics became an established and indispensable part of physics curricula. The first text, which both reflected the new reality and helped to found and define it, was Charles Kittel's *Solid State Physics*, first published in 1953. It has seen many editions, each longer and better than its predecessor, reflecting the growth of the subject. Many more texts have been written since this first pioneering effort and the subject has grown out of all recognition in every conceivable dimension. The number of scientists engaged in this field has grown from a handful to many thousands, the number of publications, sub-topics, meetings, journals and—last but certainly not least—the amount of money spent on this branch of physics, have grown by orders of magnitude. Institutions, learned societies and international conferences bear witness to its newly found identity and importance.

Despite all this growth, solid state physics has not become what has been termed 'big physics'. It does not use vast accelerators swallowing uncounted millions of whatever currency and the teams of scientists working on a single problem do not reach battalion strength. Yet in terms of contributions to technology, and thereby to real changes in the world, solid state physics is the unchallenged prime mover.

It would be quite wrong to say that solid state physics has created modern microelectronics, or modern computers and modern telecommunication systems, yet it has undoubtedly played an important part in all these technical achievements. The relationship between science and technology is a complex one and technological innovations do not result in any direct way from science, nevertheless the two activities are so intimately connected that they may be regarded as two sides of the same coin. In the case of semiconductor physics, the interdependence is particularly marked and indeed semiconductor physics has become the symbol for the symbiosis between science and technology. No wonder, therefore, that in this branch of physics industrial laboratories have played the leading role and universities have not been able to have the limelight all to themselves.

I wish this book a wide readership not only among those who were active during those turbulent years of solid state physics, but also among the younger scientists joining the ranks now. The study of the development of a scientific discipline can teach us a great deal about scientific methods. Only from knowledge of actual working methods and actual steps in the development of accepted knowledge can we abstract and generalise a methodology of science. Thus, historical research provides the raw material for philosophers of science and their abstractions feed back into the working methods of scientists. Pragmatic and opportunistic as scientists may be, they nevertheless are guided, to some extent, by their knowledge of how science is supposed to operate. When need or inspiration drives them in new directions, these in their turn influence the edifice known as 'scientific method'.

If all the lessons we can learn from the history of science fail to impress and to convince that the subject is important and deserving, then the argument of last resort must be used. It is quite simply fascinating to know how things happened, to know how we got to where we are, to know who did what and what motivated them to do it. History is there, rather like Mount Everest, and we would be less than human if we did not find fascination and satisfaction in unravelling its many strands. It is with the hope of satisfied curiosity, stimulation and enchantment with which the introduction must end and the history begin.

2 Notes on the Prehistory of Solid State Physics

Cyril Stanley Smith

2.1 INTRODUCTION: ON HISTORY

The prehistory of solid state physics is the history of practically everything that has been excluded from conventional histories of science.

Human interest in the world must precede attempts to understand it. The most superficial glance at archaeology makes it very clear that the philosophy and science of matter had a long perhistory in the technology of matter and, moreover, that technology almost always began with the selection of naturally occurring objects and phenomena to enjoy, followed by the later discovery of ways, by aiding nature, to create decorative objects. Throughout most of history, the properties of matter and the properties of the patterns of assemblies of matter were taken for granted by artisans and artists. This article will mainly deal with the structural antecedants of modern condensed matter physics, the discovery of the patterns and the interaction of subpatterns that provide the boundaries for the triumphs of mathematical physics.

Figure 2.1 shows a mosaic wall assembled from little cones in about 3500 BC, almost as if to illustrate an elementary text on crystallography. Figures 2.2 and 2.3 show objects made of metal and ceramic, the former dating from nearly a hundred human generations before the Greek philosophers began to write about the nature of matter, and the latter from some eight centuries before scientists were able to tackle problems of nucleation and interface energy. Indeed, until the very end of the last century practically all materials known and used for any purpose, including advanced engineering, had been discovered two millennia earlier; there had been great developments only in the social organisation of production and distribution. In a strange inversion, social changes are today being nucleated by the appreciation of imperfection in material structure at a different scale.

Human beings find things to be most interesting when they feel that a new distinction is imminent, and they lose interest when they have found it. The

Figure 2.1 Inlaid mosaic decoration on columns at palace at Uruk, *c.* 3500 BC, in the National Museum, Berlin, destroyed in World War II. Composed of conical units embedded in triangular close-packed array with colour segregation defining different levels of substructure. (Photograph courtesy of Bildarchiv Foto-Marburg.)

particles or units of which any structure is formed, once found, are less interesting. History is the history only of replication and aggregation, i.e. the formation, structure, and movement of boundaries as they interact on increasing or decreasing scales. History is change, and the study of history is the study of phase changes. In human knowledge, just as in matter, the period of growth of a new phase can be quantified rather easily, while nucleation is difficult to detect and impossible to understand with exactness. A historian or a scientist can trace connections only when distinctions can be made. The success of the quantum theory shows that there is a level of structure below which time-reversal between matter and antimatter (or is it space and antispace?) makes distinction impossible. Above this level, aggregates form, having various degrees of resistance to environmental change, a property that might be called structural inertia.

History must be concerned with *both* nucleation and growth. Professional historians have mainly studied the history of society at levels that submerge the role of the nucleating event. Physicists have only very recently come to see that the antisocial misfit in atomic structures is of critical importance to both the origin of phases and their properties, and since professional historians tend to be somewhat uninterested in this level it is necessary for practitioners

Figure 2.2 Gold bowl and cup made by cold working from sheet gold with frequent annealing. The rings on the bowl are soldered in place with a gold–silver alloy. From the royal graves at Ur, *c*. 2600 BC. (University Museum, University of Philadelphia.)

Figure 2.3 Bowl with 'oil-spot' temmoku glaze. Chinese, *c.* twelfth century AD. The spots are segregated crystals of Fe_2O_3 and of white silicate in distribution and shape resulting from subtle interaction effects of oxidation, diffusion, viscosity, shear, nucleation, and growth, controlled by interface energy—atoms doing what they must to adjust to the environment established by a human being adjusting to his social environment. (Fogg Art Museum. Author's photograph.)

in any field to make their own evaluations of history. Their story will be a narrow and selfish one, but history is only that of movements of boundaries between specialised areas within which certain structural forms have developed appropriate to their own scale of structure, whether in space or in time.

The history of science has largely revolved around the discovery of new levels of structure followed by periods in which the new level was explored in detail and for a time overemphasised. Because of the great contributions that analytical atomism has made in the last three centuries, scientists can now properly return to the study of the larger complex structures that first aroused their curiosity and they are discovering that the nuclear atom, though underlying everything, is not the nucleus of historical change. Clearly, both scientists and historians must remember that history is more than the social history of institutions.

2.2 ON ATOMISM AND QUALITIES

It is not generally appreciated that Greek philosophy was, to a large extent, based on the study of real materials. A sense of structure was present at the beginning, with the suggestion of Anaximander (*c.* 611 to 547 BC) that matter was a kind of segregation of distinguishable substructure from a uniform featureless *apeiron*. Most subsequent discussion has revolved around the recognition and resolution of details within the many interlacing and interacting levels of hierarchy in time and space. One of the principal schools of thought, started by Democritus, sought to explain the properties of matter in terms of component parts, believed to be atomic (indivisible) but capable of forming associations of varying density, shape and orientation. The other more popular one, championed by Aristotle and his followers, emphasised the association of elemental qualities. Both views were based on the sensual and intellectual response of the human body reacting with the properties of materials: but it required two millennia to see that both were in a way correct and to develop a view that includes the material structure of the biological body as a preliminary to understanding human thought. The structure of matter turns out to be less important than the structure of materials—throughout most of history the structure of materials has been of far less concern to philosophers and to the more rigorous scientists than to artisans.

Both atomists and Aristotelians used their theories to explain what they had seen in the workshop. Every potter, carpenter, sculptor, textile worker, and smith knew that his or her material had to be treated in certain ways if it was to be shaped, coloured, or used as intended; they knew that mixing and heating transformed the properties of materials almost without limit; and knew that visible texture, revealed by fracture or otherwise, provided a good guide both to the quality of raw materials and to the success of its treatment. Though materials, properties, and treatments are vastly more diverse today, they still depend on the basic relationship between structure and properties, with composition and treatment controlled together to produce an object having whatever combinations of shape and properties might be desired.

Pythagoras, Plato and Euclid emphasised the geometry that externally characterises the structure of matter at different scales of aggregation and segregation. Democritus saw a level of unalterable building blocks, atomic units that had no properties beyond size, shape and motion and that these remained unchanged as they formed aggregates differing only in local density and orientation. Against this was Aristotle's postulate of elementary qualities, typified by their embodiment as earth, water, air, and fire. Essentially this was an appreciation of the different degrees of rigidity, plasticity and fluidity that are today associated with the three states of matter, together with fire, i.e. energy. He realised that matter of all kinds was to a considerable extent divisible without loss of the properties that, combined, constituted the diverse nature of individual materials. He claimed nevertheless that the properties derived from the association, so that division beyond a certain minimum level must be accompanied by change in internal form and hence in the very nature of the material.

Aristotle's concepts of miscible, mixt and mixtion, in which the modern eye can detect a feeling toward the thermodynamicist's constituent, solid solution, compound and phases, were much debated and developed in medieval times. Particularly interesting was Averroës' clear realisation of a lower limit of form, more like the modern quantum than the ultimate atom. There was a *minimum naturale* or minimum natural part, a unit that embodied the minimum form that underlies the externally observed identity and properties of matter. A body could not be defined in terms of individual particles alone (a view that modern particle physicists might note). Gradually the corpuscular view became more definite. It was well summarised by J C Scaliger in 1557. 'Our corpuscles do not make contact with each other in the same way as Epicurean atoms, but as a continuous body, so that they make a unity .... The form is the act by which the parts, whose nature and aptitude make them capable of uniting, actually become one.' Note that it is the act, not the fact, of their uniting that is important. Giordono Bruno, in his *De la causa principio e uno* (*On cause, principle and unity*) published in 1584, writes perceptively of structural complimentarity, of the continuity of discontinuity, and of the central importance of relationships between the form of parts rather than absolutes. 'There is not a change which seeks another being, but a change which seeks another mode of being.'

These corpuscles owed much to earlier theories, in particular to Albertus Magnus' attempt to apply theory to definite mineral forms, to the Averroists' view of the subordination or absorption of lower forms by higher ones, the Thomists' recognition that some numerical constant persisted through qualitative change and especially to the Aristotelian ideas of minima and mixtion. In these one can see, with hindsight, perceptive theories of chemical combination, based upon distinction between the various levels of organisation that in today's chemical thermodynamics are called atom, molecule, component and phase, and at least some sense of biological complexity. However, on the whole the medieval arguments were mainly between verbal systems and definitions of problems set a millennium earlier and the philosophers neither gave nor received much help from contemporary developments in the artisan's knowledge or even from the precise gravimetric methods of separation and combination being used by the assayers, which eventually provided the quantitative basis for a scientific chemistry. Moreover, they were largely qualitative and virtually disappeared from science after the beginning of the eighteenth century.

The infinitely hard and indivisible 'atom' was revived early in the seventeenth century, and from the beginning of the eighteenth century to the end of the nineteenth Aristotelian form and qualities were derided, if not forgotten: as a consequence, physicists were utterly unable to deal with any structure-sensitive properties and even seem to have not been interested in them. Solids had mass and density and were ideally elastic, never plastic, and crystals had symmetry but not imperfections.

Of course, the new mathematics which followed Newton was enormously useful, but its utility came from the fact that its infinitesimal curvatures completely disregarded the real substructure of matter: a cynic could say that it was nothing but curve-fitting, for the necessary constants were empirically determined and required no understanding of their origin and dependence on substructure.

There was, however, plenty of activity in the seventeenth century that was of deep significance to the later physics of condensed matter ....

2.3 ALCHEMY

Partly because of the general success of Aristotelian philosophy the lowest, chemical, level of substructure was elucidated long before the nature of larger-scale structures became part of respectable science. Herbert Butterfield once commented on what he saw as the delayed revolution in chemistry compared with that in physics: this is only true of physics perceived as mechanics, for the revolution in the physics of solids was far more delayed and actually came out of the despised chemistry. We tend to think of alchemy as a backwater because it failed to uncover the possibility of transmutation at the level of the atomic nucleus. This is a narrow nineteenth century view that should be discarded: the alchemists were true precursors of today's solid state physicists. A central aim of the alchemists was to understand the manifest transformations in properties that artisans had discovered. The changes in properties that they produced were real, even if they did not depend on changes at the subatomic level. The alchemists were in fact trying to unravel the *physics* of condensed matter. Their combination of experimental, theoretical, and aesthetic approaches utilised a good deal more of the rich capacity of the human mind than anything that replaced it. Perhaps it was more art than science. Many of their mystic signs and diagrams, often combined in a fascinating way with views of practical laboratory equipment and operations (figure 2.4), reflect an (unstated) sense of the relation between the nature of chemical change and the principles of topology and geometry.

In the late sixteenth century the elemental qualities of the alchemists began to be supplanted by the three principles (salt, sulphur and mercury) that were introduced by Paracelsus. This was another attempt to base the classification of materials on the intuitive feeling of how they responded to human contact, but it had moved beyond a simple recognition of the states of matter to a sensitivity to the different kinds of behaviour associated with the types of interatomic bonding that today we call ionic, Van der Waals, and metallic: the homopolar bonding of diamond was not separately recognised because of superficial similarity to hard transparent ionic crystals.

The combustibility of organic materials allowed the Paracelsians first to include them under 'sulphur', which came to denote some reducing principle. The last stage of Aristotelian qualitative theory was the elevation of this (following suggestions by two metallurgists intimate with the spectacular change in properties resulting from the reduction of oxides by carbonaceous materials) into the two principles that were central to eighteenth century

Figure 2.4 Frontispiece of an alchemical work showing both practical mining and assaying operations in association with mystic diagrams illustrating the separation and junction of qualities and structure in ideas and in matter. Stephen Michaelspacher, Augsberg 1617. (From Ploss *et al* 1970.)

thought; phlogiston and caloric. These were almost, but not quite, intangible principles transferable during chemical reaction and responsible for physical properties. By 1781, the great chemist Tobern Bergman (Bergman 1781) quantitatively measured the amounts of what he called reducing and combining phlogiston in different alloys of iron. The former was deduced in three ways: by weighing the amount of lead produced by reaction of hot iron with fused lead oxide; by weighing the silver replaced (electrochemically) from a solution of silver sulphate; and by measuring the volume of hydrogen emitted on dissolving in acid. Combining phlogiston was released by dissolving in acid and measuring thermometrically.

Obviously Bergman was close to an electron theory of chemical reaction, but phlogiston was not weighable at the time (it takes only an ounce of phlogiston to reduce a ton of aluminium) and the new chemistry of Lavoisier was in the main so obviously correct that both phlogiston and caloric disappeared from chemistry. For about a century, the general insistence on exact measurement and the successful redefinition of the chemical element in Lavoisian terms banished phlogiston in favour of larger structural units. The electron could only be reintroduced into science at the end of a long journey that began with some observations on the twitching of a frog's leg.

The discovery of voltaic electricity at last opened up the properties of matter to the precise measurement of properties that related to the smallest levels of structure. Indeed, the physics of solids in the nineteenth century consisted mainly of the measurement of elastic, magnetic, electrical and optical properties and the empirical matching of experimentally determined constants to empirical equations. However, the groundwork for the structural revolution was being laid in studies that metallurgists and others were pursuing on microscopy and in the relation of age-old properties to the level of structure that was accessible to optical microscopy. It was eventually realised that the essence of crystallinity did not lie in the external shape that had delighted artists and crystallographers, that boundaries of misfit between crystals of different orientation and composition could be measured, and their shapes and junctions related to properties.

Though a few nineteenth century physicists—notably Felix Savart, Jean Biot, John F Daniell and Otto Lehmann—were fully alert to the importance of microscopic polycrystallinity, and Claude Berthollet insisted that there were many compounds the composition of which did not follow Daltonian rules, virtually all chemists and physicists took the existence of Daltonian stoichiometric molecules for granted. Chemists did marvellous work in establishing the relations of phases and of the hierarchy of molecular constitution, especially in gases and organic substances, but it was principally metallurgists who observed and speculated about real structure in the range between the molecule and visible matter (Smith 1960, 1968). We will return to this after a glance at another side of seventeenth century science.

2.4 THE SEVENTEENTH CENTURY

The beginning of the modern scientific era is commonly taken to be the rapid rise of mathematical astronomy in the sixteenth and seventeenth centuries and its fall-out in terrestrial mechanics. Gains are usually accompanied by some loss: in this case the adverse result was the discouragement of the corpuscular philosophers who were beginning to develop a good structural basis for the explanation of properties. Nevertheless, many things happened in the seventeenth century that pointed toward later understanding of the real structure of solids.

Johann Kepler's playful letter *Strena, seu de Nive Sexangula* (On the

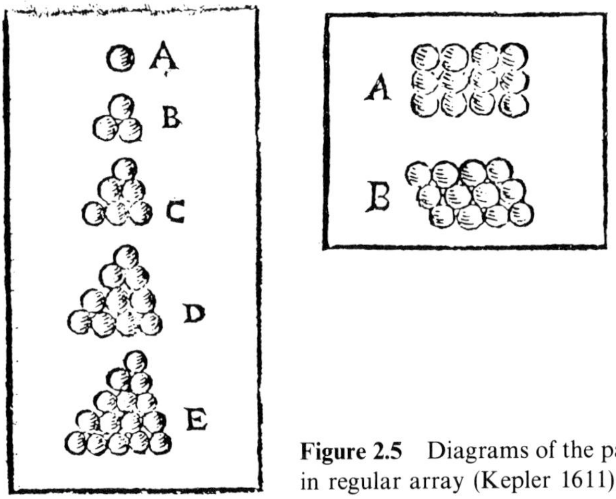

Figure 2.5 Diagrams of the packing of uniform spheres in regular array (Kepler 1611).

Hexagonality of the Snowflake, 1611) is the first published attempt to put into scientific dress the age-old knowledge of the mosaicist on the association of units and he correctly describes and illustrates the stacking of spheres into cubic and hexagonal lattices (figure 2.5). The Englishman Thomas Harriot had earlier made actual calculations of the volumes of three-dimensional piles of spheres in diverse arrays but this remained unpublished until very recently. Robert Hooke—to my mind a scientist on a level with his more famous compatriot Isaac Newton—made many observations on the structure and properties of things and he specifically showed how the multitude of anisotropic shapes taken by a crystallising substance could result from the manner of stacking of isotropic units (figure 2.6).

The introduction of the microscope early in the seventeenth century had a profound effect on biological science but relatively little on the understanding of inorganic matter. When Robert Hooke turned his new microscope on a steel razor he found nothing but a few inclusions of slag and superficial scratches due to grinding and burnishing (Hooke 1665). The actual microstructure of steel was not seen until more than two centuries later when methods of polishing and etching were devised to do for metals what sectioning and staining had earlier done for biological tissue. It is interesting to contrast the effective use of mathematics when applied to the inorganic world on the levels of astronomy and homogeneous elasticity which ignored microstructure, with sciences such as metallurgy and biology which advanced largely through the observation and classification of real structure at all accessible levels.

Corpuscular theory was far from dead. Its leading seventeenth century proponent was René Descartes, whose structural theory is the very basis of his influential *Principia Philosophia*, published in 1644. Practically all his

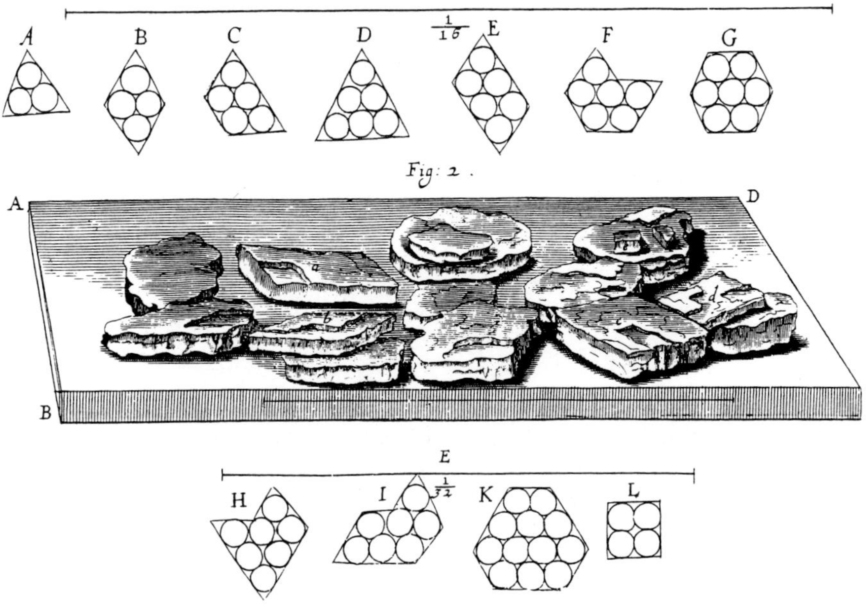

Figure 2.6 The generation of geometric crystalline shapes by the packing of spherical units (Hooke 1665).

explanations of the nature of things derive from his keen sense of the importance of structural division and aggregation. He did not believe in the atomists' ultimate particles that were without internal space and were infinitely hard, but he emphasised the incomparable diversity of association and movement. In the main, Descartes conceived of the universe as built up of three kinds of particles, two of which, by sharing faces or only points of contact, formed aggregates containing channels through which still smaller ones could circulate. With different sizes of aggregate these could qualitatively suggest a basis for any property. Today we talk of the substructure and quantised energy associated with the three states of matter. The ancestry of both views can be traced to the four elemental qualities that the Aristotelians had sensed to be important.

Descartes' structural ideas are mainly cosmological, though they are essentially scale-independent. He did not hesitate to use two-dimensional diagrams to show the nature of contact, shear and segregation. His discussions of the direction and conservation of motion and of the relations between internal and external space are an unappreciated part of the prehistory of relativity theory and quantum mechanics, but they are intuitive and were presented in diagrammatic, not elegantly mathematical form. In discussing the entry of comets from outer space into the terrestrial realm he

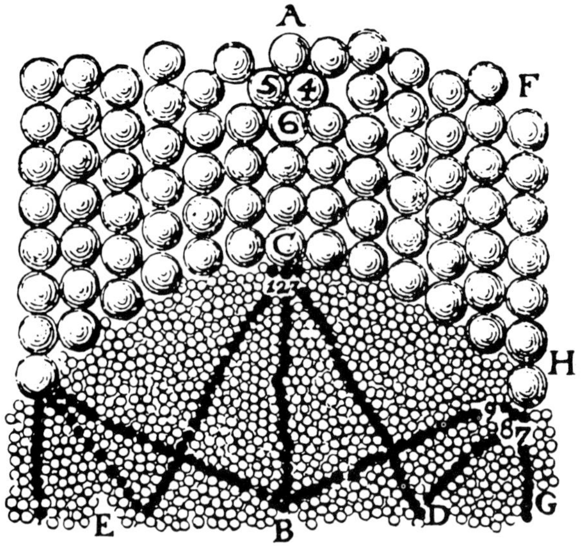

Figure 2.7 Diagram of the aggregation and interaction of corpuscles of different sizes and motions into bounded regions. (From Descartes 1644.) Though used here to explain the motion of comets between the heavenly and terrestrial regions, the diagram reveals Descartes' awareness of the nature of structural hierarchy and the formation of 'imperfections' and interphase boundaries from the interplay between different scales of order and disorder.

offers a drawing (figure 2.7) that shows a lattice with imperfections and the transfer of shear across a boundary between two 'phases' which comes close to representing modern concepts of the structure of interphase interfaces and dislocations in solids. Curiously, Descartes did not discuss the ideal crystal lattice, but he does show the formation of imperfections during the growth of a salt crystal (figure 2.8), and his discussion of the refraction of light is illustrated by a diagram (figure 2.9) of a dislocation in a hexagonal lattice.

The followers of Descartes were much less restrained and imagined particles of a vast variety of shapes to account for anything they were interested in. Cartesian corpuscular philosophy flourished mightily for some time, especially on the European continent. The 1671 *Traité de Physique* by Jacques Rohault was very popular. It was promptly translated into Latin and English and it still provides a good introduction to Cartesian physics. Its influence can be detected in the first scientific study of steel, published in 1722 by R A F de Réaumur (de Réaumur 1722). His theory, that the differences between soft iron and soft or hardened steel were a result of the diffusion, segregation and distribution of some smaller particles of something within

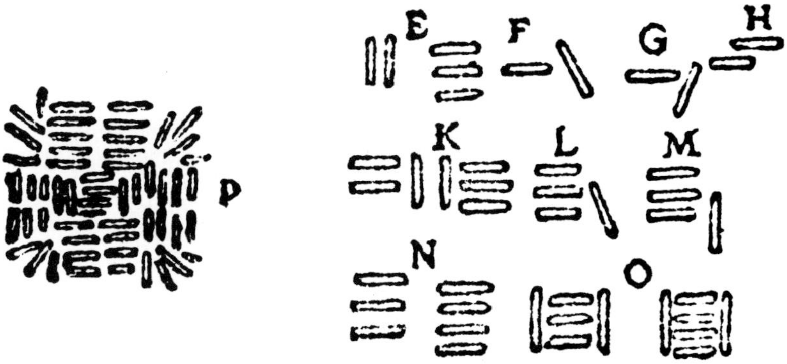

Figure 2.8 Descartes' diagram (1644) of the formation of cubic crystals of common salt.

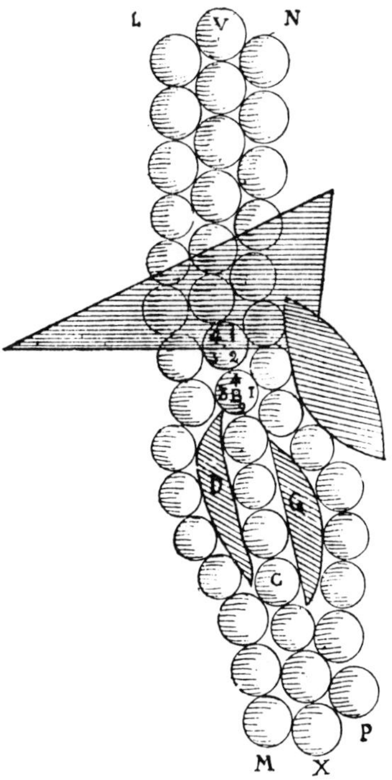

Figure 2.9 Diagram showing a dislocation at the junction of two areas of close-packed spheres in slightly different orientations. From a letter of Descartes to Ciermans, March 1638, discussing the refraction of light. (Descartes 1638.)

aggregates of larger iron particles, was illustrated by an engraving (figure 2.10) showing grains not much different from those in a modern textbook of metallography.

Despite Descartes' great contributions to algebra, his structural approach to physics was not in accord with the rigorously mathematical new(tonian)

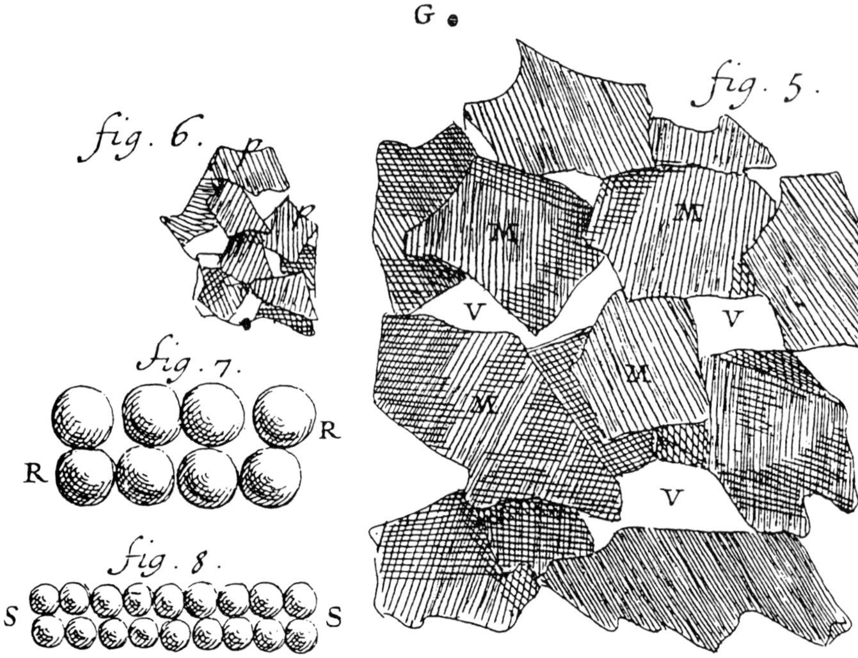

Figure 2.10 Réaumur's conception of the microstructure of steel, composed of grains, in turn composed of 'molecules', having space between them through which finer particles could diffuse and produce different properties depending on their aggregation and location. The spot G is the actual size of the structure shown enlarged in fig. 5. The particles p in fig. 6 are the supposed substructure of the regions M. (Réaumur 1722.)

fashion: it was derided and forgotten by the middle of the eighteenth century. Even the very perceptive extension by Swedenborg, to which we will return, had virtually no influence on mainstream science.

Although Isaac Newton was scornful of the Cartesians' attempts to imagine specific structures to account for the diversity of properties of materials, in the second edition of *Opticks* published in 1718 he provides a succinct summary of corpuscular views:

> The Parts of all homogeneal hard bodies which fully touch one other, stick together very strongly. And for explaining how this may be, some have invented hooked atoms, which is begging the question, and others tell us that bodies are glued together by rest, that is, by an occult quality, or rather by nothing; and others, that they stick together by conspiring motions, that is, by relative rest amongst themselves. I had rather infer from their cohesion, that their particles attract one another by some force, which in immediate contact is exceeding strong, at small distances performs the chymical operations above-mention'd, and reaches not far from the particles with any sensible effect....
>
> Now the smallest particles of matter may cohere by the strongest attractions,

> and compose bigger particles of weaker virtue and many of these may cohere and compose bigger particles whose virtue is still weaker, and so on for divers successions, until the progression end in the biggest particles on which the operations in chymistry, and the colours of natural bodies depend, and which by cohering compose bodies of a sensible magnitude. If the body is compact, and bends or yields inward to pression without any sliding of its parts, it is hard and elastick, returning to its figure with a force arising from the mutual attraction of its parts. If the parts slide upon one other, the body is malleable or soft. If they slip easily, and are of a fit size to be agitated by heat, and the heat is big enough to keep them in agitation, the body is fluid . . . .

The experimental approach to a physics of solids continued in the eighteenth century but most developments were in chemistry. Measurements of strength, hardness, plasticity, internal friction and other properties of real materials had been described by Mersenne and others early in the seventeenth century, and some fine studies of mechanical properties were published by van Musschenbroek in 1729 (van Musschenbroek 1729), but they were not amenable to the new mathematical treatment and the beauty of ideal homogeneous elastic behaviour came to pre-empt the entire field until late in the nineteenth century. The eighteenth century also saw the abandonment of the stacked-ball model of the crystal lattice which Kepler, Hooke, and Huygens had earlier put to good use. Suggested by the homeomorphic shape of cleavage fragments, crystals came to be considered as mere assemblies of mathematical cells, not anything directly appreciable to the human senses, and it became bad form even to attempt to define what might be in them. The fine mathematical treatment of the angles of crystal faces by Rene Haüy in 1782 based upon this approach (figure 2.11) was so successful and *à la mode* that when the stacked-ball model was reintroduced by W H Wollaston in 1812 it was completely ignored.

There was a strong aesthetic component underlying the alchemists' approach to matter. The decorative arts also prompted a perceptive study of combinatorial principles by a Dominican priest, Sebastien Truchet, published in 1704 (Truchet 1704). This was an analysis of the simplest tiling patterns used in the decoration of the church, expressing in diagrammatic form the ways in which regions of different density can be associated to give regions of hierarchical continuity and discontinuity of direction and density on any scale. Starting with a single tile divided diagonally into white and coloured triangles, he shows how there are 64 possible arrangements within a larger array of pairs differing only in position and orientation and that these may be combined to produce an infinity of patterns. Some of his diagrams are reproduced in figure 2.12. By excising regions of differing shapes and sizes from his figures C, D, and F and recombining them in different orientations virtually all the types of regularities and irregularities in the structure of condensed matter can be represented and, if white tiles are allowed, quantised diagrams of interaction at any scale of uniformity and divergence can be

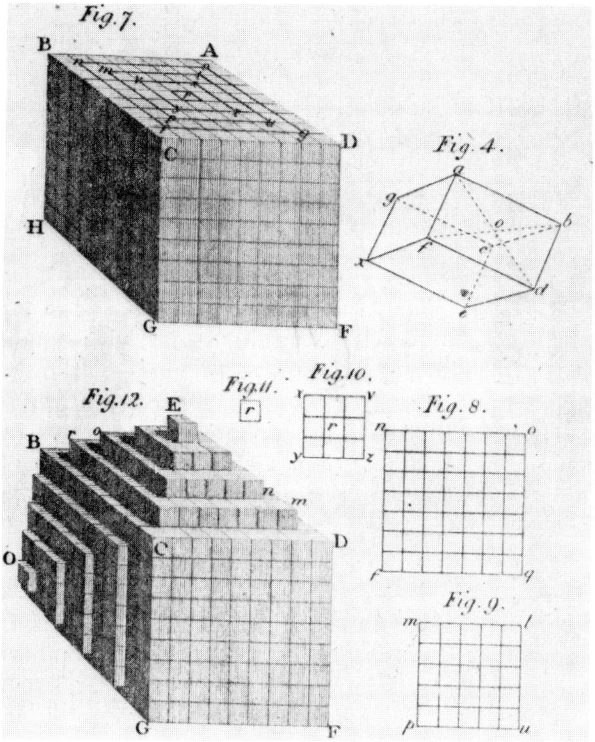

Figure 2.11 The generation of crystal faces at various angles by the assembly of integrant molecules, which were initially the smallest possible cleavage fragments (Haüy 1784). For earlier approaches to this idea and its later development see Burke (1966).

generated. The two-directional replication of any pattern of differences generates a quincunxial lattice of interpenetrating rectangles.

Truchet's idea was extended in a book published in 1722 by another Dominican named Douat (Douat 1722) which listed all 256 permutations of 4 tiles in differing orientations and explains how to use these in generating larger patterns, generally symmetrical. He illustrates 72 of these, some containing 576 tiles. This work had some influence on later decorative art, but it seems to have been unnoticed by scientists and mathematicians, who had also ignored della Porta's revival in the sixteenth century of the Babylonian gardener's quincunx—that arch diagram of the association of connected differences into uniformity.

The Cartesian approach to the structure of matter with its fundamental dependance on diagrams reached its apex in the work of Emanuel Swedenborg. Though mainly remembered as a religious mystic, Swedenborg spent his early years working for the Royal Swedish Board of Mines and

Figure 2.12 Diagrams showing the possibility of generating an infinity of symmetrical and/or asymmetrical patterns at different scales by changing only the orientation of square tiles divided diagonally into dark and light areas (Truchet 1704). (Courtesy of the Burndy Library.) The pattern at the lower right-hand side, newly drawn by the present author in accord with Truchet's principles, shows a rectilinear spiral superimposed on a background of randomly oriented tiles.

published several works on natural philosophy. In a 1721 treatise (Swedenborg 1721) he states his belief that the principle of nature was that of geometry derived from mathematical points. He shows how the cubic crystals of common salt could arise from the interlocking of acid and alkaline particles of differing shapes (figure 2.13). The same idea but with spherical or spheroidal atoms reappears in Wollaston's famous paper of 1813: combined with Haüy's close stacking of what are essentially hexahedral primitive forms, today's two-dimensional reciprocal lattice projections, and Patterson plots of electron density, they show how many different ways there are to study structure, all with some validity and utility.

In 1729 Swedenborg completed a draft of his philosophy that remained unpublished until much later (Swedenborg 1734). In this, among many other prescient remarks, he points out the apparent paradox that parallel lines in vortical circulation terminate in a triangle, suggests that the starry heaven visible to us is perhaps but a point in respect to the universe, and that undulatory pressure, which extends uninterruptedly from the sun to our earth, is the cause of light, sight and colour. It is mainly an examination of the

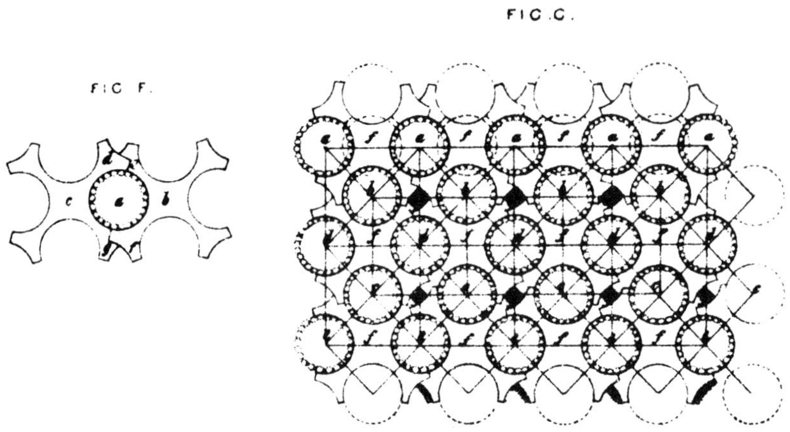

Figure 2.13 Superimposed layers of interlocking corpuscles of different shapes generating a cubic crystal of common salt (Swedenborg 1721).

interlocking of motion and density to produce stability, the idea that forms the basis of his *Principia Rerum Naturalium*, published in 1734 as one of the three fine folio volumes forming Swedenborg's *Opera Philosophica et Mineralia* (Swedenborg 1734). (The others provide comprehensive accounts of the metallurgy of copper and of iron and steel, based on visits to mines and smelters throughout Europe, interspersed with extensive extracts from the literature.) The *Principia* is an extremely perceptive study of the nature of motion, aggregation, and change. Quoting from his concluding chapter:

> Indulgent reader, I have now placed before you a sketch of my philosophical principles, which reach from the first simple of our world to the ultimate compound, from the smallest invisible to the first visible thing; ... principles connected throughout, as I think, from one end to the other by intermediates. Whoever aims at laying down principles and yet does not begin from the simple and proceed in order to the last cannot ... perceive any just connection between them. For he who stops short in intermediates only does not perceive the end of the series either on one side or the other much less does he see whether these ends have relation to each other or whether they are connected by intermediates ....
>
> In the simple itself is that one sole cause; it is latent in the first derived entity, or in what I have called the finite; indeed in this one finite are the two principles of nature, the active and the passive, from both of which the composite elementary exists.

By this time his thinking had lost its emphasis on geometry and was perceptively topological. His 'finites' (Latin *finiti*) are closures within which and between which there can be movement—in fact when studying the junction and disjunction of spiracules and vortices, he specifically says that the cause of anything being bounded is motion alone. Some of his diagrams would

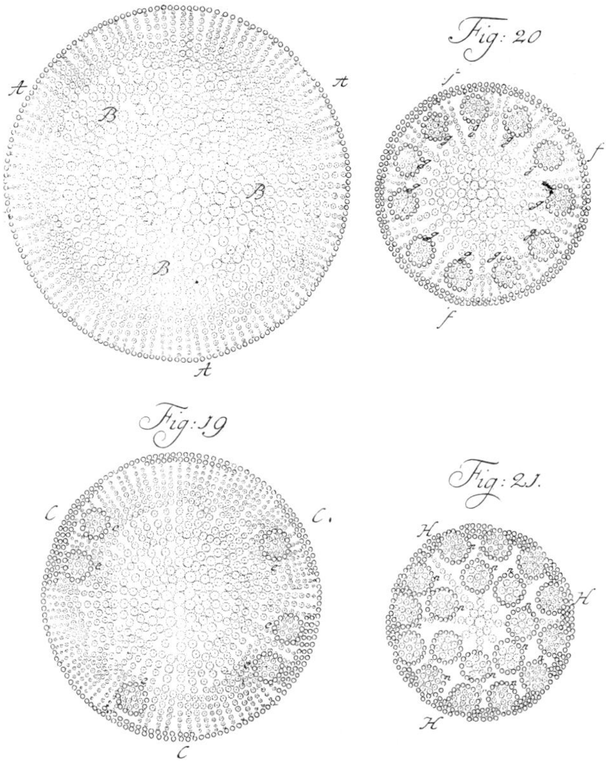

Figure 2.14 Diagram showing the interrelation between density and freedom of motion of elementary particles in hierarchical array of 'finites' on different scales (Swedenborg 1734).

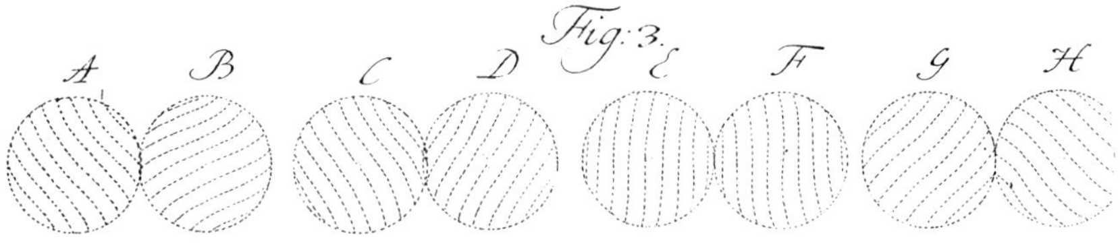

Figure 2.15 Diagrams to illustrate the interaction of magnetic lines of force in regions of different orientations. The lines represent the axes of *vorticli* (strings with opposable spins). (Swedenborg 1734.)

serve to illustrate the structural basis of what later became the degrees of freedom in Gibbs' thermodynamics (figure 2.14) and he illustrates orientation conflicts between otherwise identical arrays as at an intercrystalline boundary (figure 2.15). In his analysis of Musschenbroek's work on magnetism (which he reprints in full) he gives a beautiful diagram showing magnetic domains in aligned and random array and the resulting external fields (figure 2.16). Many

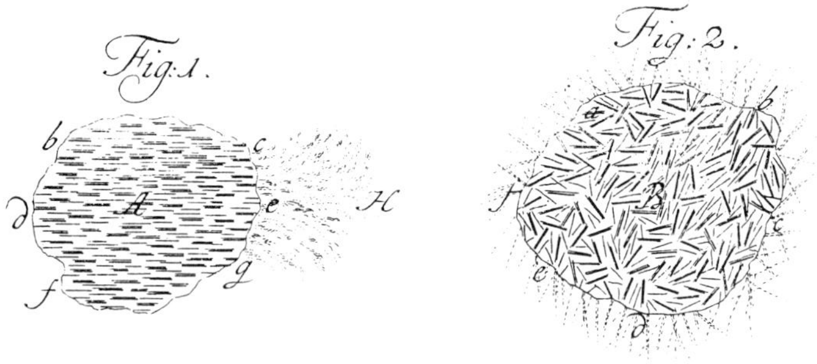

Figure 2.16 Magnetic domains in ordered and disordered array with their resulting external fields (Swedenborg 1734).

of his examples do not specifically relate to condensed matter but deal with astronomical and cosmological problems. Partly because his topological viewpoint is expressed more in diagrammatic terms than in mathematical equations and partly because he himself soon turned to mystic philosophy and religion rather than science, his work was largely ignored by scientists.

Swedenborg's approach was not that of modern physics; nevertheless both he and Descartes before him had a deep understanding of the importance of interaction between structural levels, of polarity, of time as pulsation and circulation, and of micro-movements interacting in such a way as to confer internal stability on larger clumpings and to generate boundary structures that would deflect environmental attack. The qualitative form of such ideas can be detected beneath much of twentieth century solid state physics as well as in recent attempts to combine particle physics, quantum theory, and cosmology within the unifying concepts of string theory†.

2.5 THE ROLE OF METALLURGY

During most of the eighteenth and nineteenth centuries studies of the structure and physical properties of materials were mainly work for mineralogists and metallurgists. The study of the polycrystalline nature of rocks began quite early. John Hill in 1748 used etching to identify microconstituents and has very perceptive remarks about nucleation, crystal growth and interference, as well as on the structure of aggregates in general; his work had no response (Hill 1748). But foundrymen and medical men did observe interesting shapes

†The above discussion was written after a hasty study of Swedenborg's *Principia* inspired by Norma Emerton's recent book *The Scientific Reinterpretation of Form* (Cornell University Press, 1984). Swedenborg's entire work deserves far more attention from scientists and historians of science than it has yet received.

Figure 2.17 The earliest known drawing of a metallic dendrite. (From Zannichelli 1713.)

(figures 2.17 and 2.18). Metallurgical studies of the visible texture of Damascus steel triggered studies leading to the discovery of the role of carbon in steel and, via the crystalline structures in meteorites, to those in metals generally (Smith 1960. See also Smith 1981 for a general view of the history of human interest in the shape of things.) H C Sorby of Sheffield did magnificent work on the microstructure of both rocks and steel: he founded both petrography and metallography and introduced the differential staining of biological tissues (Higham 1963). Curiously his work on steel, briefly reported in 1864, was not taken up by others for almost two decades, but the field then burgeoned. The work of the two generations of metallographers following the 1880s was, I believe, one of the most important among the many threads that combined to given modern physics. It owed very little to the physics of the time and a great deal to the chemistry. Many of those who made the significant observations were employed in industry and spent much of their time doing chemical analyses of raw materials and products. Since, however, they were intimately in contact with materials, they could not help being aware of the limitations of the chemical approach that had earlier made such great contributions to their industry. (The *Reports to the Alloys Research Committee* published between 1891 and 1904 in the *Proceedings of the Institution of Mechanical Engineers* and the *Contribution a l'Etude des Alliages* (1901) of the French *Société d'Encouragement pour l'Industrie Nationale* give a vivid picture of the excitement of the period. See also Smith (1960).)

It was in 1878, a little more than a century ago, that Adolf Martens

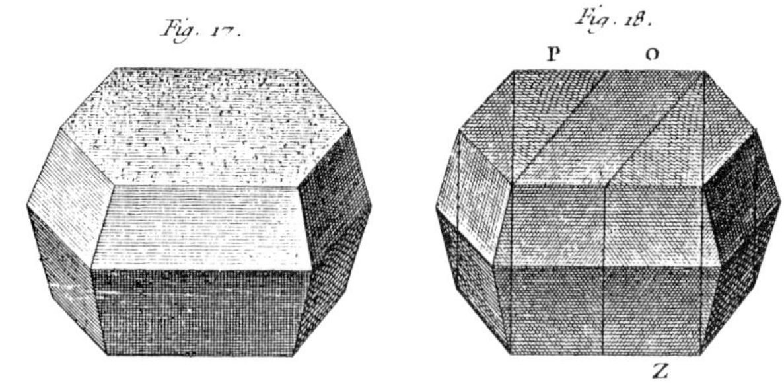

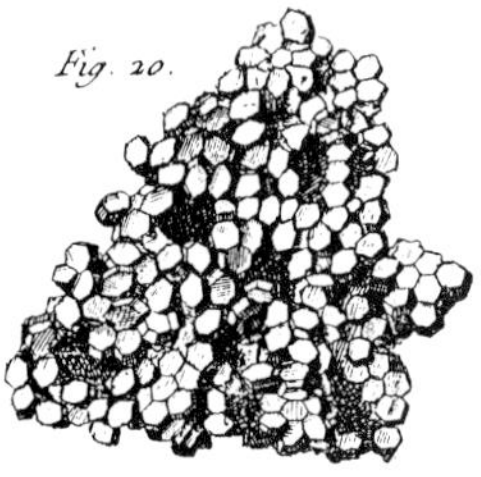

Figure 2.18 A cluster of grains of iron, revealed by intergranular fracture at a high temperature, together with an idealised grain having 14 sides and its geometric substructure of rhombs and sections of rhombs (Grignon 1775).

published the first paper on the microstructure of iron and steel to attract wide notice. A decade earlier the Russian metallurgist Dimitri K Tschernoff had discovered the thermal critical points in the heating and cooling of gun steel and their effect on grain size and hardness. In France between 1885 and 1887 Felix Osmond (who was associated both with the Creusot steelworks and with the leading thermochemist Louis Troost at the Sorbonne) examined the 'cellular' structure of steel and, more importantly, related carbon content and microstructure to the evolution of heat accompanying the phase change during cooling. The laboratory approach that he started, combining thermal analysis and microscopic metallography, gave rise to practically every subsequent advance in scientific metallurgy until the advent of x-ray diffraction forty years later. Metallographers were soon able to relate mechanical properties to microstructure and established a route for very fertile communication between physics, chemistry and engineering practice. Within two decades, Felix Osmond and W C Roberts-Austen had published several papers intimately relating the structure of steel and other alloys to their compositional, thermal, and mechanical treatment; Albert Sauveur had made the first significant measurements of grain size and volume fractions of the micro-constituents in steel; and H W Bakhuis Roozeboom and Henri Le Chatelier had shown how the experimental observations of structure on the new level conformed to thermodynamic principles that had been developed to account for macroscopic observations of change. (Le Chatelier's contributions to the techniques of the metallographer were almost as important as his famous principle.) Note how international all this is—the story involved England, Germany, France, Holland and, very early, both

Russia and the United States.

For the first three decades of the twentieth century the phase rule was the cornerstone of metallurgical science. Careful studies of phase diagrams soon made it clear that conventional ideas of chemical valence were useless when applied to alloys. This was an essential preliminary to the generation of electronic concepts which, though now seen to be physics, were first formulated in the mind of a chemist, William Hume-Rothery (Hume-Rothery 1926, 1931). Another preliminary, of course, was the much earlier discovery of voltaic electricity and the easy measurements that became possible of the most fundamental property that distinguishes materials. The fact that the electrical conductivity of a metal was highly sensitive to both composition and treatment was not predicted by theorists and was unsuspected until 1865 when it was made dramatically evident by the behaviour of different lots of copper for the first Atlantic cable: good studies were made immediately thereafter—again by a chemist, Augustus Matthiessen.

The changes of microstructure produced by mechanical deformation were observed in some detail in the period around 1900. The mechanical properties of metals had been mainly responsible for their use since their first discovery. The texture of fractured surfaces was usefully related to serviceability, but the structural reasons for strength, ductility and brittleness were almost inaccessible to science until the 1890s when metallographers saw grain boundaries and slip bands and began to study the distribution of minor phases. Intense arguments on the 'crystallisation' of metals by fatigue and on the reasons for the hardness of quenched steel are recorded in considerable detail in the transactions of the engineering societies at the time and they make fascinating reading.

In 1896 Albert Sauveur made a list of the various opinions that were then held to account for the hardness of quenched steel. Martensite, by that time a well known microstructural feature, was variously thought to be an allotropic form of iron (beta) stabilised by carbon, or an aggregate of iron with diamond, or compounds such as Fe_3C, or $Fe_{24}C$, in solution or in colloidal dispersion in either alpha or gamma iron. All this followed the lively interest in colloids and allotropy by chemists at the time and the century-old belief in simple combining proportions. And martensite was variously said to be hard because beta iron itself was hard; because the compounds were hard; because it was a supersaturated solid solution; or because its crystallites were extremely small, shattered, strained or otherwise disturbed.

Prior to the introduction of the electron microscope, microscopy was limited to magnifications of about 1000, and, in common with scientific speculation at any time, the leading ideas were of things that could not be seen. As in the seventeenth century, there was failure to distinguish between atomic, molecular, crystalline, and cellular levels of structure. Although the molecules, so important in chemistry, was inevitably overemphasised, the importance of structure was fully realised.

2.6 CHANGING VIEWS ON THE NATURE OF CRYSTALLINITY

Crystallography had begun millennia earlier with the enjoyment of gemstones and observation of external symmetry. The generation of anisotropy by the packing of isotropic spherical parts had a long history in decorative art and was brought into science by Kepler in 1611, yet for the following two centuries the respectable scientific view was that the constituent parts were inaccessible. Haüy and others who developed the mathematical analysis of the external symmetries of crystal shape specifically denied that their unit cells were ultimate units of structure.

Many different views of the nature of crystals have been held by scientists in the past. Though first noticed simply for the beauty of their polyhedral form, their substructure has at various times been regarded as being that of an aggregate of little spheres or as small molecular clumps of them; as polyhedra that filled space by translation without change in orientation; as bimodally orientable space-filling parallelopipedra (Dirichlet cells); as the mathematics of point group symmetry; or simply as statistical fields of electron density. Every one of these is useful and in a way correct, but none is complete by itself. The real basis of crystallinity lies in the *qualitative* nature of the patterns that transcends all these different views. Substructure cannot exist without superstructure and vice versa, at least above the quantum level.

In 1813 William Wollaston, quite unaware of his seventeenth century precursors, made crystal models by packing together atom balls of different sizes, and shortly afterwards chemist J F Daniell furnished the balls with interlocking auras (perhaps suggested by Boscovich's mathematical models) that both joined and separated them. These models were forgotten through most of the nineteenth century as lattice theory was being developed, but Wollaston's ideas were revived in 1883 by another Englishman of empirical bent, William Barlow, and they were widely accepted by 1912 when x-ray diffraction, as developed by W H and W L Bragg in Cambridge, at last permitted experimental measurement of structural regularities on the atomic level. It was not until then that the Daltonian paradigm of all structural change being at the molecular level could be reconsidered. (Note, incidentally, that x-ray crystallography did not develop in Munich, where not only had diffraction been discovered and its theory developed, but the leading school of chemical crystallography at the time was flourishing.)

A new kind of structural thinking could be brought to the age-old problems of alloy behaviour as the new x-ray techniques slowly merged with the findings of the older metallurgical microscopy. In the 1890s the various 'theories' of steel hardening were indeed structural, but it was structure on a molecular level with scant concern for crystalline regularity. Did not the malleability of metals dissociate them from the brittle crystals of the mineralogist? The slip bands microscopically visible on surfaces of deformed metals (figure 2.19)

Figure 2.19 Photomicrograph of slip bands on polished surface of lead after deformation (× 1000). (Ewing and Rosenhain 1900.) Although slip had been observed earlier, particularly by George Charpy whose work on recrystallisation was also important, the Ewing–Rosenhain papers were more comprehensive and were widely noticed. Rosenhain, a graduate student in 1899, went on to become the leading scientific metallurgist of the time and gave metallurgy a strong nudge toward physics.

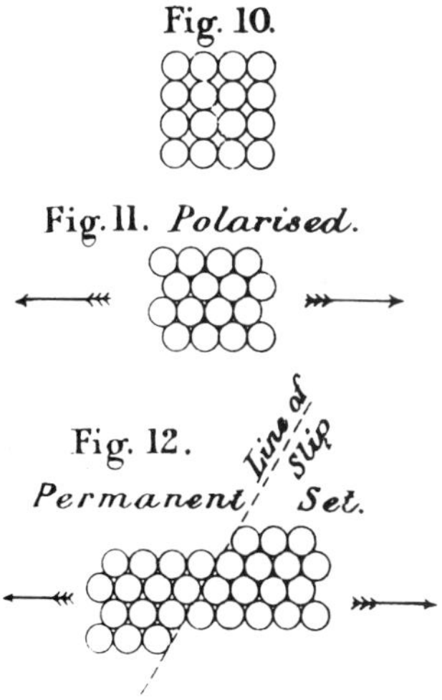

Figure 2.20 Diagram showing the movement of atoms by block slip (Roberts-Austen 1897). Compare with figure 2.24. Roberts-Austen emphasised the effect of extremely small amounts of impurities or added elements on the properties of alloys and suggested that the hardening effect of small amounts of carbon in steel arose from a longitudinally propagated disturbance in the habitual movements of the atoms of iron introduced by the added atoms.

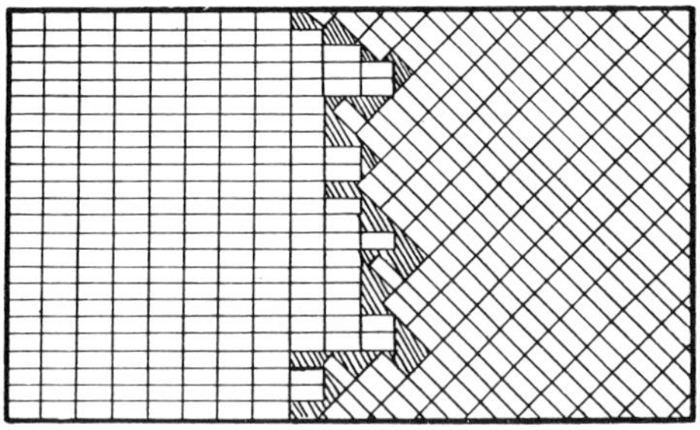

Figure 2.21 Diagram showing interstitial regions of 'amorphous' metal between crystals of differing orientation (Rosenhain 1912). Rosenhain still held the Haüyian–Daltonian view that molecules were a necessary basis for aggregation and says 'The liquid residue left in these interstices will be unable to crystallise for the simple reason that there will be no space for the aggregation of the "liquid molecules" to form a single "crystal molecule" and thus the portions shaded black [hatched] in the diagram will remain in the amorphous condition.'

suggested a way out. In 1897 Roberts-Austen drew diagrams (figure 2.20) showing the displacement accompanying slip in an ordered array of atoms that are not unlike the diagrams in introductory physics textbooks of the 1940s. In 1916 H M Howe (whose fine summaries of advanced metallurgical thinking are a boon to the historian) showed how orientation conflicts would hinder slip in polycrystalline aggregates, a view that was formalised soon after as the slip–interference theory of Zay Jeffries and R S Archer (1921). These three approaches were direct ancestors of today's dislocation theory.

In the early 1920s, the hottest debate among metallurgists was on the amorphous metal theory advanced by Sir George Beilby to account for work-hardening. Amorphous metal was next adopted by Walter Rosenhain to explain the properties of grain boundaries. (How delighted both men would have been to hear of today's metallic glasses!) The ancient polyhedral units of structure were still alive. Rosenhain thought of boundaries as being a relatively thick region of structureless disorder, almost a second phase, between arrays of polyhedral blocks (figure 2.21). A quite different model showing unitary misfits between lattices formed of spherical units (figure 2.22) was presented by F B Foley at an American Institute of Mining and Metallurgical Engineers meeting in 1925—about twenty years before J M Burger's famous dislocation model of the grain boundary eventually led physicists to accept the reality and importance of dislocations. In the discussion of Foley's paper Ernst Thum described the bubble raft model of grain boundaries later made popular by Sir Lawrence Bragg, but no one seems to have paid attention to either the theoretical idea or the demonstration.

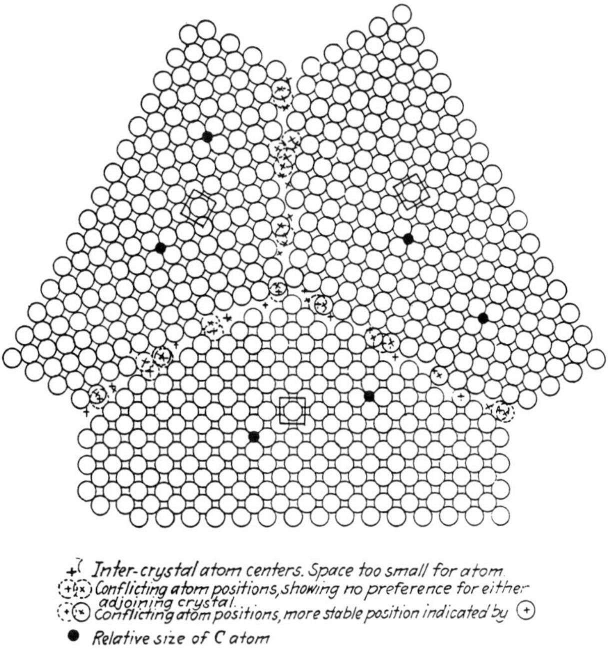

Figure 2.22 Diagram of an intercrystalline boundary showing that an amorphous region is not necessary (Foley 1925). Note the similarity of this to the well known dislocation boundary proposed by J M Burgers in 1939.

A critical event in metallurgical history occurred in 1905 with the unpredicted empirical discovery of precipitation-hardening in aluminium alloys (Hunsicker and Stumpf 1965). This slowly served to focus attention on what is probably the most interesting level of structure in inorganic materials. The new alloys were microscopically homogeneous at the stage of greatest hardness, but the particles of a second phase that became visible after over-annealing made it obvious that there were changes of solubility in solid solutions as well as in liquid ones. (Most earlier constitution diagrams had been drawn with vertical lines defining both 'intermetallic compounds' and solid solution regions.) Paul Merica in 1919 (Merica *et al* 1919) definitely established the fact that such changes did occur in solids, and proposed that hardening was a direct consequence of the presence of precipitated submicroscopic particles which interfered with slip.

Jeffries and Archer (1921) extended the theory to include the notion of critical precipitate size and orientation effects at grain boundaries. Then, in 1932, Merica (Merica 1932) had the brilliant insight that the composition gradient produced by segregation accompanying the early stages of precipitation was not necessarily associated with lattice discontinuity, but

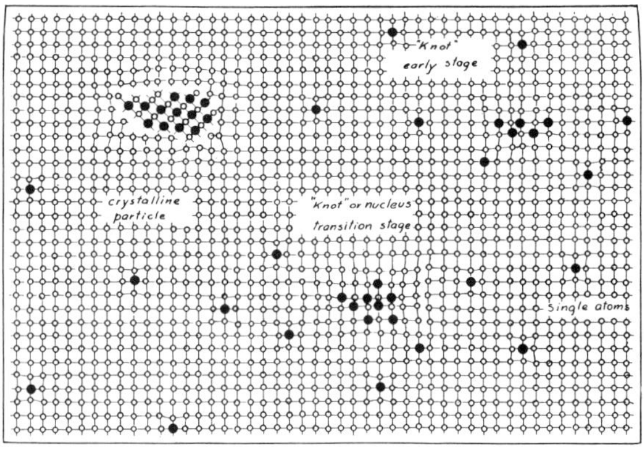

Figure 2.23 Diagram illustrating Paul Merica's concept of segregation of solute atoms with and without coherence with the parent lattice (Merica 1932).

that there could be larger diffusely bounded regions of coherent lattice strain which he termed 'knots' (figure 2.23). Not long after, Borelius (1934, 1935, 1937a and b) emphasised the thermodynamics of such spinodal transformation structures in solids generally, and in 1938 the existence of the hypothesised structures was confirmed by x-ray diffraction studies and they became known as Guinier–Preston zones. All these studies combined to make the perfect crystal lattice as obsolete in theory as it was non-existent in nature.

The 1920s were a truly revolutionary period for the development of solid state physics for it was a time of the detection and general acceptance of the importance of crystal imperfections. The detection of imperfections at last enabled theorists to tackle real materials and not just ideal simplified matter. Before then, those few physicists concerned with solids had been mainly occupied with electrical and magnetic properties or with the elaboration of classical mechanics and elasticity theory. They would have been rather scornful of the metallurgists' qualitative models. However, the work of metallurgist H C H Carpenter on the deformation of large crystals of aluminium caught the eye of G I Taylor at a time when he was thinking of A A Griffiths' 1920 theory of stress concentration around microcracks and gave birth to the idea of the dislocation as the key to plasticity in metals (Taylor 1934). Somewhat earlier, Masing and Polanyi had seen the possibility of vernier-like structures in bent crystals and in 1928–9 others, including especially Dehlinger and Orowan were, apparently, independently working on similar concepts (Orowan 1934). (See also Orowan's fine historical account as part of a symposium (1965).) The illustrations, reproduced in figures 2.24 and 2.25, are of great interest, as are pictures of any new child later to become famous.

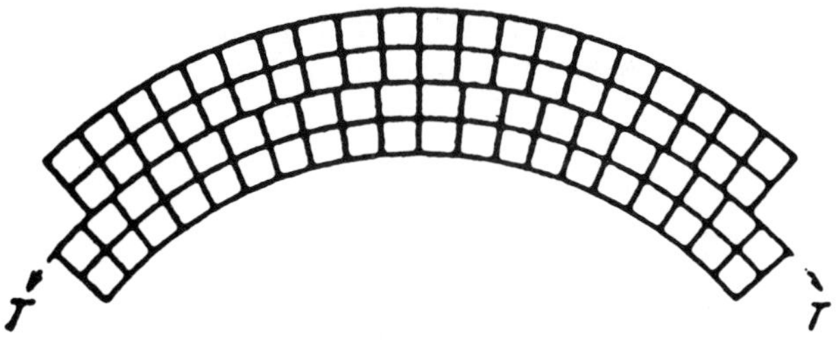

Figure 2.24 The vernier-like disregistry of lattice positions in a bent crystal lattice (Masing and Polanyi 1923).

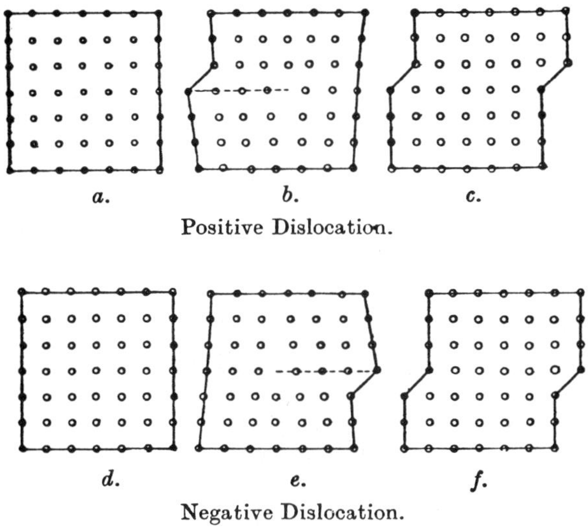

Figure 2.25 This diagram of dislocation movement was the first to be widely noticed by physicists. It illustrated the seminal paper by G I Taylor that appeared in 1934 (Taylor 1934).

It was not metals, but studies of colour changes in crystals of the alkali halides that first generated ideas of other types of imperfection. Adolf Smekal, who in 1929 advanced the idea of mosaic substructure, seems to have been the first to have pointed out the important distinction between classes of properties that were sensitive and insensitive to structure.

It will be obvious from this very superficial survey that solid state physics was a consolidation of many different developments of observation and thought. (An excellent study of the chemist's contributions is in the book by Cecil H Desch, *The Chemistry of Solids* (1934).) The period of its prehistory can be said to end in 1929, which marked not only the beginning of dislocation theory but also the publication of the first comprehensive treatment by a physicist of the properties of real matter. This was in the article by Paul Ewald and colleagues in the Muller–Pouillet *Lehrbuch der Physik* (1929).

2.7 SPECULATIVE CONCLUSION

Though we think of solid state physics as one of the most recent branches of science, it is in many ways a return to the earliest. Its renaissance marks, I believe, a watershed in human intellectual history. With its concern for the relation between real structure and observable properties it unites the Aristotelian, Democritan, and Platonic schools of philosophy, each with an overemphasis on one aspect of the nature of things. Solid state physics at last can deal with the qualitative nature of substances, and it joins with the biological sciences in accepting the fact that sensual information is as valid as intellectual, that observation and something like enjoyment of complexity must precede the mental construction of calculable models. For too long historians of science, being intellectuals, have overemphasised the mathematical part of science. I contend that while Newtonian mathematics has a beauty of its own and is of immense value in setting the limits of relationships, it does so only by avoiding substructure and hence the very nature of nature, which combines the quantitative and qualitative at all levels. It may be, as Schmandt-Besserat has suggested, that mathematics itself, the manipulation of numbers, began with distinction between shapes—using little clay spheres, tetrahedra, cylinders and prisms as counters. A real concern with the interaction between shape and space underlay the verbal disputes of early philosophers.

The recent development of Boolean algebra and topology, combined with ideas of complementarity, indeterminacy, and the integers of quantum theory, require a very different view of structure from that held by physicists of the last two centuries. Interest has shifted from the symmetry of crystals to the role of lattice vacancies, dislocations and other defects, and a viewpoint is emerging that has much in common with medieval thought. When the properties of matter are seen to arise from the interplay between internal and external form, between the possibilities of fit and misfit, on *levels* of communication, contact, comparison, connection and cohesion, one can sense a parallel with the way in which the human mind works; indeed the way in which it must work, for it too depends on the same principles of structural inertia, of freedom and fixity of electron interlock that form the inorganic world. Science, after all, is a human activity and it consists of comparing mental patterns with those provided by the senses, either directly or indirectly via instrumentation, and in real, past or future *time*.

The above may strike the reader as being closer to metaphysics than to modern physics and to deal more with intuitive concepts than with mathematical precision. Nevertheless, it is not an idea that is capable of diagrammatic representation an acceptable part of physics—perhaps even an essential preliminary to exact treatment as well as an aid in explanation? Indeed, the properties of matter and of diagrams and of the human body and human mind all depend on the same principles of pattern generation implicit in the junction and directional resolution of electron interaction. The world is far too complex to be understood even by the total of human intellect, just as it is too extensive for any local assembly of atoms to be influenced immediately by distant changes. Curiously, art is more closely dependent on the properties of matter than is science, but both depend on a human being's sense of what is uniform enough to be omitted for the time being; it is the omission that confers individuality or specificity. Some choice between 'inside', 'outside' and 'change' is necessarily made by electrons at any boundary in matter or in mind.

Condensed matter physics is concerned with those structures that currently exist because of the earlier search by the component atoms for some stability amid changing associations. Most early natural philosophers felt the need for something like it. Retrospectively one can see that the great advances that came from Newtonian emphasis on precision, on Euclidean geometry and on atomic or infinitesimal calculability, were, perhaps necessarily, accompanied by a temporary blindness to the essential contribution of substructural boundaries to the nature of matter and materials.

REFERENCES

Bergman T 1781 *Dissertatio chemica de analysi ferri* (Uppsala) (Engl. Transl. 1968 in C S Smith *Sources for the History of the Science of Steel* (Cambridge, MA: MIT Press))

Burgers J M 1939 *Proc. K. Ned. Akad. Wet.* **42** 278, 294

Burke J G 1966 *Origin of the Science of Crystals* (Berkeley: University of California Press)

Borelius G 1934 *Ann. Phys., Lpz* **20** 57–74, 650–2

—— 1935 *Ann. Phys., Lpz* **24** 489–506

—— 1937a *Ann. Phys., Lpz* **28** 507–20

—— 1937b *Ann. Phys., Lpz* **38** 517–38

Descartes R 1638 *Oeuvres* ... ed. Charles Adam and Paul Tannery (Paris) vol. 2 p76

—— 1644 *Principia philosophia* (Paris) (Engl. Transl. 1984 by V R Miller and R P Miller (Dordrecht: Reidel))

Desch C H 1934 *The Chemistry of Solids* (Ithaca: Cornell University Press)

Douat D 1722 *Méthode pour faire une infinité de designes differens avec des carreaux mipartis de deux couleurs par une ligne diagonale* ... (Paris)

Ewald P *et al* 1929 in J Muller and C Pouillet ed. *Lehrbuch der Physik* (Brunswick) vol. 1 part 2 (Engl. Transl. 1930 by F Dougall and W M Dean (London))

Ewing J A and Rosenhain W 1900 *Phil. Trans.* A193 353–77

Haüy R J 1784 *Essai d'une théorie sur la structure des crystaux* (Paris)
Foley F B 1925 *Trans. AIME* **71** 736
Grignon P C 1775 *Memoires de physique sur l'art de fabriquer le fer* (Paris)
Higham N 1963 *A Very Scientific Gentleman: The Major Achievements of Henry Clifton Sorby* (Oxford: Pergamon)
Hill J 1748 *A History of Fossils* (London)
Hooke R 1665 *Micrographia* (London)
Hume-Rothery W 1926 *J. Inst. Metals* **35** 295
—— 1931 *The Metallic State* (Oxford: Clarendon)
—— 1962 *Atomic Theory for Students of Metallurgy* (London)
Hunsicker H Y and Stumpf H C 1965 History of precipitation hardening in *The Sorby Centennial Symposium on the History of Metallurgy 1965* (New York: Gordon and Breach) pp 271–311
Jeffries Z and Archer R S 1921 The slip interference theory of hardening of metals in *Chemical and Metallurgical Engineering* **24** 1057–67
Kepler J 1611 *Strena seu de nive sexangula* (Frankfurt) (Engl. Transl. by C Hardie 1966 (Oxford: Oxford University Press))
Masing G and Polanyi M 1923 *Engebn. der exacten Naturwissenschaften* **2** 224
Merica P D 1932 The age hardening of metals *Trans. AIME* **99** 43–54
Merica P D, Waltenburg R G and Scott H 1919 *Scientific Papers of the US Bureau of Standards* No 347
van Musschenbroek P 1729 *Introductio ad cohaerentium corporum firmorum* in *Physica experimentales et geometricae* ... (Leyden) pp 421–672 and plates 17–28
Orowan E 1934 *Z. Phys.* **89** 605, 614, 634
—— 1965 in *The Sorby Centennial Symposium on the History of Metallurgy* ed. C S Smith (New York: Gordon and Breach) pp 359–76
Ploss E E *et al* 1970 *Alchemia; Ideologie und Technologie* (Munich: Heinz Moos Verlag) p129
de Réaumur R A F 1722 *L'Art de convertir le fer forgé en acier* (Paris) (Engl. Transl. by A G Sisco 1956 (Chicago: University of Chicago Press))
Roberts-Austen W C 1897 Fourth Report to the Alloys Research Committee *Proc. I. Mech. E.* 31–100
Rosenhain W 1912 *J. Inst. Metals* **8** 149–73
Smith C S 1960 *A History of Metallography* (Chicago: University of Chicago Press)
——1968 ed. *Sources for the History of the Science of Steel* (Cambridge, MA: MIT Press)
——1981 A search for structure (Cambridge, MA: MIT Press)
Swedenborg E 1721 *Prodomus principiorum rerum naturalium* (Amsterdam) (Engl. Transl. 1847 by C Strutt (London))
—— 1734 *Principia rerum naturalium* ... (Dresden and Leipzig) (Engl. Transl. 1912 by J R Rendell and I Tansley, 2 vols (London). Contains in addition the text of both earlier and later unpublished treatises on the same subject)
Taylor G I 1934 The mechanism of plastic deformation of crystals *Proc. R. Soc.* A **145** 362–454
Truchet S 1704 Memoire sue les combinaisons *Mem. Acad. Royale des Sciences* (Paris) pp 363–72 with plates 13–20
Zannichelli J H 1713 *De ferro ejusque nivis preparatione* ... (*On iron snow*) (Venice)

M T F von Laue (1879–1960)

3 Diffraction

M M Woolfson

3.1 EARLY DAYS

The discovery of x-rays by Röntgen in 1895 provoked great excitement and interest in the general public—how curious it was that one could see through objects opaque to visible light. Yet, important as that aspect of x-rays has been for medical diagnosis, it pales to insignificance compared with their application in studying the structure of matter at the atomic level and even exploring the processes of life itself. In 1905 Barkla carried out experiments which showed that x-rays were scattered in the same way as visible light and in 1907 he established the existence of characteristic x-rays. However, there was doubt for some time, particularly in the mind of W H Bragg, as to whether x-rays were waves or particles. We now know that x-rays consist of electromagnetic radiation and have both wave-like and particle-like properties.

In 1912 Ewald, a young research student in Munich, was completing his doctoral dissertation on the passage of light through crystals. This work came to the attention of von Laue who realised that the repeat distances in crystals were of the same order as the wavelengths of x-rays, which had previously been estimated by rather crude diffraction experiments to be in the range 10^{-10} to 10^{-11} m. Von Laue suggested to Friedrich and Knipping that they should try an experiment to see whether crystals would act as a diffraction grating for x-rays. The photograph they obtained, shown in figure 3.1, was no more than a collection of smudges but it was the key that would open a treasure house of scientific discovery.

The paper announcing this work appeared in the *Proceedings of the Royal Bavarian Academy of Science* in June 1912 and in the following year W L Bragg (1913) gave a simple explanation of the Munich photographs. He interpreted each spot on the film as due to specular reflection of the incident beam from equispaced planes in the crystal which were richly populated with atoms. The condition that radiation scattered from each plane should all be in phase in the reflected beam direction gave rise to what is now known as

Bragg's law

$$\lambda = 2d \sin \theta \qquad (3.1)$$

where d is the spacing of the planes and θ the angle made by both the incident and reflected beam to the normal to the planes.

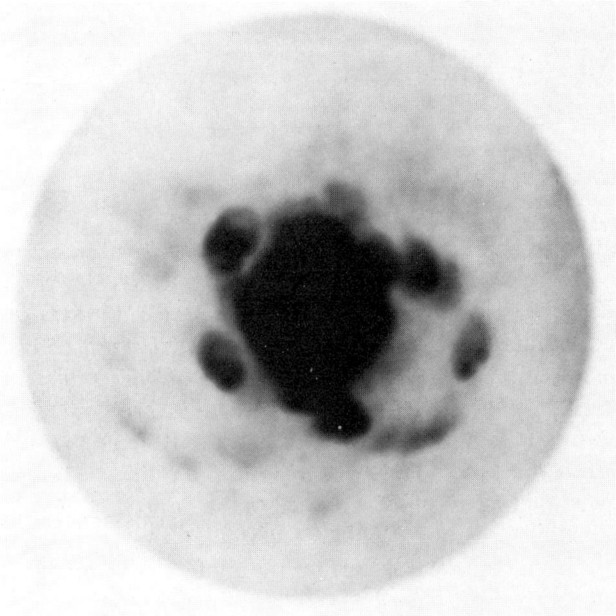

Figure 3.1 The first x-ray diffraction pattern from a crystal.

A significant step in the development of x-ray crystallography as an experimental subject was made in Leeds by W H Bragg, the father of W L Bragg, who built the first x-ray spectrometer. With this instrument spectra were measured from the alkali halides NaCl, KCl and KBr and diamond. W L Bragg (1913) and W H and W L Bragg (1913) published papers on the determination of these structures. This was the beginning of x-ray crystallography; the theoretical basis was still rather primitive and only very simple structures could be solved by semi-quantitative methods. However, the exciting prospect of being able to find the positions of atoms in crystals attracted some of the best scientific minds of the time and within a few years a sound theoretical foundation for the subject had been established. In 1913, at a Solvay Conference devoted to the new subject, Sommerfeld effectively gave the equation for calculating the intensity of a Bragg reflection. A crystal consists of a regular three-dimensional stack of identical units, the *unit cell*. Within each unit we suppose that there are N atoms at positions $\boldsymbol{r}_j$, where the components (x_j, y_j, z_j) are normalised to be in the range 0 to 1. Each type of atom scatters differently according to its *scattering factor*, f_j, which depends on the number of electrons around the nucleus and their spatial distribution. Each Bragg reflection can be defined by its Miller indices $(h\ k\ l)$ which can be thought of as a vector in reciprocal space. The Miller indices relate to the orders of diffraction in a one-dimensional diffraction grating; for scatterers separated by a vector $\boldsymbol{a}(\boldsymbol{b}, \boldsymbol{c})$, one side of the unit cell, the path difference is $h\lambda(k\lambda, l\lambda)$.

The structure factor is given by

$$F(\boldsymbol{h}) = \sum_{j=1}^{N} f_j\, \mathrm{e}^{2\pi \mathrm{i} \boldsymbol{h}\cdot \boldsymbol{r}_j}. \tag{3.2}$$

The structure factor is a complex quantity and the intensity of an x-ray reflection is proportional to $|F(\boldsymbol{h})|^2$, although it also depends on a number of physical factors—temperature, absorption, crystal perfection etc—whose effects can usually be estimated quite well. In his Bakerian lecture to the Royal Society W H Bragg (1915) suggested the use of the Fourier series in crystal-structure analysis. This led to an equation giving the electron density at the point $\boldsymbol{r}$ in the unit cell

$$\rho(\boldsymbol{r}) = \frac{1}{V} \sum_{\boldsymbol{h}} F(\boldsymbol{h})\, \mathrm{e}^{-2\pi \mathrm{i} \boldsymbol{h}\cdot \boldsymbol{r}}. \tag{3.3}$$

Another form of this equation, in which all the terms are real, is

$$\rho(\boldsymbol{r}) = \frac{1}{V} \sum_{\boldsymbol{h}} |F(\boldsymbol{h})| \cos[2\pi \boldsymbol{h}\cdot \boldsymbol{r} - \phi(\boldsymbol{h})]. \tag{3.4}$$

We should note here that, from equation (3.2), replacing $\boldsymbol{h}$ by $-\boldsymbol{h}$ gives the complex conjugate on the right-hand side so that

$$F(\bar{\boldsymbol{h}}) = |F(\boldsymbol{h})|\, \mathrm{e}^{-\mathrm{i}\phi(\boldsymbol{h})}. \tag{3.5}$$

Here we use the crystallographic convention $\bar{\boldsymbol{h}} \equiv -\boldsymbol{h}$.

The quantity $\phi(\boldsymbol{h})$, the phase of the structure factor, is of central importance in the solution of crystal structures. In general it can lie anywhere in the range 0 to 2π but for structures containing a crystallographic centre of symmetry phases are restricted to the values 0 and π. Since $\exp(\mathrm{i}0) = 1$ and $\exp(\mathrm{i}\pi) = -1$ for centrosymmetric structures, reference is often made to the *sign* of the structure factor rather than its phase. It should also be noted that even for non-centrosymmetric structures, where the origin of the unit cell is conventionally related to symmetry elements in a particular way, structure factors whose indices are special in some way (e.g. one zero index) may have restricted values—for example $\pi/2$ or $-\pi/2$.

With the availability of equation (3.4) the way was now clear for solving rather more complicated structures than had been previously solved by trial-and-error methods. If a trial structure could be found in some way—for

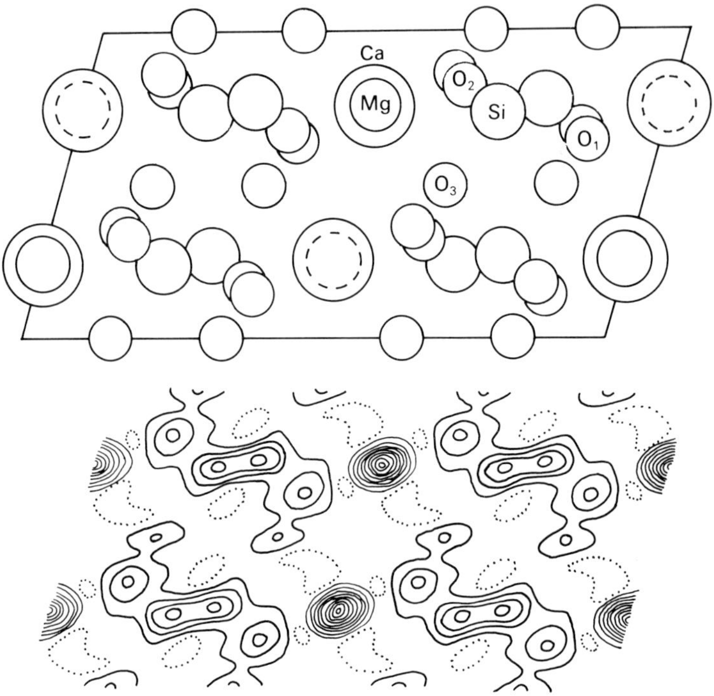

Figure 3.2 Projected electron density for diopside and its interpretation.

example from packing considerations in the unit cell—then a comparison of the magnitudes of structure factors calculated from the trial coordinates by equation (3.2), $|F_c(\boldsymbol{h})|$, and those deduced from the observed intensities, $|F_o(\boldsymbol{h})|$, would confirm, or otherwise, the essential correctness of the trial structure. Once a reasonable trial structure was obtained then it could be refined by an iterative procedure. The calculated structure factors would give an estimate of the phases $\phi(\boldsymbol{h})$. By the use of equation (3.4) an electron-density map could be calculated where observed structure amplitudes, $|F_o(\boldsymbol{h})|$, and calculated phases were used in the summation. At the usual resolution given by the x-rays employed, atoms would appear as approximately spherical blobs of electron density with the maximum of the blob giving the atomic position. With new estimates for the coordinates found in this way the cycle could be recommenced and repeated until the process converged. Most early work was done in projection. Thus a map calculated from

$$\rho(x, y)=\frac{1}{A}\sum_h\sum_k|F(hk0)|\cos[2\pi(hx+ky)-\phi(hk0)] \tag{3.6}$$

would show electron density projected on to a face of the unit cell of area A. Figure 3.2 shows projected electron density for diopside, $CaMg(SiO_3)_2$, and its interpretation as given by W L Bragg (1939).

This was how structure determination had to be done prior to 1934. Experimental measurements give $|F(\boldsymbol{h})|$ but not the phases, $\phi(\boldsymbol{h})$. If phases were available then one could immediately calculate an electron-density map and the structure would be solved. This constitutes the *phase problem in crystallography* and much of the development of x-ray crystallography has been concerned with the solution of that problem.

3.2 THE PATTERSON FUNCTION

A way of circumventing the phase problem, at least in some cases, was suggested by Patterson (1934). His function, now referred to as the Patterson function, is given by

$$P(\boldsymbol{r}) = \frac{1}{V} \sum_{\boldsymbol{h}} |F(\boldsymbol{h})|^2 \cos 2\pi \boldsymbol{h} \cdot \boldsymbol{r}. \tag{3.7}$$

The Patterson function contains only the values of $|F(\boldsymbol{h})|^2$ which can be obtained from the measured intensities without knowledge of phases.

This function has the property that it shows peaks corresponding to the vectors between all pairs of atoms and the total weight of the peak will be proportional to the product of the atomic numbers of the atoms concerned. Thus it contains potentially useful information about the structure, in principle even the whole structure, but the drawback, in general, is that it presents a very confused and poorly resolved appearance and it may not be straightforward to extract information from it. There will always be a large peak at the origin corresponding to the overlapped null self-vectors between each atom and itself. The other $N(N-1)$ vectors will fall into the same volume as do the peaks corresponding to the atoms in an electron-density map. It is only in special circumstances that one can extract information from this messy and overcrowded picture.

If there are a few atoms in the structure which have a much higher atomic number than all the rest then the vectors between these atoms may be so strong that they stand out against the background of smaller peaks. An example of this is shown in figure 3.3 which shows the b-axis projection of 2-amino-4,6 dichloropyrimidine (Clews and Cochran 1948). In projection, the structure contains two symmetry-related molecules in the repeat unit and the projected electron-density map shows clearly the large density associated with the chlorine atoms. In the Patterson map all the peaks corresponding to vectors between chlorine atoms can clearly be seen. Although there are two other substantial Patterson peaks, it turns out that the set of large peaks consistent with vectors between four chlorine atoms forming a parallelogram can easily be identified. We shall see later that finding the chlorine atoms is enough to enable the whole structure to be solved.

Another circumstance which occurs quite often and enables an

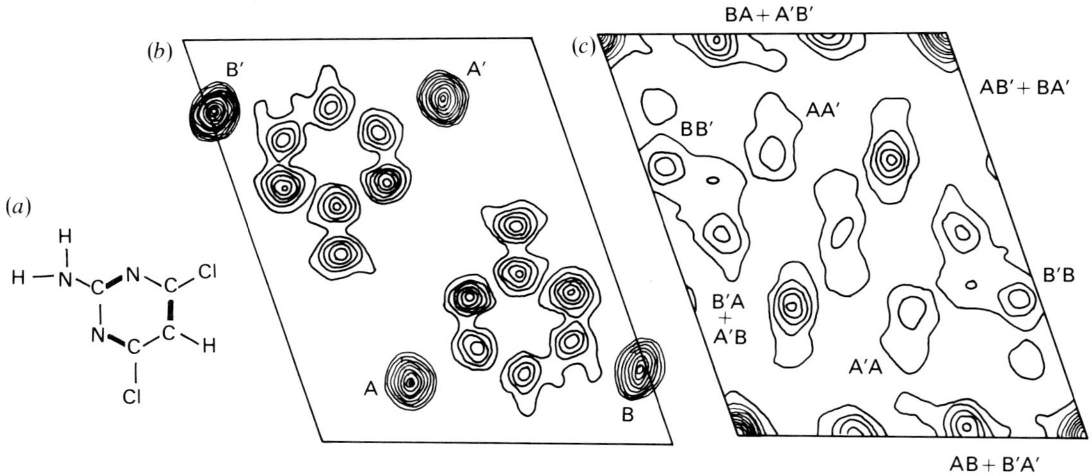

Figure 3.3 (*a*) A molecule of 2-amino-4,6 dichloropyrimidine. (*b*) The projected electron-density map. (*c*) The Patterson function for the *b*-axis projection.

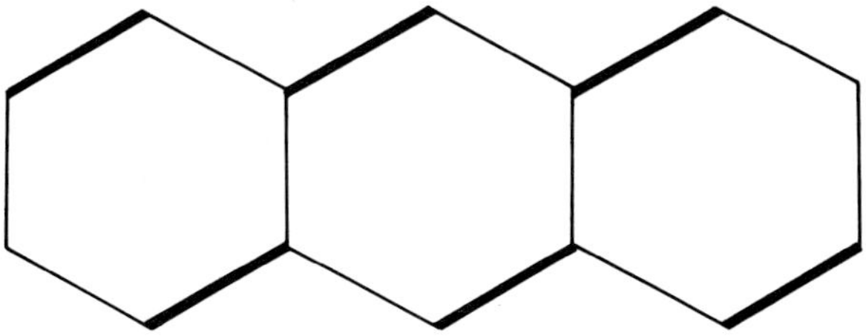

Figure 3.4 An anthracene molecule showing parallel interatomic vectors.

interpretation of the Patterson map to be made is when there are several parallel vectors in the structure. For example, in the anthracene molecule shown in figure 3.4 the interatomic vector marked boldly occurs six times. The sixfold overlap of peaks in the Patterson map may be clearly seen and indeed other multiple vectors may occur which fix or restrict the orientation of the molecule.

Most crystal structures contain a great deal of symmetry, so that each atom is related to a set of others of the same kind by symmetry elements. Patterson peaks between symmetry-related atoms may fall on special lines or special planes and so readily be identified. Again, as we saw in figure 3.3 where the chlorine atoms formed a parallelogram, the vectors between a set of symmetry-related atoms may form a characteristic pattern which can be sought in the Patterson map.

The Patterson function is still widely used to solve crystal structures, especially when heavy atoms are present. It was the first important means of solving the phase problem and methods have been devised for extracting information from Patterson maps even when they are not visually informative (e.g. Buerger 1950).

3.3 THE HEAVY ATOM AND ISOMORPHOUS REPLACEMENT METHODS

We have seen that when a structure contains a few heavy atoms it is usually possible to locate their positions by the use of the Patterson function. From the positions of the heavy atoms estimates of the phases of the structure factors can be made, these being the phases of the heavy-atom contributions alone.

The average magnitude of the contribution of a subset of m atoms to a structure factor is proportional to $(\sum_{j=1}^{m} f_j^2)^{1/2}$. Thus for a structure containing, say, 100 carbon atoms and 1 bromine atom the ratio of the average magnitudes of the contributions of bromine and carbon is $37{:}(100 \times 6^2)^{1/2}$ or, approximately, 0.6:1.

If the structure is centrosymmetric, then the sign of the structure factor will

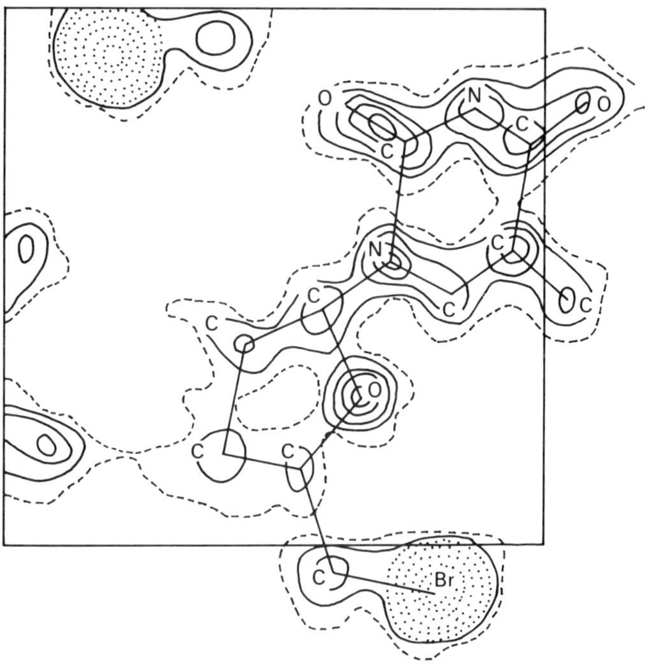

Figure 3.5 An electron-density map for 5′-Br-3′:5′-dideoxythymidine based on phases from the bromine atom.

be heavily biased towards that of the heavy-atom contribution, especially if the latter is large. It is possible, theoretically, to deduce a probability that the sign indicated is the correct one. For the non-centrosymmetric case there will be a probability distribution for the phase of the structure factor which peaks at the phase of the heavy-atom contribution.

In either the centrosymmetric or the non-centrosymmetric case an electron-density map calculated with the heavy-atom phases will give some indication of light atom positions. In figure 3.5 there is shown an electron-density map for 5′-Br-3′:5′-dideoxythymidine (Woolfson 1956a). The bromine atom was located from a Patterson map and the first trial electron-density map showed clearly the approximate position of all other atoms.

The reliability of the phase indication from the heavy atoms depends on the magnitudes of the heavy atom contribution and of the structure factor. It is possible to increase the signal-to-noise ratio in the electron-density maps by applying weights to the terms in equation (3.4) which depend on the reliability of the phase indication. Weighting schemes for the centrosymmetric and non-centrosymmetric cases were given by Woolfson (1956b) and Sim (1960) respectively.

Another method which depends on the determination of the contribution to the structure factor of a small part of the structure is that of isomorphous replacement. This occurs when there are chemically similar materials which differ only in having an atom or group of atoms in one structure replaced by similar atoms in the other—e.g. Na replaced by K or Cl by Br. The crystals of isomorphous compounds often look similar, have the same space group and similar cell dimensions. The positions of the atoms, in fractional coordinates, will be almost the same in the two structures and this can be exploited to give phase information.

Consider a structure factor in which the contribution on the right-hand side of (3.2) is split up into that of the isomorphously replaced (IR) atoms, $F(\mathrm{I})$, and the remainder of the structure, $F(\mathrm{R})$, so that

$$F(\mathrm{A}) = F(\mathrm{I}) + F(\mathrm{R}). \tag{3.8}$$

For the isomorphous partner we may write

$$F(\mathrm{B}) = F(\mathrm{I}') + F(\mathrm{R}). \tag{3.9}$$

The way in which $F(\mathrm{A})$ and $F(\mathrm{B})$ are formed from their components is shown in figure 3.6. The angle $\phi(\mathrm{I})$ is the phase of the contribution of the IR atoms and

$$F(\mathrm{I}') = kF(\mathrm{I}) \tag{3.10}$$

where k is the ratio of the scattering factors of the IR atoms and is known.

Figure 3.6(c) shows how, by a simple geometrical construction, it is possible to find the magnitude of the contribution of the remaining atoms and the phase of the contribution is restricted to two possible values. Thus the phase of the first compound is

$$\phi(\mathrm{A}) = \phi(\mathrm{I}) \pm \eta \tag{3.11}$$

and finding $\phi(\mathrm{A})$ is reduced firstly to finding $\phi(\mathrm{I})$ and then resolving the sign ambiguity.

If the IR atoms in one or other of the structures are heavy enough then it may be possible to locate them from the Patterson map and so to find all values of $F(\mathrm{I})$, both in magnitude and phase. Alternatively a Fourier summation with coefficients $|F(\mathrm{A})|^2 - |F(\mathrm{B})|^2$ (a difference Patterson map) will show peaks corresponding to vectors between the IR atoms and a background of, in general, smaller peaks which are also few in number. Such a map can often be interpreted to give the IR atom positions.

The sign ambiguity in equation (3.11) is more difficult to resolve. If for one of the isomers the two possible phases are $\phi(\boldsymbol{h})$ and $\phi'(\boldsymbol{h})$ for the structure factor of index $\boldsymbol{h}$, then Kartha (1961) has proposed calculating

$$\delta(\boldsymbol{r}) = \frac{1}{V} \sum_{\boldsymbol{h}} |F(\boldsymbol{h})| [\cos(2\pi \boldsymbol{h}\cdot\boldsymbol{r} - \phi(\boldsymbol{h})) + \cos(2\pi \boldsymbol{h}\cdot\boldsymbol{r} - \phi'(\boldsymbol{h}))]. \tag{3.12}$$

One of the terms for each value of $\boldsymbol{h}$ will systematically contribute to $\rho(\boldsymbol{r})$, the correct density, while the other term with incorrect phase will just give a noisy background. The correct structure can often be seen in such a map.

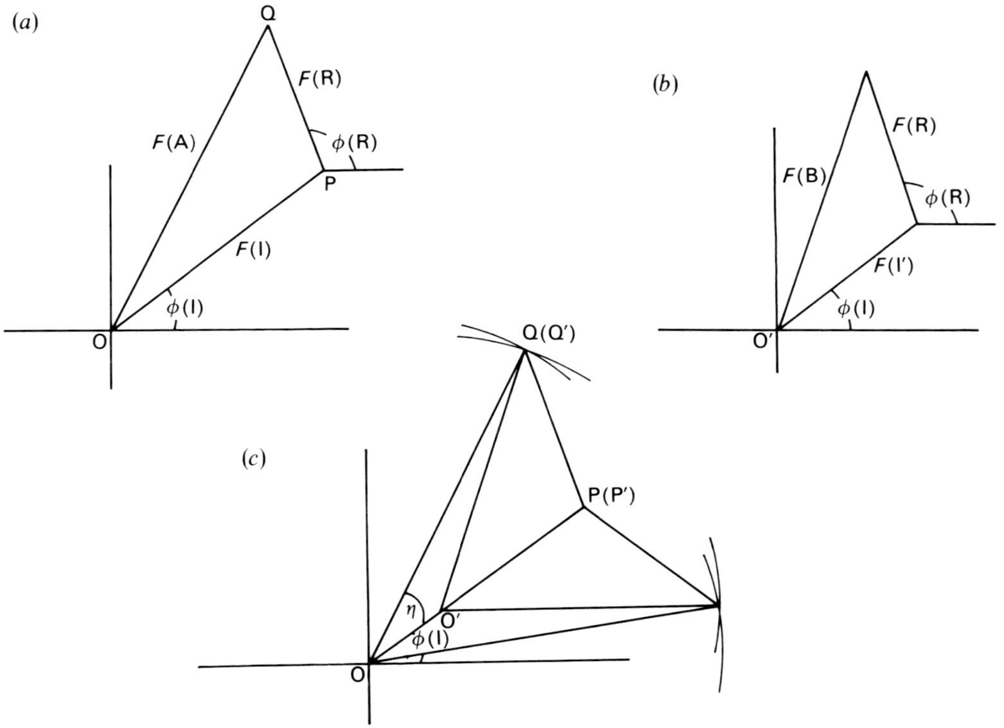

Figure 3.6 (*a*) and (*b*). The structure factors $F(\mathrm{A})$ and $F(\mathrm{B})$ with the contributions of the IR atoms, $F(\mathrm{I})$ and $F(\mathrm{I}')$ and of the residue of the structure, $F(\mathrm{R})$. (*c*) Based on the known $F(I)$, $F(\mathrm{I}')$, $|F(\mathrm{A})|$ and $|F(\mathrm{B})|$ values, a geometrical construction gives an ambiguity of phase for $F(\mathrm{A})$ or $F(\mathrm{B})$.

3.4 PROTEIN STRUCTURES

In the late 1920s and early 1930s several individuals, for example Astbury and Bernal, discovered that rather diffuse powder-like patterns could be produced from protein material which appeared to be microcrystalline in form. Eventually Crowfoot and Bernal (1934) reported some recognisable single-crystal photographs of pepsin and protein crystallography was born as a scientific discipline.

After this first success other proteins were purified and crystallised. A typical protein consists of a large number of peptide units, from several dozen to several hundred, linked together in a chain with various side groups attached. A single molecule will be folded so as to occupy some well defined region of space, typically an ellipsoid or spheroid, and several such units will be linked by symmetry elements to form a crystal. Between the protein molecules there will be solvent, e.g. water, some of this may be bound to the protein and be ordered, some may be partially ordered but the remainder will be in the form of a random liquid, filling the interstices between the molecules.

The big breakthrough in protein crystallography was essentially through chemistry. It was found that soaking protein crystals in various solutions of substances containing heavy atoms could lead to groups containing heavy atoms being bound to the protein at specific sites. The first heavy-atom derivative to be exploited was paramercuribenzoate haemoglobin, which enabled Green *et al* (1954) to produce a two-dimensional centrosymmetric projection of haemoglobin. From this starting point in the use of heavy-atom derivatives, eventually Kendrew and his co-workers (Bodo *et al* 1959) were able to produce the first three-dimensional map for myoglobin.

The way in which proteins can accommodate groups containing heavy atoms is by very local disturbances of the protein molecule. At comparatively short distances from the heavy atoms the relaxation of the molecule will be complete and the atomic positions will be very little different from those in the native protein. In essence, the heavy-atom derivative and the native protein form an isomorphous pair, although the isomorphism is somewhat imperfect.

We have seen that a single pair of isomorphous compounds leads to a phase ambiguity. However, if more than one heavy-atom derivative can be produced with the heavy atoms attached to different parts of the protein then the ambiguity can be broken. This is illustrated in figure 3.7 for the case where, in addition to the native product, two isomorphous heavy-atom derivatives have been prepared. It is assumed that the heavy-atom positions have been found, for example by a difference Patterson map, and the three magnitudes of the structure factors of the native product and two IR derivatives are known. Figures 3.7(*a*) and (*b*) show how the structure factors of the IR derivatives, $F(\mathrm{A})$ and $F(\mathrm{B})$, are formed from the contribution of the native product, $F(\mathrm{N})$, which is known only in magnitude, and that of the heavy atoms $F(\mathrm{HA})$ and $F(\mathrm{HB})$, which are known both in magnitude and phase. Figure 3.7(*c*) shows the

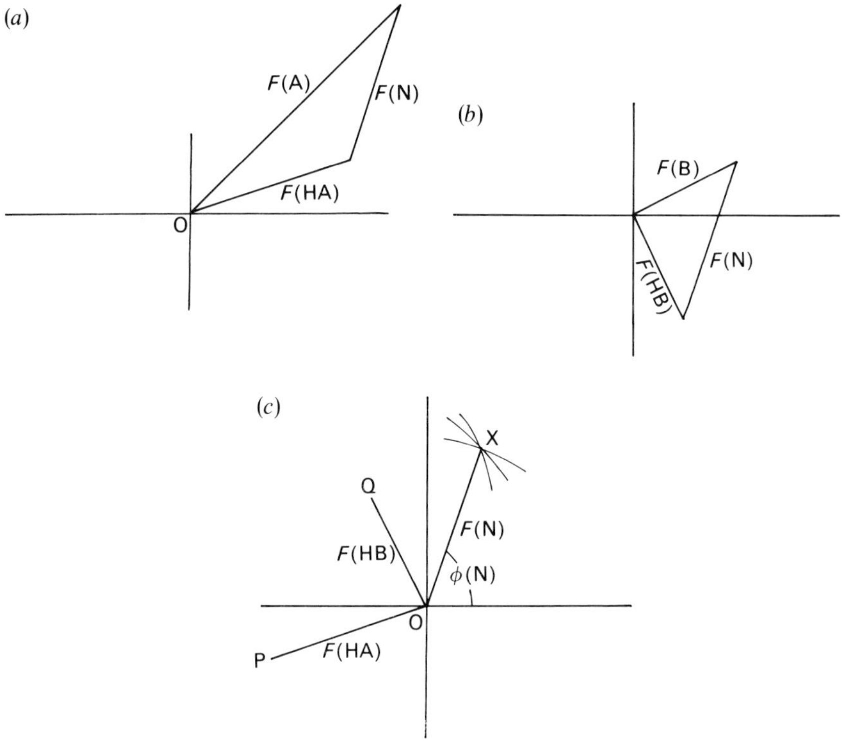

Figure 3.7 (*a*) and (*b*). The structure factors, $F(\mathrm{A})$ and $F(\mathrm{B})$, of two derivatives in terms of the contributions of the heavy-atom residues, $F(\mathrm{HA})$ and $F(\mathrm{HB})$, and the contribution of the native product, $F(\mathrm{N})$. (*c*) A reconstruction to give the phase of $F(\mathrm{N})$, $\phi(\mathrm{N})$.

construction by which $\phi(\mathrm{N})$, the phase of $F(\mathrm{N})$, may be found. First $F(\mathrm{HA})$ and $F(\mathrm{HB})$ are drawn in the way shown. A circle of radius $F(\mathrm{N})$ is drawn centred on the origin and circles of radii $F(\mathrm{A})$ and $F(\mathrm{B})$ are drawn centred on the points P and Q. A trisection of the three circles at X gives OX representing $F(\mathrm{N})$ both in magnitude and phase.

Closure to give a perfect trisection would be unusual due to the difficulty of collecting high quality protein data and the uncertain background contribution of the solvent. Another factor is the imperfection of the IR condition. Atoms in the derivative are shifted from the native protein positions; this has a greater effect the higher the scattering angle, so phase determination can only be carried out to a limited resolution. To overcome at least some of these drawbacks it is better to use as many heavy-atom derivatives as possible. One is then looking for a point in the diagram close to

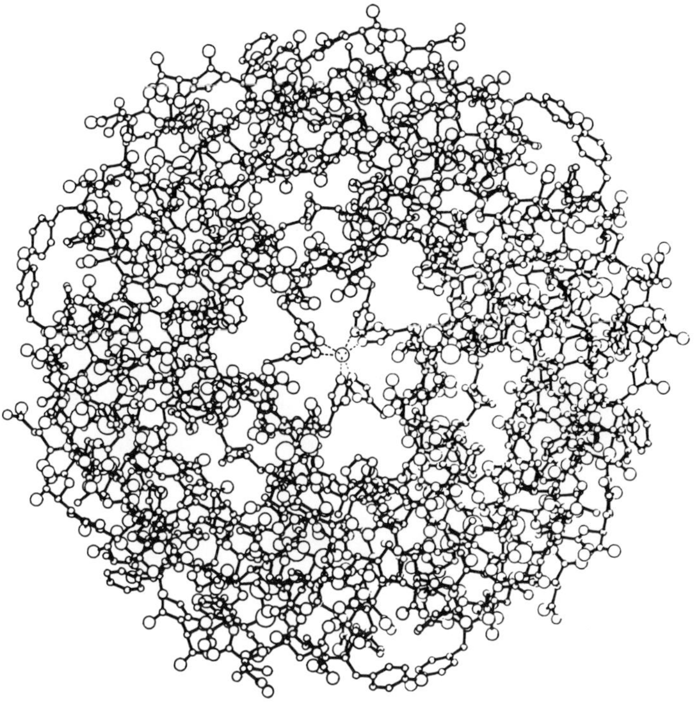

Figure 3.8 A molecule of 2 Zn-insulin.

many circles centred on many different points, although in practice computer-based analytical techniques are used rather than diagrams.

While there is still great scope for ingenuity in the solution of protein structures, and they all have their own idiosyncratic features, it is probably fair to say that at the present time if crystals of the right kind can be produced then the solution of a protein structure can be fairly routine, although requiring great effort and good experimental technique. The main interest these days is in the interpretation of protein structures in terms of their biological functions. In figure 3.8 a projection of a derivative of insulin is shown, the function of which in relation to the transport of glucose within the body has been intensively studied.

Even larger structures, bordering on the living, are now being subjected to x-ray analysis. In 1935 a crystal of Tobacco Mosaic Virus (TMV) was made and Franklin (1956) was able to produce a very low resolution image of the structure. At the present time investigation is in progress to find atomic arrangements in the protein coat of the virus.

Of all the triumphs of x-ray crystallography the ones which have justifiably

W L Bragg (1890–1971) and W H Bragg (1862–1942)

been given the greatest attention are those associated with macromolecular work. Nothing is more certain than that other triumphs are still to come.

3.5 DIRECT METHODS

1948 to 1954

In 1948 the first tentative steps were taken in the development of mathematically based methods, referred to as direct methods, for solving crystal structures. In these methods the phases (or signs) of structure factors are linked by relationships of some kind from which phases (or phase estimates) are found.

From the theoretical point of view it turns out to be better not to use the usual structure factors but to use structure factors for which the systematic fall-off in average magnitude with scattering angle has been removed. The resultant structure factors are then scaled such that the unitary structure factor satisfies the relationship

$$|U(\boldsymbol{h})| \leqslant 1. \qquad (3.13)$$

The unitary structure factor is the ratio of the actual structure factor to the maximum possible magnitude of a structure factor at that scattering angle, which would occur if all the atoms scattered in phase. Harker and Kasper (1948) derived a number of inequality relationships which could give information about the signs of structure factors for centrosymmetric structures. One example which is valid for all centrosymmetric space groups is

$$U(\boldsymbol{h})^2 < \tfrac{1}{2}[1 + U(2\boldsymbol{h})]. \qquad (3.14)$$

As an example of the use of equation (3.14) we consider $U(\boldsymbol{h}) = U(3, 2, 1) = 0.6$ and $U(2\boldsymbol{h}) = U(6, 4, 2) = 0.5$. The left-hand side of the equation equals 0.36 no matter what is the sign of $U(\boldsymbol{h})$. However, if $U(2\boldsymbol{h})$ is negative, then the right-hand side equals 0.25 which violates the inequality. Hence it is deduced that the sign of $U(2\boldsymbol{h})$, $s(2\boldsymbol{h})$, must be positive.

The drawback with the inequality relationships is that they are only useful for structures with few atoms. If there are N equal atoms in the unit cell then

$$(\overline{U^2}) = 1/N \qquad (3.15)$$

and one should not expect more than a few U values with magnitudes greater than three times the RMS value of U. Thus, with $N = 64$ the largest U will be 0.38 and inequality relationships are unlikely to be applicable.

A further important development of inequality relationships was made by Karle and Hauptman (1950) whose contributions in the field of direct methods led to the award of the Nobel prize for Chemistry in 1985.

From the condition that electron density must be positive everywhere to be physically realistic, they derived the inequality relationship

$$\begin{vmatrix} 1 & U(\boldsymbol{h}_1) & U(\boldsymbol{h}_2) & \ldots & U(\boldsymbol{h}_n) \\ U(\bar{\boldsymbol{h}}_1) & 1 & U(\boldsymbol{h}_2-\boldsymbol{h}_1) & \ldots & U(\boldsymbol{h}_n-\boldsymbol{h}_1) \\ U(\bar{\boldsymbol{h}}_2) & U(\boldsymbol{h}_1-\boldsymbol{h}_2) & 1 & & U(\boldsymbol{h}_n-\boldsymbol{h}_2) \\ \vdots & & & & \\ U(\bar{\boldsymbol{h}}_n) & U(\boldsymbol{h}_1-\boldsymbol{h}_n) & U(\boldsymbol{h}_2-\boldsymbol{h}_n) & \ldots & 1 \end{vmatrix} \geqslant 0. \qquad (3.16)$$

The determinant can be of any order but the leading column (or row) must consist of U values with different indices.

One simple inequality we may choose is

$$\begin{vmatrix} 1 & U(\boldsymbol{h}) & U(2\boldsymbol{h}) \\ U(\bar{\boldsymbol{h}}) & 1 & U(\boldsymbol{h}) \\ U(2\bar{\boldsymbol{h}}) & U(\bar{\boldsymbol{h}}) & 1 \end{vmatrix} \geqslant 0 \qquad (3.17)$$

which, for centrosymmetric structure, gives

$$U(\boldsymbol{h})^2 \leqslant \tfrac{1}{2}[1+U(2\boldsymbol{h})]$$

which we have already seen as a Harker–Kasper inequality.

An important step in the development of direct methods was the simultaneous appearance of three papers by Sayre (1952), Cochran (1952) and Zachariasen (1952). The one by Sayre introduced an exact equation between structure factors which is now referred to as Sayre's equation. The basis of this equation is that for a resolved equal-atom structure the square of the electron density, $\rho(\boldsymbol{r})^2$, will look like $\rho(\boldsymbol{r})$ but with differently shaped peaks. By the application of Fourier transform theory Sayre found

$$F(\boldsymbol{h}) = \frac{K(\boldsymbol{h})}{V} \sum_{\boldsymbol{k}} F(\boldsymbol{k}) F(\boldsymbol{h}-\boldsymbol{k}) \qquad (3.18)$$

where $K(\boldsymbol{h})$ is known. For a centrosymmetric structure it might be assumed that a large product of structure factors on the right-hand side will probably have the same sign as $F(\boldsymbol{h})$. This leads to the relationship that, for large structure factors,

$$s(\boldsymbol{h})s(\boldsymbol{k})s(\boldsymbol{h}-\boldsymbol{k}) \qquad (3.19)$$

probably equals $+1$. This was the sign relationship given by Cochran and Zachariasen in their 1952 papers. An expression for the probability of positive product of signs was given for equal-atom structures by Woolfson (1954) in the form

$$P_+(\boldsymbol{h},\boldsymbol{k}) = \tfrac{1}{2} + \tfrac{1}{2}\tanh(N|U(\boldsymbol{h})U(\boldsymbol{k})U(\boldsymbol{h}-\boldsymbol{k})|). \qquad (3.20)$$

More general probability expressions for the non-equal atom cases have been given by Cochran and Woolfson (1955), Bertaut (1955) and Klug (1958).

The sign relationship (3.19) was used by Zachariasen (1952) to solve the structure of metaboric acid. By the use of inequality relationships he was able to express the signs of forty structure factors in terms of five sign symbols, a, b, c, d, e, each of which was either $+1$ or -1. He then applied sign relationships to extend sign information and to find relationships between the symbols. He used the expression

$$s(\boldsymbol{h}) \approx s \sum_{\boldsymbol{k}} s(\boldsymbol{k}) s(\boldsymbol{h}-\boldsymbol{k}) \tag{3.21}$$

to combine indications from several pairs of signs to find a sign for $U(\boldsymbol{h})$. Thus, if there were seven contributors to the right-hand side of (3.21)

$$a \quad a \quad a \quad a \quad ab \quad ab \quad ab$$

then values of $b=+1$ and $s(\boldsymbol{h})=a$ would be indicated. If there were independent indications that $b=+1$ then these conclusions could be accepted with some confidence. In this way a sufficient body of sign information was established to compute a trial electron-density map from which the positions of the atoms could be found.

1955 to 1964

The first important tool for dealing with non-centrosymmetric structures was developed by Cochran (1955). This was in the form of a probability distribution for the combination of three phases

$$\Phi(3,\boldsymbol{h},\boldsymbol{k}) = \phi(\boldsymbol{h}) - \phi(\boldsymbol{k})\phi(\boldsymbol{h}-\boldsymbol{k}). \tag{3.22}$$

The general form of the distribution is

$$P[\Phi(3, \boldsymbol{h}, \boldsymbol{k})] = \frac{1}{2\pi I_0(\kappa)} \exp\{-\kappa \cos[\Phi(3, \boldsymbol{h}, \boldsymbol{k})]\} \tag{3.23}$$

where for the equal-atom case,

$$\kappa = 2N|U(\boldsymbol{h})U(\boldsymbol{k})U(\boldsymbol{h}-\boldsymbol{k})|. \tag{3.24}$$

The form of the distribution is shown in figure 3.9. If κ is large then the distribution can be sharply peaked at 0 (modulo 2π). The sum of three phases in (3.22) is probably close to 0 (modulo 2π) leading to the result

$$\phi(\boldsymbol{h}) \approx \phi(\boldsymbol{k}) + \phi(\boldsymbol{h}-\boldsymbol{k}) \tag{3.25}$$

which means that if the two phases on the right-hand side are known then an estimate for $\phi(\boldsymbol{h})$ may be found.

Before describing the next step in direct methods it should be said that during the late 1950s, when several important contributions were being made by J Karle and H Hauptman, the unitary structure factor, which is convenient in dealing with inequality relationships, was deposed by the normalised

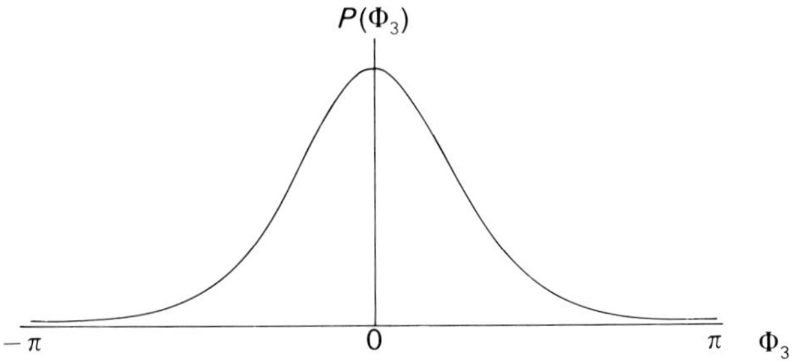

Figure 3.9 The Cochran distribution for $\kappa=2$.

structure factor, which is better for dealing with probabalistic relationships. This latter is given for the equal-atom case, by

$$E(\boldsymbol{h})=N^{1/2}U(\boldsymbol{h}) \tag{3.26}$$

and satisfies the condition

$$\langle|E|^2\rangle=1. \tag{3.27}$$

For example, in terms of normalised structure factors (3.24) may be rewritten

$$\kappa=2N^{-1/2}|E(\boldsymbol{h})E(\boldsymbol{k})E(\boldsymbol{h}-\boldsymbol{k})|. \tag{3.28}$$

The advantage of the normalised structure factor is that the theoretical distribution of E is independent of N, the number of atoms in the unit cell. From the form of (3.28) it will be seen that large values of κ are increasingly unlikely as N increases, so that, for large structures, relationships such as (3.25) will be unreliable.

If there are several pairs of phases whose sum gives an indication of $\phi(\boldsymbol{h})$, then the aggregated information should be more reliable than from any one pair alone. Karle and Hauptman (1956) developed the tangent formula which shows how to combine several indications. This is

$$\tan\phi(\boldsymbol{h})=\frac{\sum_{\boldsymbol{k}}|E(\boldsymbol{k})E(\boldsymbol{h}-\boldsymbol{k})|\sin[\phi(\boldsymbol{k})+\phi(\boldsymbol{h}-\boldsymbol{k})]}{\sum_{\boldsymbol{k}}|E(\boldsymbol{k})E(\boldsymbol{h}-\boldsymbol{k})|\cos[\phi(\boldsymbol{k})+\phi(\boldsymbol{h}-\boldsymbol{k})]} \tag{3.29}$$

where $\phi(\boldsymbol{h})$ is the best estimate of the phase of $E(\boldsymbol{h})$. By multiplying the top and bottom of (3.29) by $2N^{-1/2}|E(\boldsymbol{h})|$ it can be rewritten as

$$\tan\phi(\boldsymbol{h})=\frac{\sum_{\boldsymbol{k}}\kappa(\boldsymbol{h},\boldsymbol{k})\sin[\phi(\boldsymbol{k})+\phi(\boldsymbol{h}-\boldsymbol{k})]}{\sum_{\boldsymbol{k}}\kappa(\boldsymbol{h},\boldsymbol{k})\cos[\phi(\boldsymbol{k})+\phi(\boldsymbol{h}-\boldsymbol{k})]} \tag{3.30}$$

or

$$\phi(\boldsymbol{h})=\text{phase of}\left\{\sum_{\boldsymbol{k}}\kappa(\boldsymbol{h},\boldsymbol{k})\exp\mathrm{i}[\phi(\boldsymbol{k})+\phi(\boldsymbol{h}-\boldsymbol{k})]\right\}$$

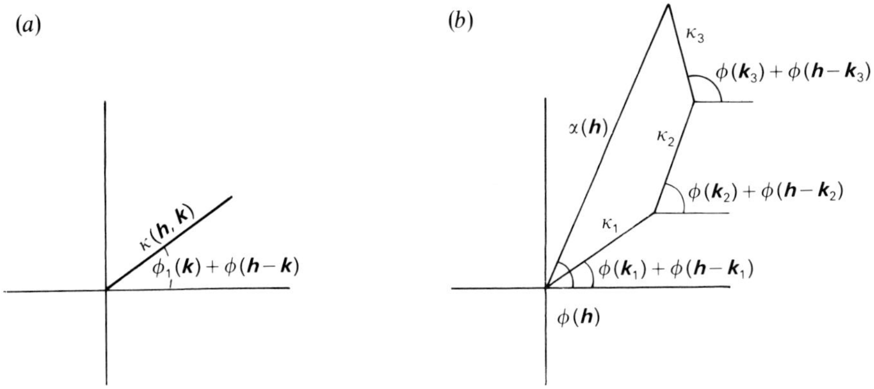

Figure 3.10 (*a*) The indication of $\phi(\boldsymbol{h})$ from a single phase relationship. (*b*) The indications of $\phi(\boldsymbol{h})$ from several phase relationships.

where

$$\kappa(\boldsymbol{h},\boldsymbol{k})=2N^{-1/2}\left|E(\boldsymbol{h})E(\boldsymbol{k})E(\boldsymbol{h}-\boldsymbol{k})\right|. \tag{3.31}$$

The form (3.31) enables us to envisage the tangent formula in a graphical way. In figure 3.10(*a*) the relationship (3.25) is illustrated as a complex quantity with argument $\phi(\boldsymbol{k})+\phi(\boldsymbol{h}-\boldsymbol{k})$ and modulus $\kappa(\boldsymbol{h},\boldsymbol{k})$. Similarly, in figure 3.10(*b*), expression (3.31) is seen as the sum of several such complex quantities and the resultant, with modulus $\alpha(\boldsymbol{h})$, has an argument which is the best estimate of $\phi(\boldsymbol{h})$.

At this stage the means were at hand for solving non-centrosymmetric structures but, surprisingly, it was eight years before the first such structure was solved using phase relationships. A development of the Zachariasen method using symbols had been made by I L Karle and J Karle; in this the necessity of an inequality-derived base was removed. Shortly afterwards they extended the technique to non-centrosymmetric structures and reported their first success in an historic paper (Karle and Karle 1964). The structure they solved was L-arginine dihydrate with space group $P2_12_12_1$. In table 3.1 the starting point of their solution is shown.

The first three reflections, whose phases are restricted if the origin of the unit cell is chosen according to the normal conventions, are given specific values and this corresponds to a choice of one or other of the allowed origins. The other four reflections are given symbolic values; for the first three the possible phases are 0 or π but the last one can have any value in the range 0 to 2π.

The first few steps of the symbolic procedure are illustrated in table 3.2. The first three steps each give a new phase indication from a single pair of 'known' phases. Step 4 gives two consistent indications but step 7 gives indications $\pi+m$ and 0. The stronger of the two indications is the first one, and this is carried forward, but there is also an indication that $m=\pi$. An accumulation of

Table 3.1 Starting point for the solution of L-arginine dihydrate by symbolic addition.

h	k	l	$\|E(h\ k\ l)\|$	Symbol or phase	
3	0	10	3.46	0	Fix origin
3	3	0	2.17	$-\pi/2$	Fix origin
3	0	1	2.77	$\pi/2$	Fix origin
2	12	0	3.21	p	0, π
2	10	0	2.31	s	0, π
4	0	14	2.56	m	0, π
3	8	3	2.31	a	General

Table 3.2 The first few steps of the symbolic-addition procedure for L-arginine dihydrate.

1.	3	0	1	$\pi/2$	5.	3	0	−10	π
	3	3	0	$-\pi/2$		2	3	13	$\pi+m$
	6	3	1	0		5	3	3	m
2.	−3	0	10	π	6.	6	3	1	0
	6	3	1	0		−5	−3	3	m
	3	3	11	π		1	0	4	m
3.	−4	0	14	m		−3	0	−10	0
	6	3	−1	π		4	0	14	m
	2	3	13	$\pi+m$		1	0	4	m
4.	−3	0	−1	$-\pi/2$	7.	5	3	−3	m
	3	3	11	π		−3	−3	11	π
	0	3	10	$\pi/2$		2	0	8	$\pi+m$
	3	3	0	$-\pi/2$		1	0	4	m
	−3	0	10	π		1	0	4	m
	0	3	10	$\pi/2$		2	0	8	0

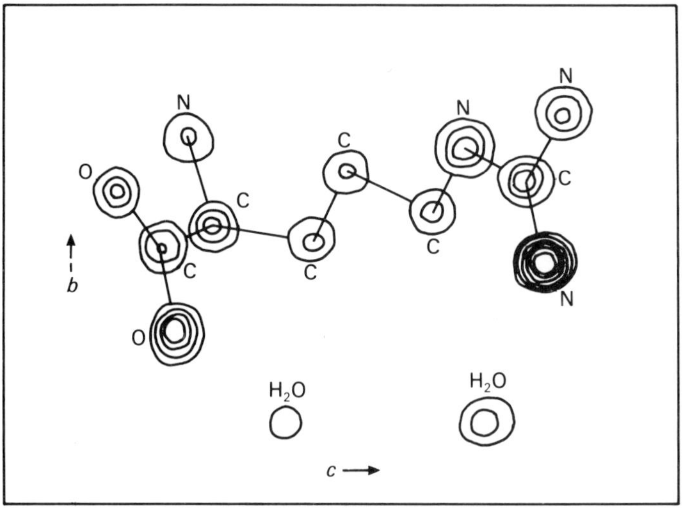

Figure 3.11 The first *E*-map for L-arginine dihydrate.

indications $m=\pi$ would eventually lead to the acceptance of that value for m.

At the end of the symbolic addition process several hundred phases may either be expressed as symbols or have explicit values. If some symbols are not indicated as having particular values then trial values can be alloted to them. The set of phases is then inserted in the tangent formula and the phases refined to self-consistency. This will occur when each phase has the value indicated by the right-hand side of equation (3.31).

When a set of phases has been found then an *E*-map is calculated. This is a Fourier synthesis like that in (3.4) but with coefficients $E(\boldsymbol{h})$. The *E*-map obtained by Karle and Karle for L-arginine-dihydrate with 400 phases is shown in figure 3.11.

Symbolic addition is still occasionally used and has solved some very complicated crystal structures. However, improvements in computers and their greater availability have given a preference for other types of direct method which are more easily automated.

1965–1974

During the late 1960s many structures, some quite large, were solved by symbolic addition, particularly by the Karles. They also did an interesting piece of theoretical work in deriving the variance of the Cochran distribution in terms of κ (Karle and Karle 1966). Their result is illustrated in figure 3.12; the variance shown will also represent the uncertainty in the estimate of a phase derived from (3.25) although this will be on the assumption that the precise values of the phases on the right-hand side are known. Where a phase is estimated from the tangent formula, again assuming precise knowledge of

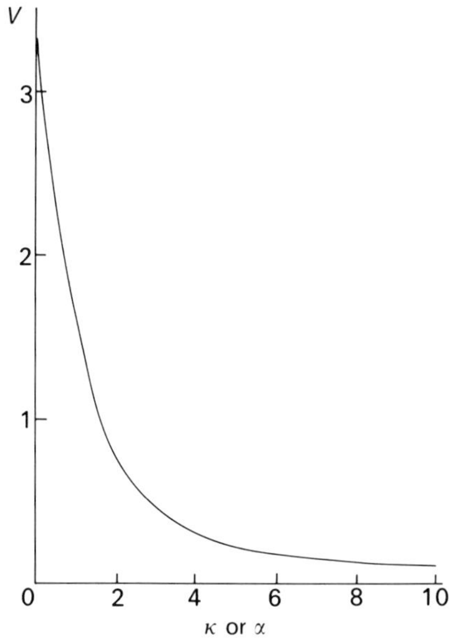

Figure 3.12 The variance of a phase indication as a function of κ or α.

phases on the right-hand side of (3.31), the variance of the estimate is given by the same function but with $\alpha(\boldsymbol{h})$, shown in figure 3.10(*b*), replacing κ.

In table 3.2 it was noted that at step 7 there were two estimates for a phase but that these could not be combined. It is not permitted to take an average of estimates given in different symbolic form. For example, if there are two estimates, $a+b$ and $c+d$, and $a=b=\pi/2$ and $c=d=-\pi/2$, then the two estimates, π and $-\pi$, are the same. However, an average, $\frac{1}{2}(a+b+c+d)$, equals zero, which is as far wrong as it possibly can be.

The early stages of phase development, starting with a small base, are very critical and as many relationships as possible should be involved. For this reason Germain and Woolfson (1968) suggested an alternative approach in which explicit phases rather than symbols were used throughout so that the tangent formula could always be used to combine estimates. The method was a multisolution one which developed and refined phases by the tangent formula from several different starting points. Applying this approach to the L-arginine dihydrate example in table 3.1, the symbols p, s, and m would be replaced by the pairs of values $(0, \pi)$ and the general reflection by 'quadrant' values $(\pi/4, 3\pi/4, 5\pi/4, 7\pi/4)$. This would give $2\times2\times2\times4=32$ starting points, each of which would be developed to give a set of several hundred estimated phases. It turns out that taking four quadrant values for the general phase is sufficient; the greatest possible error with one or other of the values is $\pi/4$ and an expected error is $\pi/8$.

This concept has been developed and extended to give the computer-based direct method MULTAN (Multiple Tangent Formula Method). The various stages in MULTAN are as follows.

(i) Raw observed intensities plus basic information about the space group, cell dimensions and cell contents (which need not be known accurately) are processed to give a complete set of E values. Taking account of the structural complexity, a number of reflections (usually 2–500), say M, are chosen in terms of which the structure is to be solved.

(ii) All the triple-phase relationships linking the M reflections are found. This number, Q, is usually in the range 3000–10 000.

(iii) The M reflections and Q relationships are analysed to find a set which will give a strong pathway for phase development by the tangent formula. The process by which this is done, known as CONVERGENCE, has been described by Germain, Main and Woolfson (1970).

(iv) For each starting set of phases, obtained by permutation as described for L-arginine dihydrate, more phase estimates are developed by a modified tangent formula (Germain *et al* 1971) which incorporates weights, namely

$$\phi(\boldsymbol{h}) = \text{phase of} \left\{ \sum_{\boldsymbol{k}} w(\boldsymbol{k}) w(\boldsymbol{h}-\boldsymbol{k}) \kappa(\boldsymbol{h}, \boldsymbol{k}) \exp \mathrm{i}[\phi(\boldsymbol{k}) + \phi(\boldsymbol{h}-\boldsymbol{k})] \right\}. \tag{3.32}$$

As each phase is estimated, so is an associated weight which indicates the reliability of the estimate. This weight increases with the magnitude of the right-hand side, reaching unity when the magnitude is five.

(v) Each set of phases is tested by being inserted in some figure of merit (FOM) functions. These are easily calculable functions such that a set of phases leading to an E-map with desirable features (e.g. equal peaks in a non-negative background) will have an extremum value. There are three main FOM functions and the values of those for a particular phase set are combined to give a combined figure of merit, CFOM. The values of CFOM for all the phase sets lie between 0 and 3 and the higher the value the more plausible is the set of phases.

(vi) The set of phases with the highest value of CFOM, or any other chosen by the user, is used to compute an E-map. The peaks of the map are found by a peak-search routine and from some simple rules of chemistry, which are included in the programme, the peaks are interpreted in terms of chemically realistic fragments.

(vii) A favourable projection, which spreads the peaks out as far as possible, is found and the projected peaks are output on the line printer in the correct configuration together with a table indicating possible bonds together with bond lengths and angles.

(viii) This process often reveals the complete structure but if only a part of it is found then MULTAN has a facility for processing the fragment until all the structure is revealed.

A typical outcome for a fairly simple molecule is shown in figure 3.13(*a*) where almost the complete structure is revealed. Figure 3.13(*b*) shows a

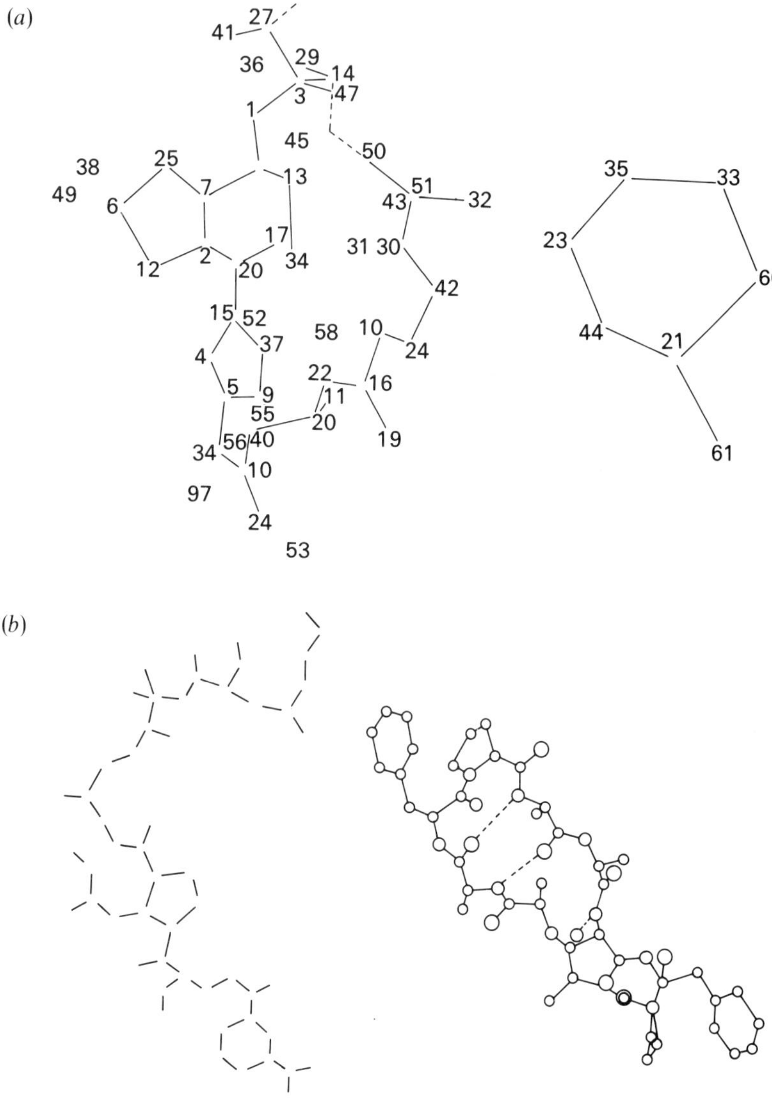

Figure 3.13 (*a*) A line printer output for Factor M, a macrocyclic lactone $C_{28}H_{35}N_7O_7$ plus one dioxane, $C_4H_8O_2$. (*b*) A fragment found for gramiciden S and a drawing of the complete molecule.

fragment found for gramiciden S (Hull *et al* 1978) and a drawing of the complete molecule.

By 1976 MULTAN had become established as the main tool for solving moderately sized structures without heavy atoms. Where heavy atoms are present then Patterson maps are often used, although MULTAN is quite effective in such cases.

1975 to 1985

The main improvements in direct methods since 1975 have been in the tactics of using the earlier theoretical results. Thus White and Woolfson (1975) introduced an idea they called magic integers which may be regarded as an extension of the symbolic-addition concept but which leads systematically to a multisolution approach.

Consider n reflections with phases, in cycles, ϕ_j ($j=1$ to n). The magic-integer approach is to say that, for some value of x in the range 0 to 1,

$$\phi_j \approx m_j x \text{ (modulo 1)}$$

where the m values are integers. An example will illustrate this principle. With $\phi_1=0.3$, $\phi_2=0.2$ and $\phi_3=0.7$ and $m_1=4$, $m_2=6$ and $m_3=7$, then for $x=0.538$

$$m_1 x \text{ (modulo 1)}=0.152$$

$$m_2 x \text{ (modulo 1)}=0.228$$

$$m_3 x \text{ (modulo 1)}=0.766$$

giving errors of representation 53°, 10° and 24°.

Magic-integer sequences may be found for more than three phases and Main (1977) has given rules for generating efficient sequences of any size. For a given integer sequence a number of variables may be used so that the elements of the matrix

$$\begin{pmatrix} 8 \\ 12 \\ 14 \\ 15 \end{pmatrix} (x\ y\ z) \tag{3.33}$$

may be used to represent twelve phases.

Since phases may be represented in magic-integer form so also may three-phase relationships. With $\phi_1=8x$, $\phi_2=14x$ and $\phi_3=12y$ a relationship as produced by MULTAN in the form

$$\phi_1-\phi_2+\phi_3+0.5\approx 0 \text{ (modulo 1)} \tag{3.34}$$

becomes

$$-6x+12y+0.5\approx 0 \text{ (modulo 1)}. \tag{3.35}$$

In general a relationship would appear in the form

$$Hx+Ky+Lz+b \approx 0 \text{ (modulo 1)} \tag{3.36}$$

and there will be several, say M, relationships linking the magic-integer-represented phases. If we look for peaks in the function

$$\psi(x, y, z) = \sum_{r=1}^{M} \kappa_r \cos[2\pi(H_r x + K_r y + L_r z + b_r z)] \tag{3.37}$$

then their values of (x, y, z) substituted in the matrix (3.33) will give trial phases for a dozen reflections. These will satisfy reasonably well the relationships which interlink them plus other reflections in the starting set—for example, those fixing the origin.

The magic-integer concept has been developed so that the number of phases represented in magic-integer form can be 40 or more. These may be linked by 150 or more relationships and some 100 peaks may be selected from the ψ map to give starting points for phase extension and refinement by the tangent formula. Programs based on this idea, MAGIC (Declercq *et al* 1975) and MAGEX (Zhang and Woolfson 1982), have proved to be very powerful and can solve structures where, because of unfavourable circumstances, MULTAN has not succeeded.

Another quite significant advance in the application of direct methods was made almost by accident. As an alternative to the tangent formula Woolfson (1977) suggested a least-squares process for refining phases which led to a cyclic refinement process with linear equations. Tests have shown that this method of refinement is always at least as good as the tangent formula and is usually much better.

With any refinement process the radius of convergence is always a matter of interest. Tests were carried out with a known structure by applying random errors with gradually increasing RMS values to the correct phases. With RMS errors $<60°$ nearly correct phase sets were recovered every time. The surprising result was that with RMS errors of 75° correct phases could be recovered six times out of ten and with 90° four times out of ten. The question naturally arose of what would have happened with random phases; tests indicated that substantially correct phases could still be recovered for a high proportion of random starting sets.

This strange and unexpected result led to the design of a new direct-methods procedure, YZARC, the steps of which are as follows.

(i) Run MULTAN up to and including the CONVERGENCE procedure.

(ii) Select a number, P (default value 100), of the reflections at the bottom of the CONVERGENCE map and the three-phase relationships which link them.

(iii) A pseudo-random generator is used to produce M sets of P random phases in cycles, which are cyclically refined by the least-squares process.

(iv) Each set of phases is extended by the tangent formula, figures of merit are calculated and the MULTAN procedure followed thereafter.

The current version of YZARC does contain features additional to those described above, in particular the inclusion of an efficient weighting scheme, but the essence of the procedure is as described.

Both MAGIC and YZARC will solve structures which cannot be solved by routine runs of MULTAN. These two newer methods deal simultaneously with a large number (40–100) of phases and the relationships which link them. This is the main way they differ from MULTAN which begins with a much smaller starting set which is then developed by a step by step process using very few relationships (even one) at each step. For complicated structures these first few steps are very uncertain and may often give a completely wrong pattern of *relative* phases no matter what are the phases in the starting set. Where this occurs MULTAN is not a multiple-solution method so much as a multiple wrong-solution method.

The process of extending the starting set and using many relationships was taken to the limit by Yao (1981) in his procedure RANTAN. In this procedure all phases to be determined are given initial random values and also assigned weights, usually 0.25, although some phases which are used to fix the origin etc are given different and higher weights. A process of tangent-formula refinement is then commenced but a phase is not changed until its calculated weight (see equation (3.32)) is greater than the initial weight. Thereafter the phase is allowed to follow its refinement path.

Tests indicate that this is a most powerful procedure. A structure solved with its assistance is shown in figure 3.14. It has the advantage of using the tangent formula and is therefore considerably faster to run on a computer than is YZARC.

3.6 ANOMALOUS SCATTERING

When the frequency of the x-rays is close to the natural frequency of some of the inner electrons, that is near an x-ray absorption edge, then resonance (anomalous) scattering takes place. The scattering factor of the atom has additional components so that

$$f = f_0 + f_1 + \mathrm{i} f_2 \qquad (3.38)$$

where f_0 is the normal scattering factor and f_1 and f_2 are the additional real and imaginary components due to the anomalous scattering.

An important consequence of anomalous scattering is that it leads to a breakdown of Friedel's law. For example, if we write $f_c = f_0 + f_1$ then

$$F(\boldsymbol{h}) = \sum_{j=1}^{N} (f_{cj} + \mathrm{i} f_{2j}) \exp(2\pi \mathrm{i} \boldsymbol{h} \cdot \boldsymbol{r}_j) \qquad (3.39)$$

and

$$F(-\boldsymbol{h})=\sum_{j=1}^{N}(f_{cj}+\mathrm{i}f_{2j})\exp(-2\pi\mathrm{i}\boldsymbol{h}\cdot\boldsymbol{r}_j). \qquad (3.40)$$

Here, $F(-\boldsymbol{h})$ is no longer the complex conjugate of $F(\boldsymbol{h})$, as it is with normal scattering, and hence in general $|F(-\boldsymbol{h})|\neq|F(\boldsymbol{h})|$.

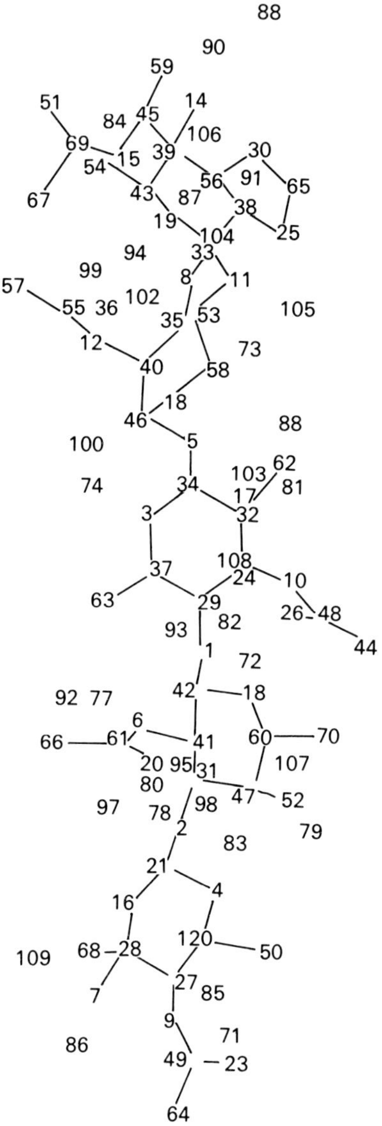

Figure 3.14 A RANTAN solution of avileurekauose, $C_{46}O_{27}H_{70}$.

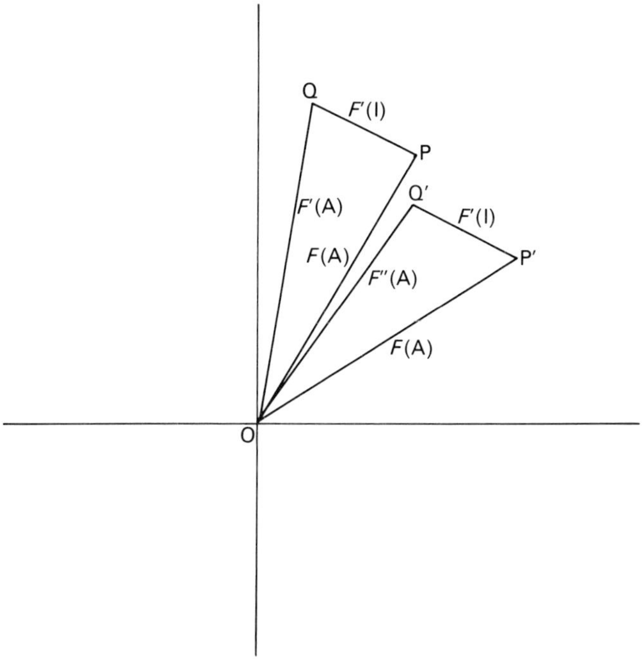

Figure 3.15 Isomorphous replacement gives the alternative structure factors OP and OP′, each of magnitude F(A). With anomalous scattering F'(I) is added to each of them giving OQ and OQ′ of different magnitudes.

Anomalous scattering information can be used in a variety of ways and new ways are constantly being found. For example, it provides a means of resolving the ambiguity of phase when the single isomorphous replacement method, described in §3.3, is used. In figure 3.15 we see reproduced that part of the phase-ambiguity diagram, figure 3.6(c), which gave the result

$$\phi(\mathrm{A}) = \phi(\mathrm{I}) \pm \eta. \tag{3.41}$$

Since we have seen that it is possible to find the positions of the IR atoms, then the contribution of these to the structure factor is known. This contribution F'(I) is added to each of the two possible complex structure factors of magnitude F(A) to give F'(A) and F''(A), which are of different magnitudes. The magnitude which agrees with that experimentally observed indicates the correct phase alternative.

Ideas have also been advanced for using anomalous scattering alone for phase determination, sometimes with data from several wavelengths. For example, Woolfson (1984) has considered the case where a structure contains several anomalous scatterers of the same kind and where data are taken close to the absorption edge and also well away from it so that anomalous scattering effects are negligible.

Measurements of $|F(\boldsymbol{h})|$, the structure factor without anomalous scattering, and $|F_{\mathrm{A}}(\boldsymbol{h})|$ and $|F_{\mathrm{A}}(-\boldsymbol{h})|$, the structure factors with anomalous scattering, are required. Two quantities are found

$$\Delta(\boldsymbol{h})=\left(\frac{|F_{\mathrm{A}}(\boldsymbol{h})|^2}{|F(\boldsymbol{h})|^2}\right)-1 \tag{3.42}$$

and

$$\Delta(-\boldsymbol{h})=\left(\frac{|F_{\mathrm{A}}(-\boldsymbol{h})|^2}{|F(\boldsymbol{h})|^2}\right)-1. \tag{3.43}$$

For the structure factor of index $\boldsymbol{h}$ the scattering factor for the anomalous scatterer is

$$f_{\mathrm{A}}=f+f'+\mathrm{i}f''$$

and we define $\tan\delta=f''/f'$, a known quantity.

One may then set up two equations

$$g^2+2g\cos(\theta+\delta)-\Delta(\boldsymbol{h})=0 \tag{3.44}$$

and

$$g^2+2g\cos(\theta-\delta)-\Delta(-\boldsymbol{h})=0 \tag{3.45}$$

and, in general, there are two pairs of values (g, θ) satisfying them. The quantity g is the ratio of the magnitude of the contribution of the anomalous scatterers to $|F(\boldsymbol{h})|$. Usually only one of the two values of g is physically realistic and, even where both are possible, one will have a much greater probability of being the true value. In this way the magnitudes of the structure factors of the anomalous scatterers alone may be found and hence the comparatively simple anomalous scatterer structure may be solved—for example by direct methods. With this structure the phase, $\eta(\boldsymbol{h})$ of the anomalous scattering contribution may be found and the phase of $F(\boldsymbol{h})$ is then given by

$$\phi(\boldsymbol{h})=\eta(\boldsymbol{h})+\theta. \tag{3.46}$$

Determining the phases $\phi(\boldsymbol{h})$ is tantamount to solving the overall crystal structure.

The importance of anomalous effects is enhanced by a technical advance; the availability of powerful synchrotron sources of radiation. It is possible to select a required wavelength and so to obtain optimised anomalous scattering effects. The power and collimation of the beams offer other advantages as well. Photographs of diffraction patterns from proteins may be taken in a very short time and record many thousands of reflections in a single exposure. Apart from reducing the sheer tedium of collecting data from proteins, it has other advantages as well. Protein crystals often suffer damage in x-ray beams and many crystals may be needed to collect a full set of data. However, it turns out that damage is not just a function of total exposure—it also has a time-dependent component since secondary damage is caused by radicals diffusing through the crystal structure. The overall effect is that one obtains more data, better data and that much more quickly. The photographs so obtained may be analysed by computer-controlled automatic densitometers so that obtaining a data set for a protein, which thirty years ago required several man-years of effort, now takes only days, at least in principle.

An extensive review on the production, properties and use of synchrotron radiation has been given by Helliwell (1984).

3.7 THE FUTURE

Since the discovery of x-ray diffraction in 1912 there have been enormous advances in the development of techniques of structure analysis. Events over the last decade or so suggest that further advances are in store. The advent of powerful radiation sources, large and fast computers and new methods enable solutions to be found both for moderately small structures, containing up to about 200 independent atoms, and also for large structures, perhaps even viruses. There is, however, a curious gap between these two categories which is still resistant to structure solution. The structures are too large to be tackled by existing methods, such as direct methods, but too small to accommodate a heavy-atom containing group and still retain isomorphism. If such structures contain up to a few atoms capable of giving anomalous scattering, then a structure solution should be possible. Failing that, there must be developed new methods capable of dealing with such structures and some progress is being made in this direction (e.g. Debaerdemaeker *et al* 1985).

Another direction in which the first tentative steps are being taken is in obtaining phase information directly from experiment. When a crystal is so positioned that two diffracted beams are produced simultaneously, then the profile of the diffracted beam can give information about the sign product (3.19) or, more generally, the phase sum (3.22). Post (1979) and Billy *et al* (1984) have reported convincing results for the centrosymmetric and non-centrosymmetric cases respectively.

The physical methods are still in their infancy and the quality of the information provided and the effort required to obtain it do not make them serious challengers to existing methods at present. Nevertheless they may well turn out to be the methods of the twenty-first century.

REFERENCES

Bertaut E F 1955 *Acta Crystallogr.* **8** 544–8
Billy H, Burzlaff H and Hümmer K 1984 *Collected Abstracts Thirteenth International Congress of Crystallography (Hamburg) 9–18 August 1984* Abstract 16.2–7
Bodo G, Dintzis H M, Kendrew J C and Wyckoff H W 1959 *Proc. R. Soc.* A **253** 70–102
Bragg W H 1915 *Phil. Trans. R. Soc.* **215** 253–74
Bragg W H and Bragg W L 1913 *Proc. R. Soc.* A **89** 277–91
Bragg W L 1913 *Proc. R. Soc.* A **89** 248–77
Bragg W L 1939 *Nature* **143** 678
Buerger M J 1950 *Acta Crystallogr.* **3** 87–97
Clews C J B and Cochran W 1948 *Acta Crystallogr.* **1** 4–11
Cochran W 1952 *Acta Crystallogr.* **5** 65–7
Cochran W 1955 *Acta Crystallogr.* **8** 473–8
Cochran W and Woolfson M M 1955 *Acta Crystallogr.* **8** 1–12
Crowfoot D and Bernal J D 1934 *Nature* **133** 794–5
Declercq J P, Germain G and Woolfson M M 1975 *Acta Crystallogr.* A **31** 367–72
Debaerdemaeker T, Tate C and Woolfson M M 1985 *Acta Crystallogr.* A **41** 286–90
Franklin R E 1956 *Nature* **177** 929–30
Germain G, Main P and Woolfson M M 1970 *Acta Crystallogr.* B **26** 274–85
—— 1971 *Acta Crystallogr.* A **27** 368–76
Germain G and Woolfson M M 1968 *Acta Crystallogr.* B **24** 91–6
Green P M, Ingram V M and Perutz M F 1964 *Proc. R. Soc.* A **225** 287–307
Harker D and Kasper J S 1948 *Acta Crystallogr.* **1** 70–5
Helliwell J R 1984 *Rep. Prog. Phys.* **47** 1403–97
Hull S E, Karlsson R, Main P, Woolfson M M and Dodson E J 1978 *Nature* **275** 206–7
Karle I L and Karle J 1964 *Acta Crystallogr.* **17** 835–41
Karle J and Hauptman H 1950 *Acta Crystallogr.* **3** 181–7
—— 1956 *Acta Crystallogr.* **9** 635–51
Kartha G 1961 *Acta Crystallogr.* **14** 680–6
Klug A 1958 *Acta Crystallogr.* **II** 515–43
Main P 1977 *Acta Crystallogr.* A **33** 750–7
Patterson A L 1934 *Phys. Rev.* **46** 372–6
Post B 1979 *Acta Crystallogr.* A **35** 17–21
Sayre D 1952 *Acta Crystallogr.* **5** 60–5
Sim G A 1960 *Acta Crystallogr.* **13** 511–12
White P and Woolfson M M 1975 *Acta Crystallogr.* A **31** 53–6
Woolfson M M 1954 *Acta Crystallogr.* **7** 61–4
—— 1956a *Acta Crystallogr.* **9** 974
—— 1956b *Acta Crystallogr.* **9** 804–10
—— 1977 *Acta Crystallogr.* A **33** 219–25
—— 1984 *Acta Crystallogr.* A **40** 32–4
Yao J-X 1981 *Acta Crystallogr.* A **37** 642–4
Zachariasen W H 1952 *Acta Crystallogr.* **5** 68–73
Zhang S-H and Woolfson M M 1982 *Acta Crystallogr.* A **38** 683–5

H C Sorby (1826–1908) and J W Gibbs (1839–1903)

4 Solid State Physics and Metallurgy

Robert W Cahn

The art of metallurgy has been practised by man for several millennia, but the science of metallurgy has barely celebrated its first centenary. The first tentative stirrings of that science grew from the mineralogy and chemistry of the time, and the involvement of physics was slow in coming. The aim of this chapter is to outline how physics came to contribute, at length, to the science of metals and alloys, how physical metallurgists in turn gave something back to the practice and theory of solid state physics, and to speculate where physical metallurgy is likely to go next.

Most of the early attempts to apply natural philosophy to metals were aimed at identifying the true nature of steels, and especially at finding how they differed from cast and wrought irons—a long and difficult task, as it proved. The precocious experiments of Réaumur in the early eighteenth century were conducted in a philosophic climate unsympathetic to his approach and so did not lead to a science of steels. Real progress was contingent on two later developments, a proper understanding of the chemical elements and an appreciation of the regularities governing the coexistence of phases.

The Russian Chernov in 1868 (following earlier indications from Réaumur's work) first showed that a simple carbon steel, during cooling, passed through several critical temperatures, though he was obliged to use rough, indirect methods to estimate those temperatures. To turn this recognition into proper science required two innovations, one experimental and one theoretical. Le Chatelier, a French industrial chemist, invented the Pt/Pt–Rh thermocouple in 1888 (it is still used today in its original form). Willard Gibbs, a shy academic tucked away at Yale University—a geometrician and mechanical engineer turned thermodynamicist—derived and published the Phase Rule in 1876 in a New England journal. This, while not a major periodical, was nevertheless widely distributed to scientific academics. European scientists, including van der Waals, Jüptner, and Le

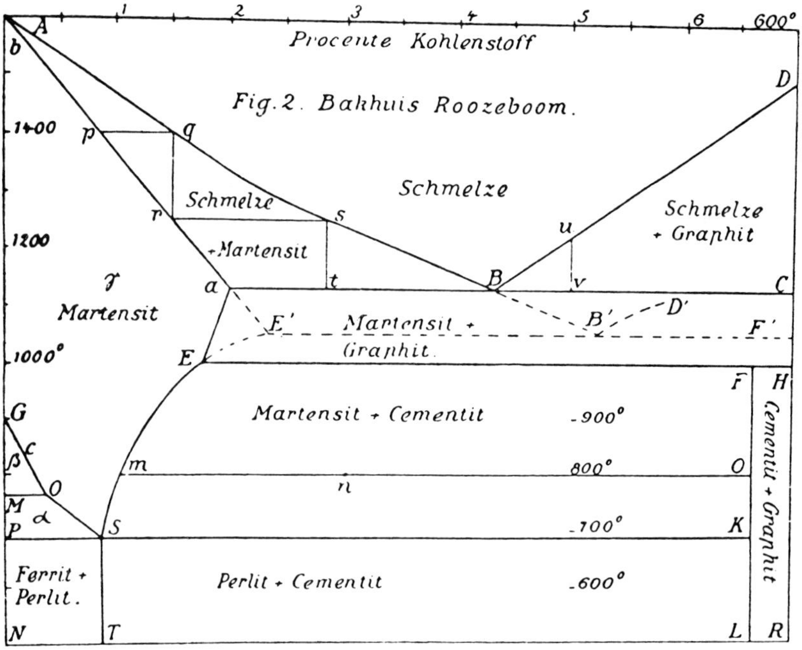

Figure 4.1 Roozeboom's 1901 phase diagram for the iron–carbon system—the first diagram properly based on Gibbs' phase rule.

Chatelier himself, who had access to this periodical, repeatedly rediscovered Gibbs' original papers. In spite of the conceptual difficulty of Gibbs' work, its crucial importance for the understanding of the constitution of alloys in general and steels in particular was eventually recognised. Sauveur, the Harvard metallurgist, measured volume fractions of phases and thereby helped to establish the Phase Rule. In France, Osmond (one of the first scientists to study metallurgy as a recognised discipline) used Le Chatelier's thermocouple with a chronograph to record accurate cooling curves in 1890. The anomalies resulting from the release or absorption of latent heat pinpointed the critical temperatures which Chernov had first shown to exist. In England, Roberts-Austen (who gave his name to austenite) extended these studies. The role of carbon was by now widely recognised.

The next stage was crucial. During the last decade of the century, Sauveur at Harvard and Roberts-Austen at Imperial College plotted the measured critical temperatures as a function of carbon content, and thus produced a rough approximation to a *phase diagram*. To turn their diagrams into the real thing required a proper application of Gibbs' Phase Rule, and this was the achievement of the Dutch physical chemist Roozeboom, who in 1901 published the first proper phase diagram of iron–carbon alloys (figure 4.1). At about the same time, working in isolation but to quite remarkable standards of experimental accuracy, Heycock and Neville in Cambridge determined the copper–tin phase diagram, which they published in 1904. In that same year, the physical chemist Tammann in Göttingen began a long series of determinations of alloy phase diagrams, very numerous, if not highly accurate. These innovations were largely the work of people trained as chemists: this accounts for the fact that metallurgy is still treated as a branch of chemistry in France today!

The phase diagram had now arrived, and with it the first of three vital preconditions of scientific metallurgy.

4.1 THE THREE VITAL INGREDIENTS

It is one of the more remarkable features of the history of physics that a comprehensive, quantitative kinetic theory of gases preceded by several decades the study of crystal structures, in spite of the conceptual subtlety of the methods required to handle the chaotic molecular constitution of a gas. Even the purely mathematical theory of space groups (not dependent on a knowledge of specific crystal structures) was derived some twenty years after Maxwell's kinetic theory of gases. Thus, order was plucked out of gaseous chaos at a time when the understanding of solid order was still in a somewhat chaotic state. Mineralogists, who were a long way ahead of metallurgists in scientific understanding in the second half of the nineteenth century, were still restricted to a study of the outer appearance of minerals. It is thus hardly surprising that scientific metallurgists could not make much progress until they could find out more about the precise nature of crystals. This is well illustrated by the persistent beta-iron controversy, which concerned a particular hypothesis advanced to interpret the hardening of carbon steel on quenching. It was supposed by many that beta-iron, existing above what we now know to be the ferromagnetic Curie temperature of iron, was a crystallographically distinct phase. Some thought that it was a glass! The supposed phase transformation of beta-iron into alpha-iron (existing below the Curie point) was thought to play a major part in the hardening of steel. This controversy absorbed a great deal of energy that might have been better expended elsewhere. There is an important lesson to be learned from the fact that, as long as there was no possibility of directly testing the central hypothesis, great passion was engendered by the controversy! There is also a piquant irony in the fact that the paramagnetic–ferromagnetic transition would indeed now be categorised as a phase transformation by solid state theorists.

The discovery of the diffraction of x-rays by crystals by Laue and Knipping in 1912, leading on to the art of crystal structure determination, was the most crucial contribution by physics to scientific metallurgy in the first half of this century. If we date the origins of this innovation back to Röntgen's discovery of x-rays in 1895, then one can see that the roots of crystal structure determination go back to the same magnificent decade (1895–1905) that included so many other major innovations in physics, and also (as we have seen) the determination of the first alloy phase diagrams. Since 1912 it has been a simple matter to identify the phases that occupy the various fields of a phase diagram.

From 1912 onwards, physics replaced chemistry as the crucial source of scientific insight into the behaviour of metals and alloys. Not at all coincidentally, it was in 1914 that the Australian engineer-turned-metallurgist Walter Rosenhain published the first textbook of scientific metallurgy (he was then working at the National Physical Laboratory near London) and called it *An Introduction to the Study of Physical Metallurgy*. The term 'physical metallurgy' dates from that year.

The third vital ingredient was skill in the microscopy of opaque materials. The undisputed father of this skill was Henry Sorby, an independent scientific investigator, who between 1849 and 1887 devoted himself to the micrographic study of rocks and minerals, marine micro-organisms, and metals and alloys, always working in his native city of Sheffield. He is regarded as the originator of the micrographic study of rock sections by transmitted polarised light, a technique which met great scepticism at first from traditional geologists. De Saussure, the Swiss geologist, cast ridicule on the notion that one could 'look at mountains through a microscope'. Having demonstrated that this could indeed be done, Sorby next showed metallurgists that, as N Higham has put it 'it made sense to look at railway lines through a microscope'.

An important step in the transition from transparent minerals to opaque metals was made via the study of polished sections of metallic meteorites (a kind of cosmic mineral). Nearly a century earlier, the Austrian amateur scientist Alois von Widmanstätten had observed geometrical patterns on polished and etched surfaces prepared on metallic meteorites (figure 4.2). However, such patterns were not published until 1804. Von Widmanstätten had seen them with the naked eye, while Sorby looked at polished meteorites under a microscope arranged to operate by reflected light. (As we now know, some large meteorites have cooled so exceedingly slowly during their space travels that they can consist of single crystals weighing several tons, and the pattern of two phases arranged epitaxially—plates of one phase parallel to specific crystal planes of the other—is then coarse enough to see with the naked eye; in other, smaller meteorites, such Widmanstätten patterns are very much finer and need a microscope for their resolution.) Sorby's meteorites in due course gave way to specimens of heat-treated steels and other alloys, and modern metallography had been born.

Optical metallography came of age in 1887, when Sorby published a detailed account of metallographic researches spanning more than 20 years. This was thus just in time to play a part in the determination of the first phase diagrams. In particular, Heycock and Neville in Cambridge were early

Figure 4.2 Widmanstätten pattern: reproduction of a direct typographical imprint from a polished and etched section of a metallic meteorite (reduced from natural size). (Courtesy of C S Smith.)

masters of this application, which has continued unabated ever since. They were the first to recognise the importance of examining micrographically sections of samples of different compositions which had been quenched from known temperatures (see, for example, figure 4.3); this was the decisive innovation that made possible the accurate determination of complex phase diagrams.

It would be difficult indeed to exaggerate the central position of microscopy in the development of physical metallurgy. To understand this, one can regard the study of metals and alloys as being subject to a tug-of-war, tied at one extreme to chemistry with its concern for composition and chemical bonding, and at the other extreme to physics, involved with atomic structure and with the interplay of energy and matter. Midway between these extremes lie two other subject areas—crystallography, itself ever undecided whether it belongs to chemistry or physics, and *microstructure*.

Figure 4.3 Optical micrographs of Cu–Sn alloys of various compositions quenched from heat-treatment temperatures, polished and etched. (From a paper by Heycock and Neville 1904.)

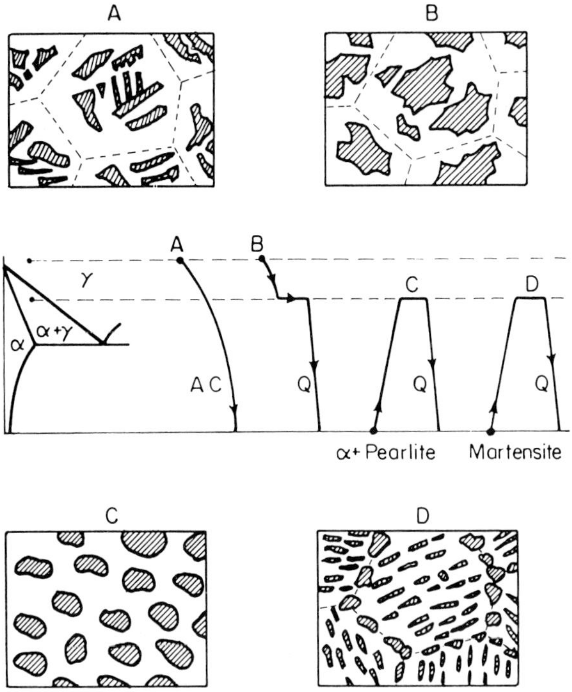

Figure 4.4 Different kinds of microstructure (schematic) attainable by four different heat-treatments of a Fe–0.1 wt% C–0.5 wt% Mn steel. All microstructures are 'dual-phase' and consist of martensite (hatched) dispersed in a ferrite matrix. (AC = air-cooled, Q = water-quenched.) (Courtesy of G Thomas and J Y Koo.)

Microstructure is a scientific category which is the peculiar contribution of physical metallurgy to the study of the solid state. Students of mineralogy, petrology and solid state chemistry are also concerned peripherally with microstructure, but for metallurgists it is central to their understanding of their subject-matter. By extension in recent years, materials science, which incorporates the study not only of metals but also of ceramics, semiconductors and polymers, is likewise built around the study of microstructure, a concept which denotes the geometrical and crystallographic features that define the mutual disposition of phases, including the arrangements of crystal grains of a single phase. A Widmanstätten pattern (figure 4.2)—a pattern which is found in many different alloy systems, not just in meteorites—is a perfect example of what is meant by a microstructure.

To get some feel for the importance of microstructure in the metallurgist's scheme of things, consider figure 4.4. The diagram refers to a novel category of high-strength steels called 'dual-phase steels', first investigated about 10 years ago. Such steels consist of a dispersion of martensite in a matrix of ferrite. Ferrite is the soft, ductile, body-centred cubic form of iron (α) generated by slow diffusion-controlled transformation of the face-centred cubic austenite phase (γ), stable at high temperature. Martensite is the very hard and brittle, crystallographically distorted, form of alpha-iron generated by quenching austenite so that it transforms by a form of shear, far too rapidly for diffusional control. Slow cooling from the γ single-phase field to the $(\gamma+\alpha)$ field creates a limited amount of ferrite, and subsequent quenching turns the residual austenite into martensite. The diagram shows various ways of bringing about this dual-phase microstructure and indicates, schematically, what the resultant microstructures look like.

The resulting mechanical properties depend not only on the relative proportion of the two phases (which are much the same in the four cases shown in figure 4.4) but also on the mean size of the ferrite (matrix) grains, on the size and mean spacing of the martensite dispersoid, and on the 'connectivity' of the martensite particles. It turns out, for instance, that the ductility (limiting plastic deformation before fracture) is greater for microstructure B than for microstructures C or D, because the mean free path for dislocations involved in slip is greater in microstructure B.

It is important to emphasise that a phase diagram only specifies the crystal structures and relative proportions of one, two or more phases that coexist *in equilibrium* for an alloy of specified composition held at a specified temperature. It leaves open the scale and geometry of the microstructure, which depends on the complete thermal history of the specimen. Neither can it predict whether a specimen which has reached a particular point in a phase diagram by a particular thermal route will be in equilibrium or in a *metastable* condition, in the sense that the proportions of the phases may not correspond to those required by equilibrium, or even that the crystal structures present may correspond to some other part of the phase diagram, or to none. In that sense, martensite in a simple iron–carbon steel is a metastable structure which does not appear at all in the iron–carbon diagram, and reheating to a sufficient temperature will decompose the martensite into a mixture of ferrite and the compound Fe_3C. (The old phase diagram (figure 4.1) includes a martensite field but the modern version does not.)

The all-important geometry of a microstructure can be measured and expressed in terms of variables which have a precise, statistically defined meaning. Such quantities are the mean grain diameter, grain-shape anisotropy, mean free path between dispersoid particles, volume fractions of phases, mean diameter of dispersoid particles, alignment of anisotropic particles along, say, a tensile direction, etc. They can be related to properties of interest, such as elastic modulus, flow stress, ductility, thermal conductivity and ferromagnetic coercive field. This kind of measurement is the subject-matter of *stereology*, a discipline which is important in anatomy as well as metallurgy and the science of materials more generally.

4.2 PHASE TRANSFORMATIONS

Microstructure determines properties. This lapidary phrase has become the metallurgist's expression of scientific faith. Phase transformations determine microstructure, so it is not surprising that the study of phase transformations—their mechanism, crystallography and kinetics—has become as central for metallurgists as the study of microstructure itself. This category includes not only transformations in the solid state but also crystallisation from the melt. Indeed, the understanding of solidification has been revolutionised since 1950.

The solid state transformations in simple carbon steels, responsible for hardening, were the first to be studied, and this study is still in progress (except that many steels today are no longer simple). At the one extreme such work generates essential kinetic information exemplified by the time–temperature–transformation (TTT) curves shown in figure 4.5; this information is vital for

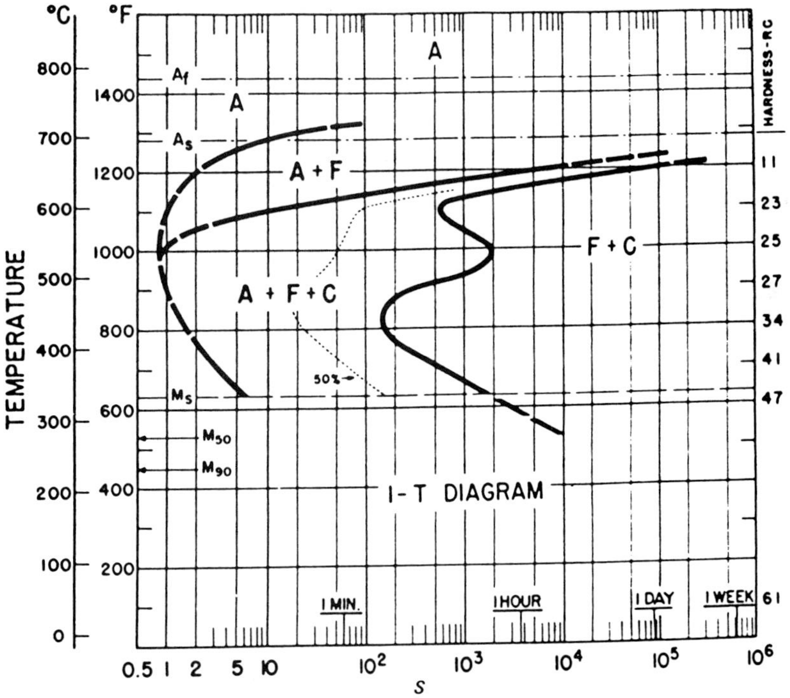

Figure 4.5 Time–temperature–transformation curves for a 0.4 wt% C–1.0 wt% Mn steel. A = austenite, F = ferrite, C = cementite (Fe_3C). The three curves show the beginning of ferrite precipitation, beginning of cementite precipitation and end of precipitation. M_s is the temperature at which the martensite transformation begins on cooling; M_{90} indicates 90% completion of the transformation.

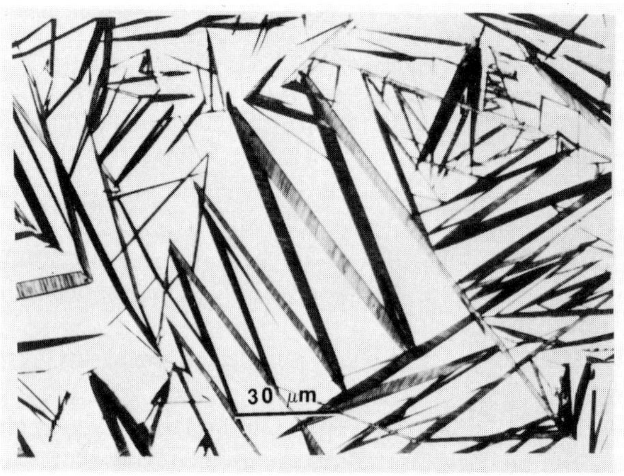

Figure 4.6 Optical micrograph of martensitic microstructure in a Fe–31 wt% Ni–29 wt% C steel isothermally transformed at −160 °C. (Courtesy of C M Wayman.)

correct heat-treatment of high-strength steels in engineering practice. At the other extreme, there has been extensive research into the precise mechanical and crystallographic characteristics of the category of phase transformations termed 'martensitic', which involve a kind of shear on irrational crystal planes. It has now become possible to understand the criteria by which a transforming system selects the habit planes and shear vectors leading to a microstructure such as that shown in figure 4.6. Likewise, in the 1930s, Mehl and Barrett in Pittsburgh, two of the leading physical metallurgists of their day, systematically studied the epitaxial relationships between matrix and product phases in Widmanstätten structures (where one phase has precipitated slowly from another); their early observations have enabled their present-day successors to interpret the choice of planes and the kinetics of the transformations in quantitative detail.

Following the early studies on the hardening of steels, the next major development could be said to be the real herald of scientifically based development of new alloys. This was the discovery of *age-hardening* of aluminium-based alloys. Just as every schoolchild knows the tale of Newton and the gravitationally accelerated apple, so every undergraduate student of metallurgy learns the story of how Wilm discovered age-hardening. Alfred Wilm was a chemist who headed the newly established metallurgical section of a German government research institute near Berlin. He set about seeking a high-strength alloy of aluminium by examining different combinations (the time-honoured approach of pure empiricism first recorded in 1788 by his countryman Karl Achard, who studied some 900 alloys of numerous metals). In September 1906 he was examining an alloy of aluminium containing about

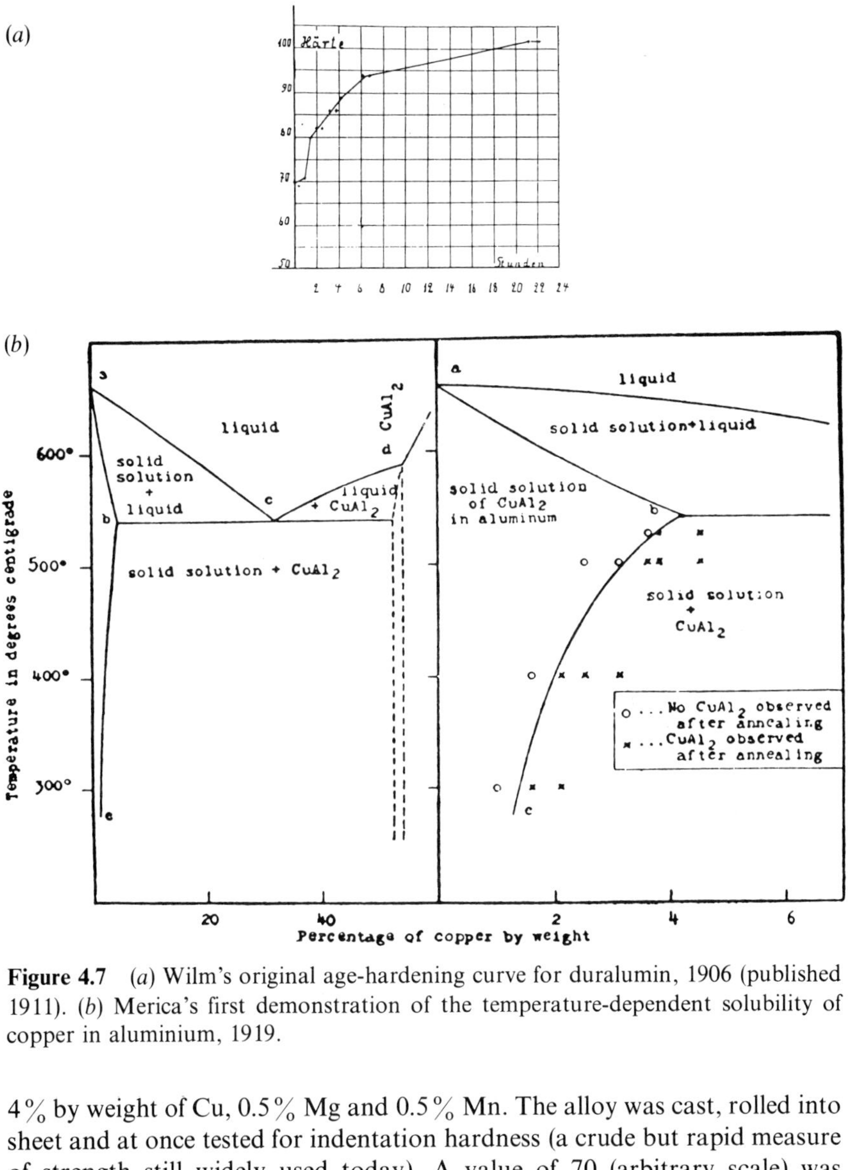

Figure 4.7 (*a*) Wilm's original age-hardening curve for duralumin, 1906 (published 1911). (*b*) Merica's first demonstration of the temperature-dependent solubility of copper in aluminium, 1919.

4% by weight of Cu, 0.5% Mg and 0.5% Mn. The alloy was cast, rolled into sheet and at once tested for indentation hardness (a crude but rapid measure of strength still widely used today). A value of 70 (arbitrary scale) was recorded at 1 PM on a Saturday. The measurement was hasty even by the undemanding standards of hardness measurements and Wilm was conscientious enough to insist on a repeat measurement on Monday: the value was about 100. He then set about repeating the experiment with frequent measurements, and obtained the curve shown in the figure 4.7(*a*). He had discovered age-hardening. By 1914 the alloy, now named 'duralumin', was being manufactured to build German Zeppelins.

The real lesson lies in the sequel. No one knew which was the 'active principle'; Wilm himself felt sure that only the magnesium was important. A number of desultory experiments failed to resolve the relative importance of the various constituents and the original form of duralumin was used for the next few years: it was of course wonderfully helpful to have an alloy which when freshly made was soft and easy to fabricate, and spontaneously hardened afterwards. No one knew how to set about finding other, perhaps better, age-hardening alloys. The Harvard metallurgist Albert Sauveur reported to the US Air Service in 1918 that he did not know of any other metal or alloy exhibiting a similar phenomenon so that analogy did not help. Thereupon Paul Merica at the National Bureau of Standards in Washington micrographically established the variation of solubility of copper in aluminium as a function of temperature and published his results in 1919 (figure 4.7(*b*)). The *sine qua non* of age-hardening was now manifest: it evidently depended upon the precipitation of the equilibrium phase from a supersaturated solid solution. The interesting thing about this insight was its very partial truth: it was later found that the hardening is due to a metastable *precursor* of the equilibrium phase, here Cu_2Al; in duralumin specifically, much of the hardening is in fact due to the inchoate precipitation of the compound Mg_2Si, which incorporates an unintended impurity species. (Wilm was not far from the truth.) No matter: once there was a working hypothesis, metallurgists had a basis for seeking other age-hardening systems, using known phase diagrams, and progress was rapid in the development of new, improved alloys of aluminium and other metals, whereas between the years 1906 and 1919 no new age-hardening systems had been discovered.

A good parallel in the history of physics was the blockage on the exploitation of semiconductors until the recognition of the role of dopants in defining p and n regions; after that, the crucial categories had become clear and researchers had a conceptual basis on which they could proceed.

In 1937, two independent investigators, Guinier in France and Preston in Scotland, established by x-ray diffraction applied to single-alloy crystals that the age-hardening of Al–Cu alloy was due to the strain fields surrounding minute zones in the solid solution crystal which were locally enriched in copper; by the time annealing had progressed far enough to deposit the equilibrium phase, all the hardening had once again disappeared. Since that time, all the minutiae of the age-hardening process have been mapped out in many alloys, first by means of x-rays (using advanced techniques which were all borrowed from solid state physicists) and later, when electron microscopes were available, by their use. Age-hardening is now a mature topic attracting relatively little attention in research.

The first systematic use of x-ray diffraction to study the nature and mechanism of a phase transformation was in connection with an atomic ordering reaction. The history of the investigations of the copper–gold system is most instructive in tracing the gradual shift from a chemical to a physical view of alloy equilibria. The first approach was by the chemist Tammann, who discovered that at certain critical compositions the resistance of these alloys to specific acids underwent a sharp variation, and he sought to interpret these purely chemical discontinuities in terms of postulated intermetallic compounds. (He did, however, also speculate about the possible conversion of a random into an ordered atomic arrangement.) A further protagonist was the Russian chemist Kurnakov, who in 1916 published a major investigation of the entire binary system, comparing the same alloys in the slowly cooled and the quenched conditions. He examined the electrical conductivity of a range of Cu–Au alloys and found maxima at the CuAu and Cu_3Au compositions; he also found hardness singularities at these compositions, moreover, only if they were in the slowly cooled state. He supposed these results to prove the sluggish formation of stoichiometric intermetallic compounds and persisted in this view into the thirties. Indeed, for years to come there was much rather fruitless argument as to whether the concept of the intermetallic compound was applicable to solid solutions with a substantial homogeneity range. Old habits die hard.

In 1925, the Swedes Johansson and Linde proved in a classic study by x-ray powder diffraction that the slowly cooled alloys at the CuAu and Cu_3Au compositions had undergone an atomic ordering reaction below their critical temperatures. Kurnakov's physical results on alloy resistivity at once fell into place. It is noteworthy that the Swedes were the first to apply x-ray diffraction systematically to the study of alloys: Phragmén, Westgren and Hägg were three early Swedish metal crystallographers, and the Swedish metallographer Borelius also contributed a great deal to the understanding of ordering reactions.

It gradually became clear that the perfection of the order increased gradually with falling temperature, just like the magnetisation of a ferromagnet. It fell to Bragg and Williams, in their 1934 paper, to show how these facts could be interpreted in terms of a very simple model with nearest-neighbour interactions. Ever since the publication of that paper, solid state theorists have been fascinated by the theory of collective phenomena in general and of the ordering reaction in particular, in both its long-range and short-range manifestations. The theory has now become exceedingly involved, and unfortunately the physicists' and the metallurgists' approaches to these alloy systems have diverged to the point where there is now limited comprehension between the two groups. The metallurgists, together with a few physicists, study the crystallography, microstructure and especially the mechanical behaviour of ordered alloys. Theoretical physicists think about the reasons why particular alloy systems undergo ordering and study the nature of the transitions in great detail in terms of advanced statistical mechanics. One theorist of such topics, Kenneth Wilson, has recently been awarded the Nobel Prize for studies of such *critical* phenomena (covered by D J Yeomans elsewhere in this volume). This is an area where physicists and metallurgists have been able to contribute to each other's interests, even though the approaches and terminologies of the two communities are at variance.

The understanding of solidification mechanisms, as already mentioned, has advanced very rapidly since the crucial researches of the 1950s led by Bruce Chalmers at Toronto and later Harvard. All these developments have centred on a study of the behaviour of the solid–liquid interface. This has permitted an understanding of microsegregation in crystals grown from the melt, and led directly to the invention of the zone-refining process by William Pfann, just when it was needed by the emerging semiconductor industry. The theory of the stability of the solid–liquid interface, which (like ordering theory) has now become very complex, is an essential tool in understanding what happens when liquid alloys are quenched at ultra-rapid rates so that metastable phases of various kinds are formed and great solute supersaturations can be achieved. This has become one of the major fields of metallurgical research in the 1980s.

Here again, the concepts created by students of solidification have had a major impact on the theory of phase transformations in general, which is increasingly centred on an understanding of processes at the interface between the old and the new phase.

4.3 DEFECTS AND INSTRUMENTS

It has become a truism of solid state science in general that all real crystals contain defects, and that these defects govern properties, especially mechanical ones. In so far as the concept of microstructure has been extended to encompass isolated dislocations, dislocation arrays, point defects and point defect assemblies, the governing metallurgical principle that microstructure determines properties remains intact!

The recognition of the importance of defects is one of the most important contributions of solid state physics to metallurgy. The blend of physics and metallurgy which has grown out of the incorporation of defect theory in both subjects has played a crucial part in understanding and bypassing the problems associated with radiation damage in the materials used in nuclear reactors.

The dislocation concept was postulated in 1934, but for the next 15 years it was barely respectable among either physicists or metallurgists. Only slowly did respectable theorists apply elasticity theory to deduce the precise interactions that dislocations might be expected to have. The reason was certainly that until micrographic evidence was offered—until, to put it crudely, dislocations could be seen—their very existence was in doubt, and therefore investment of intellectual capital in the field was too risky. For metallurgists, it has always been true that seeing is believing, and in this case solid state physicists were the same. Various forms of micrographic evidence for the reality of dislocations became available from 1947 onwards (polygonisation boundaries, decorated dislocations in transparent crystals, growth spirals, etc). By the mid-1950s the evidence had become compelling.

The list in brackets in the penultimate sentence omits the most important evidence of all, that obtained by Peter Hirsch and his colleagues by thin-film electron microscopy. This technique grew out of exasperation with microbeam x-ray diffraction, which had reached its limits. The study of post-war physical metallurgy in general, and understanding of the behaviour and consequences of defects in particular, largely consists of such advances in instrumentation, leading to insights previously unthinkable.

Recent advances in physical metallurgy are closely linked with improvements in resolution, with the fact that defects can be seen and characterised *in situ*, and with the equally important ability to analyse chemical composition as it varies from point to point, by means of x-ray fluorescence and other related methods. There can be little doubt that the combination of electron microscope with electron microprobe analyser has played the same role in the period following the Second World War as x-ray diffraction played in the inter-war period.

Moreover, chemical methods of analysis have, since the Second World War, been replaced by physical methods, and metallurgy students no longer practise chemical analysis, as they did until about 30 years ago.

Not all physical methods used in metallurgy depend upon elaborate instrumentation. The Snoek pendulum is an exceedingly simple technique for measuring minute concentrations of interstitial impurities such as carbon in iron. It measures the spontaneous damping of free torsional oscillations of a specimen wire at the resonant frequency, as a function of temperature (figure 4.8). The peak damping is proportional to the solute concentration, because at this peak temperature each solute atom, by jumping in synchrony but suitably out of phase with the imposed mechanical oscillation, exerts a maximum braking action on the oscillation. At this maximum, the angular frequency ω is related to the carbon atom relaxation time τ by $\omega = 2\pi/\tau$. The variation of peak temperature with frequency allows the diffusion coefficient to be determined (figure 4.6): $D = a^2/36\tau$, where a is the iron lattice constant. Here we have something special: a simple method whose sophistication lies in the mode of interpretation, not in the hardware. The same thing is true of the simple melt-spinning technique used nowadays to quench metallic melts at such high cooling rates as $10^6\ \mathrm{K\,s^{-1}}$.

The understanding of *point* defects has come about almost entirely through solid state physics. In the 1930s it was recognised that certain solid solutions (notably those in the Ni–Al system near the NiAl composition) contain excess vacancies, so-called 'constitutional vacancies'. At that time the combination of density and x-ray measurements was introduced for the study of vacancy concentrations. The later development of the field was greatly stimulated by the needs of radiation damage research.

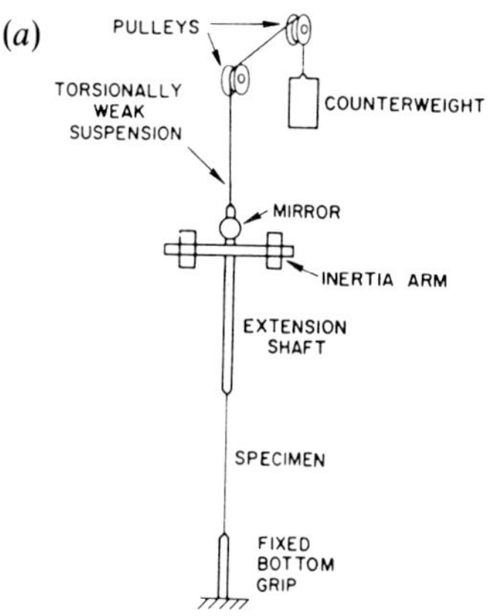

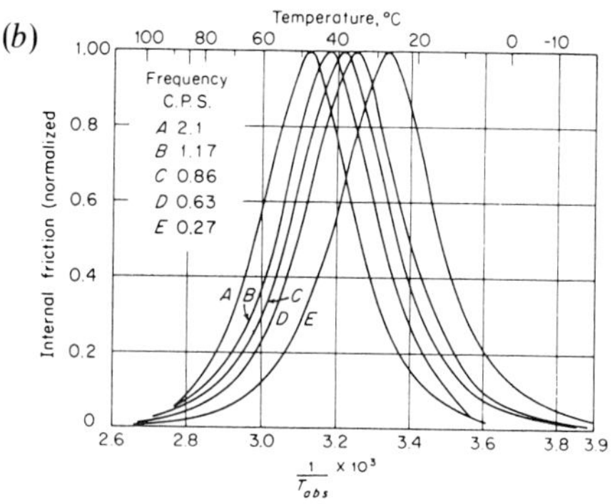

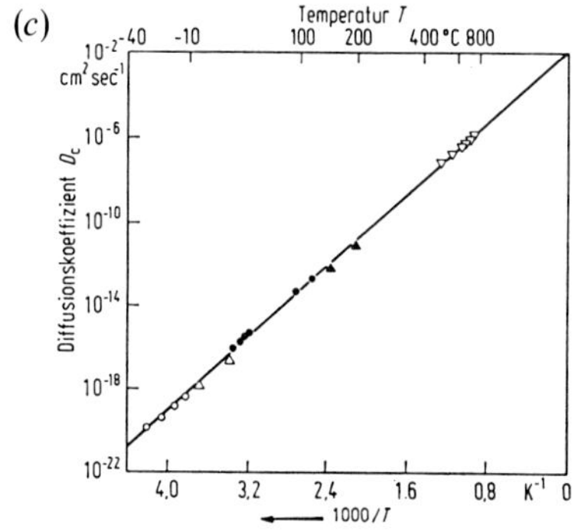

Figure 4.8 (*a*) Arrangement of an inverted torsion pendulum used for analysis of carbon in iron by means of the Snoek effect. (*b*) Internal friction as a function of temperature for a solid solution of carbon in iron at five different pendulum frequencies (Wert and Zener 1949). (*c*) Diffusion coefficient of carbon in iron over 14 orders of magnitude, using Snoek effect (−30–200 °C) and a conventional radio-isotope method (400–700 °C).

Around 1950, it was first recognised that the rate of age-hardening of light alloys at ambient temperature was many orders of magnitude faster than would have been predicted by extrapolating the measured diffusivities of solute atoms in aluminium down to room temperature. This was attributed either to diffusion along dislocation short circuits ('pipe diffusion') or to anomalously rapid lattice diffusion made possible by the presence of a massive supersaturation of vacancies. In fact, both forms of anomalous diffusion are involved.

This was the beginning of what is now a considerable body of theory and experiment, important particularly in interpreting forms of radiation damage such as anomalous specimen swelling during neutron irradiation. Point defects, both vacancies and interstitials, are created as a result of neutron impact and they are differentially absorbed at various kinds of sinks, including precipitates, grain boundaries, single dislocations, grouped dislocations and grouped point defects (so that point defects can be seen to indulge in cannibalism), apart from mutual recombination of the two kinds of point defect. First tiny voids form. How they do so is still the subject of extensive theorising. The voids absorb excess vacancies; they can do this because the more rapidly migrating interstitials are preferentially absorbed by dislocation loops, and so more vacancies than interstitials are absorbed by voids. In modelling this exceedingly complex process, the 'strengths' of both sinks and sources and the jump rates of both kinds of point defects (not to mention such chimeras as divacancies) all have to be taken into account. Equally complex calculations apply to the 'stress-induced preferential absorption' of interstitials at dislocations in the course of irradiation. This can result in very easy motion of dislocations under a small stress—the phenomenon of 'irradiation creep'.

4.4 MECHANICAL PROPERTIES

The most important single objective of metallurgists has always been to optimise the mechanical properties of metals and alloys: high strength, creep or fatigue resistance, ductility or toughness are the usual desiderata.

This century has produced alloys to withstand high stresses at white heat in jet engines, resist embrittlement in arctic gas pipelines, combine the ability to be press-formed with great strength in thin sheet form (for motorcar bodies), allow shaping at low stresses at moderate temperatures while being very strong after cooling to ambient temperature (superplasticity), withstand high stresses while undergoing intense neutron bombardment, permit shaping into wires a few microns thick which then stand up to temperatures of 3000 °C or more, capable of being turned, by rapid quenching, into metallic glasses which excel long-established steels for transformer laminations ... to quote a few examples only. The early stages of most of these developments were indeed empirical (the development of tungsten lamp filaments is a good example) but the later stages of improvement undoubtedly were fed by modern understanding of dislocation behaviour, heat-treatment of steels or age-hardening alloys, microstructural control and, most recently, by the development of a proper quantitative science of fracture mechanics.

A very simple instance of the role of such understanding was the interpretation of discontinuous yielding of steels through the work of Cottrell and Bilby in 1949. Some steels, when loaded in some states of heat-treatment and after certain mechanical pretreatments, begin plastic deformation suddenly and then suffer transient weakening. In consequence, plastic strain is then very non-uniform. This leads to unacceptable defects in steel sheets on press-forming. The origin of this behaviour was argued about for several decades and, in some quarters, mythical amorphous particles were held responsible. Cottrell and Bilby conceived the idea that 'atmospheres' of carbon or nitrogen atoms might form, by room-temperature diffusion, in the stress field surrounding dislocation lines, and then hold the dislocations back when these were subjected to a force as a result of an applied stress, because of the elastic interaction between the carbon atoms and the dislocation lines.

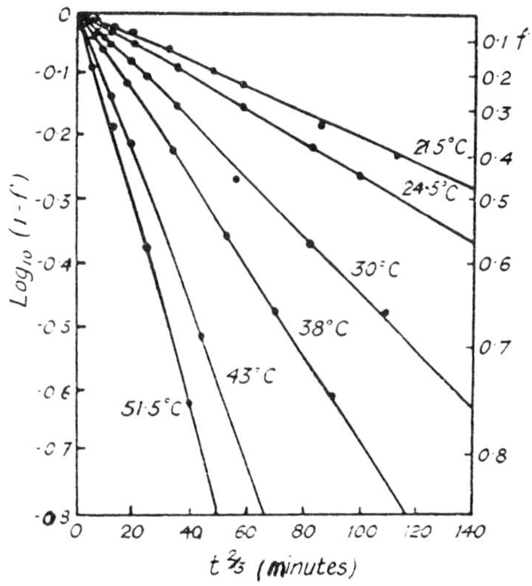

Figure 4.9 The effect of temperature on the kinetics of strain-aging of a dilute iron–carbon alloy, i.e. the gradual return of carbon atoms into dislocation atmospheres after plastic deformation. Here, f is the fraction of carbon atoms which have segregated into atmospheres and are effectively locked (being thus incapable of contributing to the Snoek internal friction peak). Elastic theory predicts that f should be proportional to the 2/3 power of the aging time. (Harper 1951.)

Rigorous elastic calculations, combined with equally exact study of the mobility of carbon atoms by S Harper, proved this hypothesis beyond possible doubt, and also predicted the rate at which this metallurgical condition would return after the steel had been induced to yield. The model was accurately confirmed by experiments using the simple Snoek torsion pendulum, already mentioned, which was able to distinguish between solute atoms in free solution and those captive around a dislocation (figure 4.9).

Since that early and now classic study, an enormous amount of research has been done to understand and improve various mechanical properties. A good example of this is the development of a family of superplastic alloys: these alloys depend upon the creation and maintenance of an ultra-fine-grained microstructure. To ensure that the fine grains do not coarsen at the high temperature at which superplastic forming is done requires all the subtleties of microstructural control, particularly the precipitation of the right volume fraction of very fine second-phase particles.

Another modern family of alloys, the superalloys, are used for the hottest parts of jet engines. They are based on a dispersion of the ordered γ' phase, of composition $Ni_3(Al, Ti)$, within a disordered nickel–base matrix. To prevent the growth of these particles requires an understanding of what metallurgists call 'Ostwald ripening', that is to say, the growth of large particles at the expense of small via diffusion of a constituent between particles through the matrix lattice. It is now known what microstructural features minimise the rate of such ripening: misfit at grain boundaries, solute solubility and solute diffusivities all enter into the theory. A good superalloy has almost perfect fit between matrix and particles.

This example carries a lesson: if metallurgists are to be able to optimise some microstructure for a particular task, then they must have an understanding of many subsidiary processes and features, based on a long series of experiments, often on model substances which simplify the necessary understanding. Lattice diffusion, grain-boundary diffusion, grain-boundary fine structure, the characteristics of dislocation arrays, interface structure, nucleation mechanisms of precipitates, excess point defect concentrations, irregularities in eutectic structures . . . these are a few examples of 'subsidiary processes and features' which have to be understood in their own right.

The conclusion is that to optimise realistic and therefore unavoidably complex alloys, it is first necessary that other metallurgists should have carried out numerous experiments on intentionally simplified model systems in order to gain understanding of subsidiary processes. Metallurgy differs from physics in that useful systems are structurally very complex (whereas a theoretical physicist wishing to achieve the maximum of rigour may shy away from the helium nucleus as being too complex and settle for a proton instead.) *The optimisation of a complex material system presupposes the understanding of simple constituent subsystems.*

4.5 PHASE STABILITY

Even though this is a chapter in a book on solid state physics, up to now nothing at all has been said about quantum mechanics and it is time to rectify the omission.

The quantum theory of metals was thoroughly established in the 1930s, and became widely known among physicists through the classic books by N F Mott and H Jones and by A H Wilson. However, metallurgists did not at that time have the educational background to understand such books—and even today, many would have great conceptual difficulties with them. The translation of quantum-mechanical ideas into terms that metallurgists could follow and exploit was due to three pioneers: William Hume-Rothery in Oxford and Alan Cottrell and Geoffrey Raynor in Birmingham.

Hume-Rothery was a brilliant chemist who devoted his professional life to establishing the regularities that govern the composition ranges within which particular intermetallic phases are stable. His studies were based on the scrupulously precise determination of phase diagrams, and this enabled him both to establish the exact solubility limits of primary solid solutions of various solutes predominantly in copper, silver and gold, and also the homogeneity ranges of intermetallic phases which had different structures from either constituent element, such as the body-centred cubic phase in the copper–aluminium system. In the early 1930s he established a series of empirical rules that came to be known as the 'Hume-Rothery rules', which specified the regularities of phase stability limits in terms of the concepts of atomic radius, electronegativity and, in particular, electron-to-atom ratio (i.e. the ratio of the number of valency, or free, electrons to the number of atoms in a solid solution).

These rules—especially those based on electron/atom ratios—found, or seemed to find, simple and natural interpretations in terms of the gradual filling up of Brillouin zones as the ratios became larger. The related theory was worked out by Jones just after Hume-Rothery drew up his rules. A connection had thus been established between quantum theory and alloy constitution.

Cottrell did not do research in this field: his contribution was to write a slender volume, *Theoretical Structural Metallurgy*, which appeared in 1948 and for the first time presented, in terms accessible to undergraduates without an advanced physics background, the quantum-mechanical ideas underlying the Hume-Rothery rules, the statistical mechanics underlying the general forms of phase diagrams, and the fundamental ideas of dislocation and point defect theory. Raynor taught in the same department of physical metallurgy (the first with such a name), and the innovations jointly wrought by Cottrell and Raynor, under the stimulus and encouragement of one of the great metallurgical teachers of the century, Daniel Hanson, brought about a sea of change in metallurgical education which subsequently affected the syllabus for the teaching of metallurgy throughout the world.

Post-war research, described in Chapter 5, showed that the form of Fermi surfaces within Brillouin zones was more complex than had been thought by Jones and his contemporaries, and the detailed interpretation of the Hume-Rothery rules—which retained their predictive power undiminished—was thrown back into the melting-pot. But the Jones/Hume-Rothery symbiosis had fulfilled its task of transforming the study of phase equilibria. Hume-Rothery himself, looking back on his life's work in 1965, shortly before his death, remarked that 'the theory of alloys is . . . at the stage of Kepler and not of Newton'. He was perhaps too modest. Further comments on our present state of understanding of such matters are to be found in Chapter 6.

4.6 THE ROLE OF TRACE ELEMENTS

In practical metallurgy, unlike the semiconductor industry, ultra-pure materials are not often used and alloys generally contain impurities which must be limited within specified margins but which it would be uneconomic or even impossible to remove almost completely. An impurity, by definition, is tolerated but not wanted. A trace element or dopant in a metal or semiconductor, on the other hand, is often something that is desired and fulfils a specific function. Metallurgical history is full of episodes in which trace elements (desirable) or impurities (to be suppressed) proved the key to some manufacturing crisis.

Examples of unacceptable impurities are arsenic in copper, which greatly increases electrical resistance, and phosphorus, sulphur or tin in most steels, which embrittle them. This kind of limitation in turn sets limits on the use of scrap metal in alloy production; for instance, the use of tin plate inevitably introduces tin into recycled steel which is then almost impossible to remove.

In this way, fundamental research on model systems has a direct and potent influence on manufacturing practice. The relevant research uses modern instrumental techniques of localised microanalysis, and particularly methods (notably Auger electron analysis) which give information on the composition of thin surface or interfacial layers. It not infrequently turns out that a useful trace element, established by systematic research to satisfy a specific need, turns out to have been discovered by chance perhaps a century previously. A good example of this is a research episode at Johnson and Matthey, the firm specialising in noble metals. An alloy containing roughly equal numbers of gold and copper atoms (18 carat French red gold) is widely used in the jewellery industry. It hardens substantially as a result of atomic ordering when heated and, if this goes too far, the alloy becomes embrittled. Since the alloys have to be soldered in use, it was thought desirable to find a way of slowing down the ordering reaction so that a rapid soldering operation would not embrittle the alloy. Systematic but empirical research on ordering kinetics by Svoboda and Chaston revealed that a small addition of iron had this desirable effect, whereupon it was discovered that an old melting book compiled by the same firm at the end of the last century (drawing on immemorial practice in the jewellery trade) called for the addition of two 'French nails' to each melt of the 75/25 alloy. No one knew what was special about French nails ... perhaps they feel at home in French gold!

Another famous episode is the development of sintered tungsten (using powder metallurgy, that is) specifically for drawing into the very fine coiled filaments used in electric light bulbs. This turned out to be extremely difficult and research from 1905 onwards in the General Electric laboratories in Schenectady under the direction of William Coolidge ran into endless problems because crystal grains became too large in the white-hot filaments and they fell apart (among other disasters). The use of particular crucibles for annealing the powder compact at length solved the problem, and prolonged chemical analysis of trace elements (using the relatively insensitive methods of the day) finally pin-pointed the role of silicon, potassium and aluminium picked up from the crucible. These have ever since been added intentionally to the tungsten. It is only the use of high-power electron microscopy in the 1970s which at last explained the mode of action of these trace elements. Potassium

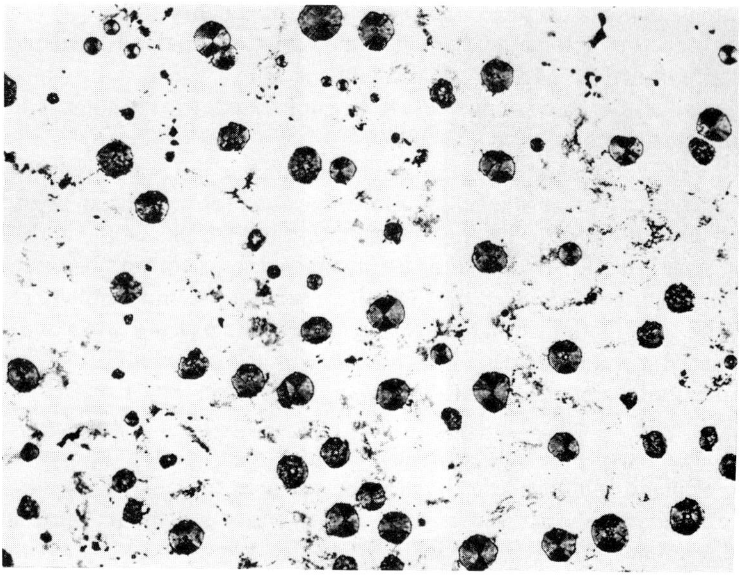

Figure 4.10 Microstructure of spherulitic cast-iron (× 100). (Courtesy of H Morrogh.)

vapourises at a certain stage of processing, the vapour pressure being held low by the other dopants until a sufficiently high temperature is reached, and the microbubbles so formed are drawn into arrays by wire-drawing and restrict the mobility of grain boundaries.

Other trace elements can counteract the deleterious action of impurities. Small additions of some rare earth elements in steels can neutralise the residual embrittling action of phosphorus, which is very difficult to reduce below a limiting concentration. The mode of this counteraction, which is reminiscent of similar effects in pharmacology, has been resolved in detail at the National Physical Laboratory by Auger electron microanalysis.

Perhaps the most industrially important trace element of all is the magnesium (0.02–0.06 %) added to cast iron to convert the graphite in the microstructure, which is normally in the form of plates, into small dispersed spherulites (figure 4.10). This converts the normally extremely brittle metal into a tough, ductile form. Cast iron can contain either cementite, Fe_3C, which is metastable, or graphite, which is the stable form of carbon, in association with iron. The magnesium is now known to 'poison' potential growth sites on the (0001) planes of incipient graphite flakes; multiple nucleation then takes place to generate a polycrystalline spherulite. Certain impurities in the melt, such as Pb, Sb, Bi, Ti and Cu, can hinder the function of the magnesium—they poison the poison—but it has been found that cerium can neutralise these 'subversives'; cerium is thus a poison three times removed! Inoculation of cast iron in this way has been practised for some 30 years (Morrogh in England had the major role in this development), but an even older inoculant is sodium, added in trace quantities to Al–Si casting alloy to refine the eutectic structure of Al and Si lamellae; sodium in some way facilitates nucleation of the solids so that the lamellae become much finer.

4.7 CURRENT AND FUTURE DEVELOPMENTS

In many parts of the world, university departments of metallurgy are changing, by degrees, into departments of materials science and engineering. The physical methods and concepts which have served to convert the old art of practical metallurgy into the metallurgical science of today are in turn converting the study of ceramics into an advanced science, and many are convinced that because of their compressive strength and temperature resistance, ceramics will assume some of the most demanding tasks performed today by alloys. For instance, uncooled internal combustion engines with very high working temperatures, made predominantly of ceramic components, are close to realisation. Some engines of this type have recently been operating in Japanese experimental cars.

Likewise, the science of polymers is now mature and is also increasingly biased towards physics: this bias, in the view of some, has gone too far in polymer science, as indeed it has in metallurgy, so that now extractive metallurgists are scarce and polymer chemists (those that design and synthesise entirely new polymer molecules) are even scarcer.

What does the future hold for metallurgy itself? The first thing to note is that improvements of alloys will increasingly be linked to improved methods of processing. A good illustration of this trend can be found in the steel industry. Because the aggregate demand for all steel products is falling, most onlookers make the mistake of treating the steel industry as technologically backward and short on innovation. This is quite erroneous: there have been more radical improvements in steel properties in the last two decades than in the previous six decades. As in so many other fields, the demand is for a smaller output with more added value per unit, and the industry has responded.

An illustration of this tendency has been in the development of steel sheet for motor cars. Ordinary mild steel—the bulk of low-grade steel, widely used in former years for pressing car bodies—is no longer strong enough: cars made from such steel require thick gauge and are thus too heavy and need too much fuel. Also such simple steel is apt to be used as pressed. One can hardly expect to heat a whole car body to red heat, quench and temper it! All that was possible was to anneal a complete coil of sheet, weighing many tons, on a batch basis and that would take a day or two for each coil. Rapid cooling was out of the question.

The industry has now developed a range of *high-strength low-alloy* and *microalloyed* steels (the two categories merge into each other) which combine ease of forming in a press with high strength and toughness after aging. The perfection of the *continuous annealing line*, in which Japan has taken the lead, has made it possible to heat newly rooled steel sheet, cool and if need be temper it according to a precisely controlled schedule, and do all of this continuously. A thin moving steel sheet can be heated and quenched in a few seconds. There is no longer any need for day-long batch treatments.

What has this to do with physics? A continuous annealing line can only repay the investment required to construct it if new families of steels are also developed at the same time. Some of the new Japanese microalloyed steels contain less than 0.5% of alloying additions in total and yet have quite exceptional strength and toughness, as well as formability, as a result of the continuous treatment. In designing such steels, TTT curves such as those shown in figure 4.5 are essential design tools, and microstructural study is needed at every stage. For instance, the composition and heating/cooling cycle can be so arranged that the hot-worked austenite recrystallises into a fine polycrystalline structure which on cooling then transforms into a fine-grained ferrite phase plus hardening martensite. This is an illustration of the fact that in metallurgical innovation one can never afford to leave the size, shape and orientation of the crystal grains out of consideration. The essential lesson is that the new compositions were necessary to exploit the new processing technique; new technique and new steel developed in mutual symbiosis, rather like semiconductor devices and zone-refining 30 years ago.

Light alloys are also at a turning point. In the design of aircraft, a battle for survival is in progress between aluminium alloys and composites reinforced with high-strength carbon fibres; titanium alloys are beginning to make inroads into the gap left between the two main rivals. The main American aircraft manufacturers appear to be convinced that the continuing use of aluminium in aircraft beyond the end of the century is contingent on the success of a new family of alloys based on aluminium–lithium.

Excellent strength has long been available to the designer who has to select light alloys for aircraft construction: this development probably reached its apogee in the advanced light alloy used for much of the skin of Concorde. The problem lies with stiffness and with density. Elastic modulus is not a structure-sensitive property and is therefore very hard to change. The only approach has been to include a large proportion of a strongly bonded (and therefore stiff) second phase and rely on the law of mixtures to create a stiff alloy. Such alloys have always been too brittle to be acceptable.

Now alloys containing some 2–3% by weight of lithium (and several other elements needed to provide age-hardening and grain refinement) offer a 10% increase in stiffness combined with a similar reduction of density. Since in many design calculations, the necessary mass of a component—say, a wing spar—is limited by elastic deflection rather than by resistance to flow or fracture, these improvements jointly should lead to an estimated 12% saving in all-up weight, and this in turn leads to a very large saving in fuel. (Every 1 lb of weight saved should save some $400 in fuel cost over the life of an aircraft at present fuel price.) The American projection is that if these alloys can overcome their numerous teething problems, notably an unimpressive level of fracture toughness, then future aircraft will contain about half their mass as aluminium; otherwise, aluminium ten years from now may well only make up one tenth of an aircraft's mass.

The most intellectually exciting single development in physical metallurgy since the invention of the electron microscope is, arguably, the introduction of rapid quenching of alloys from the melt; it also has the greatest long-term technological promise. It is hard, at first sight, to see why freezing a volume of liquid in a millisecond rather than several minutes should have such remarkable consequences. Yet, at present, publications on metallic glasses alone are running at several thousand per annum, and this excludes the rapidly increasing volume of publications on crystallisation mechanisms associated with rapid quenching. Why all the excitement?

Rapid quenching (RQ) has an impeccable physical pedigree. The technique was originated in 1959 by Pol Duwez, a metallurgist working in Pasadena, in order to solve an intellectual puzzle. He was interested in phase stability rules and was troubled by the fact that whereas the Cu–Au and Ag–Au systems form continuous ranges of solid solutions, the Cu–Ag system has limited solid solubility at each end and a eutectic mixture which crystallises in the middle. Duwez predicted that with sufficient encouragement it should be possible to force a metastable but continuous series of solid solutions in that system; in other words, he predicted that the Cu–Ag system is teetering on the thermodynamic edge of continuous mutual solubility. The encouragement, he thought, could come from cooling an alloy melt so fast that there would be no time for the long-range diffusion needed to create two solid solutions of different compositions, and the melt would then freeze to produce a metastable solid solution of the same composition. He thereupon invented the first of a series of devices intended to flatten a small volume of melt into thin fragments or ribbons in contact with an efficient heat sink; figure 4.11 shows schematically some of the more recent and effective of such devices. When the new Duwez gun was used to rapid-quench a concentrated Cu–Ag alloy, the equilibrium eutectic was indeed replaced by a metastably supersaturated, face-centred cubic solid solution.

The techniques of rapid quenching from the melt are important in metallurgy because only from the melt can cooling rates of 10^4–10^7 K s^{-1} be achieved; all available methods of quenching in the solid state are restricted to a maximum of about 10^3 K s^{-1}. Therefore, various interesting metastable structures which result from quenching a high-temperature *solid* phase exceedingly fast can only be achieved by starting from the liquid state. Solid state quenching is as old as the use of steel, but liquid state quenching is novel.

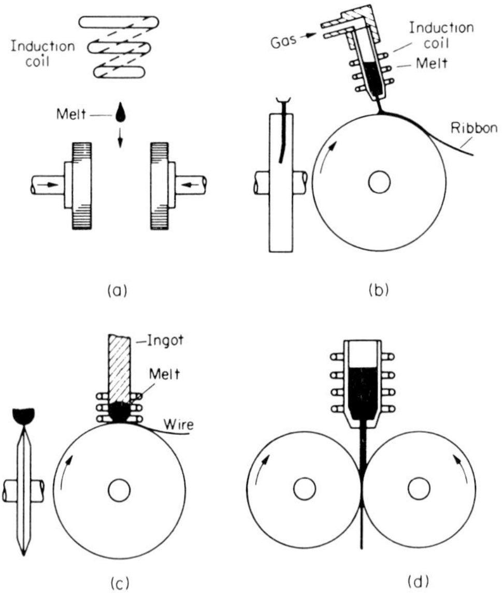

Figure 4.11 Some current methods of rapid quenching from the melt. (*a*) Drop-smasher, using levitation melting by induction; the pistons are pneumatically or electromagnetically accelerated and come into action when the falling drop breaks a light beam. (*b*) Melt-spinning. (*c*) Pendant-drop melt-extraction (there is also a version with the wheel dipping into a melt surface). (*d*) Twin-roller quenching device. (From Cahn and Haasen 1983.)

Briefly, this way of freezing an alloy can have one of three distinct consequences: greatly supersaturated solid solutions may be generated; metastable crystalline phases, hitherto unknown, may form; or, if crystallisation is suppressed altogether, a metallic glass is formed. From a scientific point of view, the glasses have proved the most irresistible of these categories, and a whole new branch of solid state physics has resulted; a number of applications based upon ferromagnetic, corrosion-resistant and catalytic properties are also on the verge of exploitation.

However, in the long term it is probable that the practical consequences of rapidly quench-casting crystalline alloys will be the more important. Tool steels, improved because the greater supersaturation of solutes created by RQ in turn enables a greater measure of age-hardening, are the first fruit of this approach. The next development seems likely to be a family of aluminium alloys, hardened by a large volume fraction of intermetallic phase and strong up to a temperature around 300 °C: these alloys are made by rapidly quenching minute drops of melt into helium and then consolidating the powder particles by techniques such as hot isostatic pressing. Titanium alloys, and in due course superalloys, are likely to be the next families of alloys to benefit from this powder-metallurgical approach.

From a physics viewpoint, the challenge of this subject resides in the theory of solidification applicable to these conditions. Here, the modern theories of interface-controlled motion of the solid–liquid interface come into their own; attention at present is concentrated on the exceedingly complicated processes which govern the relation between the compositions of the solid and liquid phases, ranging all the way from equilibrium segregation to 'partitionless solidification' in which the solid is constrained by *force majeure* to freeze at the composition of the liquid, and going on beyond this to enforced glass formation. Studies based on interface velocities up to several metres per second are complemented by the revival of a technique first employed 30 years ago by David Turnbull. This involved the cooling of a melt divided into tiny droplets, so that the majority can be relied upon to freeze homogeneously, without benefit of nucleation catalyst (because a catalyst particle only affects the droplet in which it is contained). The two approaches jointly are now casting much light on both nucleation and growth of crystals over an extremely wide range of cooling rates. There can be no doubt that these studies will lead to a whole new range of metallurgical processes. As with steels, new processes will call for new alloy compositions, only these will have to be more concentrated in solutes than hitherto, whereas the new steels are more dilute in solutes than their precursors were.

Very recently, new ways of creating metastable alloy phases have begun to be studied. One approach is ion-implantation (originally developed by people looking for new ways to construct integrated circuits in silicon but now adopted by people studying metallic systems). For instance, implantation of sufficient boron or phosphorus in a thin surface layer of pure nickel or tantalum will convert that layer into glass (with the result, among others, that the surface is now more resistant to wear and offers lower frictional resistance to sliding). Closely related is the technique of ion-mixing: here a thin layer of one metal is evaporated or sputtered on to another and the two layers are mixed by bombarding with a stream of energetic rare-gas ions.

Perhaps the most remarkable of the new techniques was pioneered by Johnson and co-workers at California Institute of Technology in 1983. They discovered that interdiffusion in various metal couples (for example, Au–La) could lead to the generation of an amorphous phase starting from two crystalline phases. This does not, as at first appears, fly in the face of thermodynamics. There is actually a hierarchy of relative stabilities: the amorphous phase is stabler than the mixture of the two starting metals, whereas a potential crystalline intermetallic phase is stabler than the amorphous phase. The latter will form if diffusion in it is much faster than in the potential crystalline compound, so that its nucleation and growth is retarded relative to the amorphous phase. By starting with an evaporated sandwich of the two precursor metals, thick layers of amorphous phase can, in

Figure 4.12 Electron microscope diffraction pattern from a rapidly melt-quenched foil of Al–Mn alloy containing 14% Mn atoms. (Courtesy W M Stobbs and A L Greer, Cambridge University.)

principle, be made, something that cannot be done by previous techniques.

The practice of laser-annealing of silicon implanted with electrically active dopants is closely related to the rapid quenching of alloys, but developed quite independently; here also, very large supersaturations of solutes such as In and Sb can be attained. The theoretical treatment of silicon resolidification after localised laser-melting has developed in parallel with that for metals and in fact metallurgists and solid state physicists have now begun to apply the technique of nanosecond and even picosecond laser pulsing to a number of alloy systems as well. As in ion-mixing, one starts with a specimen consisting of evaporated sandwich layers of two metals which are to react with each other.

In November 1984, D Shechtman and his colleagues at the National Bureau of Standards published a truly extraordinary observation: alloys of aluminium with 10–14% of atoms of Mn, Fe or Cr, by rapid melt-quenching, can be induced to solidify in a form which is intermediate between crystallinity and the amorphous state. There is *no* periodicity, but the interatomic bonds are orientationally ordered and consequently the phases give sharp diffraction patterns. That they cannot be crystalline is sufficiently attested by the five-fold symmetry of the patterns (figure 4.12). The formal mathematical nature of this packing was prefigured by R Penrose some years ago when he studied the orientational, non-periodic assemblies in two dimensions of certain curiously shaped families of tiles. Shechtman and his colleagues refer to this new state of matter, as they call it, by the term 'near-neighbour icosahedral bond orientational order' or, for ease in articulation, as 'quasicrystal'. There have been other types of solid with orientational disorder among substances which have been termed 'glassy crystals'; the point, however, of these, mostly organic, substances is that they are true crystals with periodicity (albeit with a measure of disorder within each unit cell). Since 1984 many papers have been published on quasicrystals and activity is still accelerating. The quasicrystal really is something quite new and shows the capacity of materials science to pull utter surprises out of the hat from time to time.

I cannot conclude this chapter without acknowledging my profound indebtedness to that incomparable historian of metallurgy, Cyril Stanley Smith. His writings and his personal advice have been an inspiration to me for a quarter of a century.

FURTHER READING

Barrett C S and Massalski T B 1966 *Structure of Metals*, 3rd edn (New York: McGraw-Hill)

Cahn R W 1981 Metallic Glasses *Contemporary Physics* **21** 43

Cahn R W and Haasen P ed. 1983 *Physical Metallurgy*, 3rd edn (2 vols) (Amsterdam: North-Holland)

Cottrell A H 1953 *Dislocations and Plastic Flow in Crystals* (Oxford: Clarendon Press)

Cottrell A H 1955 *Theoretical Structural Metallurgy*, 2nd edn (London: Edward Arnold)

Cottrell A H 1964 *The Mechanical Properties of Matter* (New York: Wiley)

Hume-Rothery W, Smallman R E and Haworth C W 1969 *The Structure of Metals and Alloys*, 5th edn (London: The Institute of Metals)

Martin J W 1968 *Precipitation Hardening* (Oxford: Pergamon Press) (Includes reprints of a number of classic papers)

Martin J W and Doherty R D 1976 *Stability of Microstructure in Metallic Systems* (Cambridge: Cambridge University Press)

Pickering F B 1978 *Physical Metallurgy and the Design of Steels* (London: Applied Science Publishers)

Porter D A and Easterling K E 1981 *Phase Transformations in Metals and Alloys* (New York: Van Nostrand Reinhold)

Ranganathan S, Arunachalam V S and Cahn R W ed. 1981 *Alloy Design* (Bangalore: Indian Academy of Sciences)

Smith C S 1960 *A History of Metallography* (Chicago: University of Chicago Press)

Smith C S (ed.) 1965 *The Sorby Centennial Symposium on the History of Metallurgy 1963* (New York: Gordon and Breach)

REFERENCES

Cahn R and Haasen P ed. 1983 *Physical Metallurgy*, 3rd edn (Amsterdam: North-Holland)

Heycock C T and Neville F H 1904 *Phil. Trans. R. Soc.* A **202** 1–69

Wert C and Zener C 1949 *Phys. Rev.* **76** 1169

Harper S 1951 *Phys. Rev.* **83** 709

D Shoenberg (b. 1911)

5 Electrons at the Fermi Surface

D Shoenberg

5.1 EARLY HISTORY

About 30 years ago there was a great surge forward in the detailed understanding of the properties of metals. On the experimental side, new techniques were developed and new phenomena discovered, which confirmed that the Fermi surface (FS) was a genuinely meaningful concept and made possible the precise determination of the FS of nearly all the metallic elements and many ordered alloys in the following years. On the theoretical side, advances in computing methods and a better understanding of the role of many-body interactions, together with the guidance provided by the new experimental results, made possible rapid progress in the calculation of realistic band structures. The history of this new branch of solid state physics, often loosely called Fermiology, is the subject of this chapter.

In fact this history goes back a lot further than 30 years, but we shall pass rapidly over the earlier phases. A decisive breakthrough, which might be called the 'birth' of Fermiology†, came in 1926 with the independent discovery by Fermi and Dirac that an ideal gas of a certain class of particles (including electrons) obeyed a new kind of statistics—Fermi–Dirac (FD) statistics as it came to be called. At temperatures well below a 'degeneracy temperature' T_0 (typically of order 10^4 K for electrons in metals), very nearly all states of energy below $k_B T_0$ (the 'Fermi energy') are occupied and very nearly all those above unoccupied. If the states are represented geometrically by points in a wave-vector (k) space, the states at the Fermi energy define a surface in k-space and this is the Fermi surface, which turns out to play a crucial role in determining many properties of the metals.

† The subject might with equal justice have been called 'Diracology' but Fermi presented some of the important results more explicitly than Dirac, who merely remarks on them as obvious; moreover, Fermi's paper, though appearing in the same year as Dirac's, was communicated five months earlier. For an interesting review see Peierls (1972).

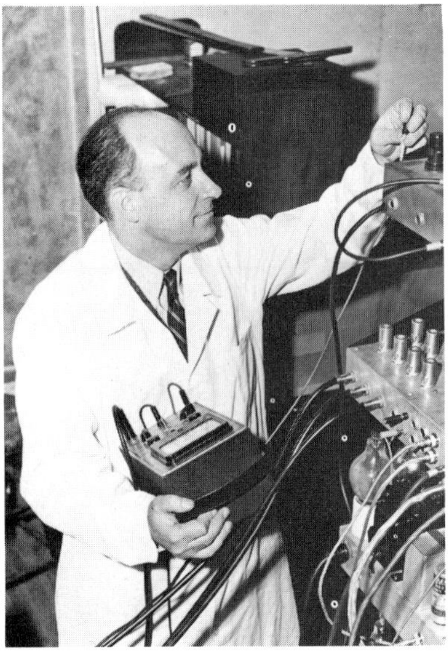

E Fermi (1901–1954) and P A M Dirac (1902–1984).

If the FD statistics marked the 'birth' of the subject, its 'conception' could be said to have come a quarter of a century earlier—an unusually long period of gestation. Thus in 1900 Drude attempted to interpret the properties of metals in terms of the electrons recently discovered by J J Thomson and his theory was refined by H A Lorentz in 1904. The great success of this Drude–Lorentz (DL) theory was that it explained an experimental generalisation discovered half a century earlier. Wiedemann and Franz had found in 1853 that the ratio of the thermal and electrical conductivities κ and σ was much the same for all metals, and L Lorenz showed in 1872 that this ratio was proportional to the absolute temperature T; the constant of proportionality L is usually called the Lorenz number. In Drude's original theory L came out as $3(k_B^2/e^2)$ which was only about 10% below the experimental value (not then very precisely known). However, some gilt came off the gingerbread with Lorentz's refinement of the theory, which spoiled the agreement by changing the 3 to a 2. Nevertheless the qualitative agreement was still a considerable achievement, suggesting that the DL theory must be on the right lines, even though it had several serious defects. Thus it could not plausibly interpret the magnitudes and temperature dependences of σ and κ separately and it incorrectly predicted that the electrons in a metal should have an extra specific heat comparable to that of the ionic lattice. Much later, when an electron spin with a magnetic moment was postulated, an additional puzzle was presented by the fact that the electrons in a metal did not contribute a large paramagnetic susceptibility obeying Curie's law.

It was only after the development of quantum mechanics, Pauli's exclusion principle and, most significantly, the FD statistics, that the difficulties in the way of the DL theory could be removed and the road opened to a realistic theory of metals. That the road was open was not immediately realised. Thus R H Fowler, who had appreciated the implications of the FD statistics for astrophysics, has been quoted (Wilson 1980) as saying 'I had the theory (of electrons in metals) right under my nose but I couldn't see it was there. I kick myself whenever I think of it'. Pauli was the first to take to the open road; in 1927 he showed that the magnetic susceptibility of an electron gas according to the FD statistics is (T/T_0) times the classical value, where $k_B T_0$ is the Fermi energy. Since T_0 is typically of the order of 10^4 K, the paramagnetic susceptibility becomes feeble and almost independent of temperature, in agreement with experiment.

A year later, Sommerfeld systematically recast the DL theory in terms of the FD, rather than classical, statistics and remarkably, in spite of the radical changes in the basic assumptions, the Wiedemann–Franz law and the Lorenz number still came out right—indeed even better than before, for Drude's original 3 now became $\pi^2/3$ in almost exact agreement with experiment. That L should be much the same as in the classical DL theory is somewhat of a fluke. It comes about because the ratio κ/σ involves the product of the energy and the specific heat of the electron gas and in FD statistics for $T \ll T_0$, the former is roughly T_0/T times the classical value, while the latter is roughly T/T_0 times the classical value, so that the product is much the same. The great reduction in the value of the specific heat was of course a considerable triumph since it removed one of the major difficulties of the classical theory. The predicted linear dependence on T is most apparent at low temperatures where the lattice specific heat (falling off as T^3) becomes negligible and it was observed in Leiden a few years later by Keesom and Kok. In fact the coefficient of the linear term subsequently became an important diagnostic feature since it proved to be a measure of the density of states at the Fermi energy.

The next big theoretical advance was made in the same year by Bloch, who took into account the fact that the electrons were not moving in a field-free space, as had for simplicity been assumed earlier, but rather in quite a strong periodic field with the periodicity of the lattice. The electrons were still treated as independent particles but it was implicitly supposed that the periodic field could to some extent take account of the strong inter-electronic forces. One important feature of Bloch's theory was that the finite mean free path of the electrons was linked with slight departures from an ideal lattice (i.e. perfect periodicity), such as caused by the thermal vibrations of the lattice. For the first time the theoretical picture of the conductivity of a metal and its temperature dependence was in qualitative agreement with experiment. But even more significant in the present context was Bloch's demonstration that the energies of the possible states of an electron fell into bands. In the 'tight binding' limit the bands were simply broadened versions of the sharp energy levels of an electron in the field of a single ion, while in the opposite limit of a very weak periodic field the bands nearly merged into a continuum, with only small gaps between them.

Although Bloch did not himself point it out, the idea of bands of states combined with that of occupation of the states according to the FD statistics (at ordinary temperatures essentially two electrons of opposite spins occupy each state up to the Fermi energy) soon led to an understanding of the distinction between metals and insulators. This was most explicitly discussed in 1931 by A H Wilson. He also pioneered the concept of semiconductors as an intermediate case, the importance of which was only realised much later.

Subsequent theoretical progress in the pre-war years lay broadly in two directions. On the one hand considerable effort went into the calculation of band structures more realistic than those based on Bloch's idealised weak and tight binding limits. On the other hand various fundamental physical properties were calculated quantum mechanically, though based on simplified models of the band structure.

As regards band structure, an important early advance was the appreciation of the relevance of the detailed crystal structure. Bloch had for simplicity considered only the simple cubic structure (rarely characteristic of any real metallic element, though occurring in some intermetallic compounds). The generalisation to other symmetries was made in 1930 by Brillouin, who showed that the wave-vectors k of the states in a given band could be represented by points within a polyhedron characteristic of the crystal symmetry—the Brillouin zone (BZ) as it is usually known. The BZ for face-centred cubic symmetry (as for the noble metals) is shown in figure 5.1. This is a reproduction of what is probably the earliest example of the useful geometrical representation of band structure mentioned earlier. It is taken from Sommerfeld and Bethe's classic review (1933) and shows a surface of constant energy in wave-vector (k) space, based essentially on Bloch's tight binding formula but extended to the face-centred cubic structure. Although this is not intended to be the Fermi surface corresponding to the highest occupied states, it is interesting that it does in fact qualitatively resemble the FS of a noble metal (see figure 5.6), determined a quarter of a century later.

During the 1930s a number of methods were developed to overcome the formidable difficulties of calculating the band structures of real metals and drastic simplifications of uncertain validity were introduced to make the problem tractable. As before, it was assumed that the actual very strong interactions between electrons could somehow be taken into account by suitable choice of a periodic potential so that the problem could be reduced to that of a single electron moving in this periodic potential. Even with this assumption, which found its justification only much later, the problem could be tackled only with considerable further simplifications. The first such simplification (in 1933) was the method of Wigner and Seitz in which the volume of the metal was divided into primitive cells centred on each ion and

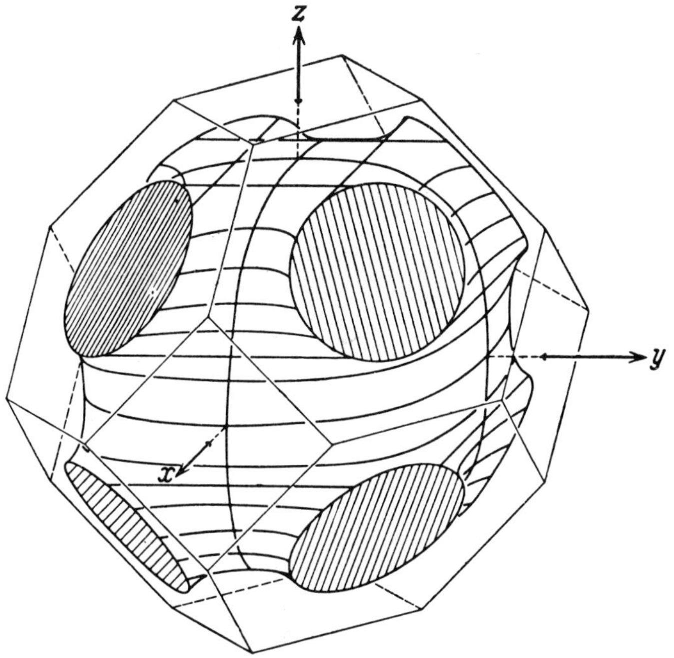

Abb. 25 c. $E = E_0 + C + 2|A|$.

Figure 5.1 A surface of constant energy within the Brillouin zone of a face-centred cubic metal (Sommerfeld and Bethe 1933).

the potential within each cell taken as something like that of the central ion alone. This type of method was subsequently modified and improved in various ways. In 1937 Slater proposed the augmented plane wave (APW) method, in which the potential was of the 'muffin tin' type: essentially ionic within a spherical core and constant (corresponding to plane waves) over the rest of the Wigner–Seitz cell. An alternative powerful method, that of orthogonal plane waves (OPW), was proposed in 1940 by Herring. In this, a combination of core wave functions was added to a plane wave in such a way that the combination was orthogonal to each core level.

Although these methods of band calculation had some qualitative success in giving approximately correct values of various physical properties, they could not be properly exploited at the time, partly for lack of adequate computing facilities and partly for lack of the kind of experimental information which could provide searching tests of the reliability of the band structure. Moreover, many solid state physics theoreticians felt that the drastic approximations which lay at the heart of such methods were likely to cause gross inaccuracies. The methods did, however, provide a semi-empirical framework for the analysis of experiments and for much of the theory which enabled band structure calculations to come into their own 10 or 20 years later. This came about with the development of modern computers and of more sophisticated methods of calculation, such as the KKR method of Korringa, Kohn and Rostoker, as well as the discovery of new experimental techniques (to be described below), which produced results serving as guides to further refinement of the calculations†.

In the other kind of theoretical activity during the pre-war years considerable progress was made in laying down the basic principles for the quantum mechanical calculation of the physical properties of metals. The results of such calculations usually turned out to depend in detail on the form of the surfaces of constant energy in the neighbourhood of the FS. But since the qualitative effect of introducing quantum mechanics did not usually depend very much on the details of the band structure and since, moreover, these details were usually not known, the calculations were made using greatly over-simplified models. Thus, as we have already seen, the free-electron model, implying spherical energy surfaces and in particular a spherical FS, was sufficient to demonstrate the striking reduction in magnitude of the electronic specific heat and of the spin paramagnetism brought about by calculating with quantum, rather than classical, mechanics. Similarly, in 1930 Landau used the free-electron model to show that with quantum mechanics a finite orbital diamagnetism should occur in metals, in contrast to the zero prediction of classical mechanics. More sophisticated calculations were based on ellipsoidal, rather than spherical, energy surfaces and introduced an effective mass parameter (or mass tensor for ellipsoids) to modify the free-electron model.

Another generalisation of the free-electron model took account of a situation that often occurs in practice, when conduction comes about partly from electrons at the top of a nearly filled band and partly from a few electrons in the next higher band. In 1929 Peierls first showed that in some respects the nearly full band was best described in terms of positively charged holes (missing electrons). This shows up particularly strikingly in the Hall effect, with a contribution from the holes which is of sign opposite to that of free electrons. Even though the electrons in the higher band may contribute a Hall effect of the normal sign, the combination of the two contributions in this so-called 'two-band model' can have the 'anomalous' sign. This explained, at least in principle, the long-standing mystery of how it was that some metals had an 'anomalous' Hall effect of sign opposite to that expected for the free-electron model. It was fortuitous that Hall's original experiments in 1879 were on copper, which has the normal sign, so that his results were qualitatively compatible with the idea of negative electrons being the carriers.

It has already been mentioned that the development of band structure

† For a more detailed account of the modern techniques of band structure calculation a standard textbook such as that of Ashcroft and Mermin (1976) should be consulted.

theory was hampered by lack of suitable experimental information. Some qualitative features of a band structure could indeed be checked against the optical properties of the metal and its soft x-ray absorption spectrum. But most of the experimental data available before the war were of the kind specified by the numerical values of only a few parameters, such as the principal components of the conductivity or the susceptibility tensor. In the general theory (not explicitly formulated at that time) such parameters could be expressed as complicated integrals over the FS, often involving other variables, such as the mean free path. At best, if a band structure was known in detail around the FS, it was possible to calculate the relevant parameter, but the reverse was not possible: any number of band structures were compatible with any given numerical value. What was needed for a more searching comparison with band theory was some kind of effect whose variation with the orientation of a single crystal of the metal had a one-to-one relation with the form of the constant energy surfaces, rather than being specified by a small number of parameters. One such phenomenon was in fact already known before the war, but its significance for Fermiology was not properly realised until much later. This was the oscillatory variation with field of the magnetisation of a single crystal of bismuth, discovered by de Haas and van Alphen (1930), which we shall now discuss.

5.2 THE DE HAAS–VAN ALPHEN EFFECT AND THE FERMI SURFACE OF BISMUTH

By a rather remarkable coincidence the oscillatory effect discovered by de Haas and van Alphen (figure 5.2(*a*)) was quite independently predicted theoretically by Landau at about the same time. The motivation behind the experiment was that Shubnikov and de Haas had very recently observed an oscillatory character in the field dependence of electrical resistance and de Haas had a long-standing hunch that there should be a close correlation between diamagnetism and magnetoresistance. The main aim of Landau's theory was to calculate the diamagnetic susceptibility of a free electron gas according to quantum mechanics and he showed that it should be just one third of Pauli's paramagnetic spin susceptibility rather than vanishing, as in classical theory. In his paper he noted that if H/T is not small enough, a 'smoothing' approximation (use of the Euler–Maclaurin formula) is no longer valid and the variation of magnetisation with field should show strong periodic oscillations about the smoothed linear variation corresponding to the steady diamagnetism. But this was somewhat of a throw-away remark, for he added: 'Because of this periodicity it would hardly be possible to observe this effect experimentally, since on account of the inhomogeneity of available fields there would always be an averaging.'

(a) (b)

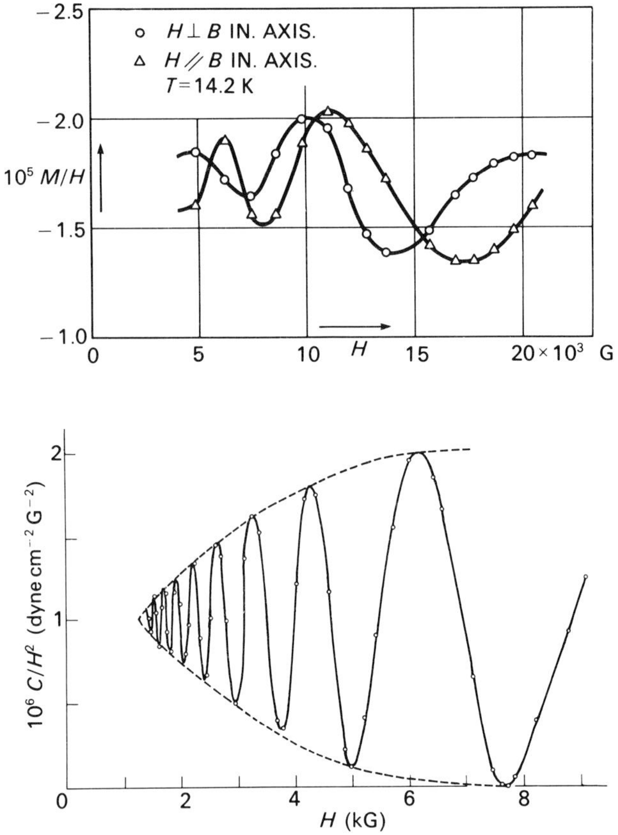

Figure 5.2 (a) First observation of oscillatory field dependence of susceptibility (M/H) in bismuth by the Faraday method for two different field directions (de Haas and van Alphen 1930). (*b*) Oscillatory field dependence of torque for bismuth at 4.2 K (Shoenberg 1939). The ordinate is C/H^2 where C is the torque per unit volume at field H. The field is oriented at about 10° from the binary axis. The improvement of (*b*) over (*a*) reflects both the absence of field inhomogeneity and the lower temperature.

In dismissing the possibility of observing the oscillatory variation Landau had, however, overlooked two things. First, that the effective mass and the effective number density of the electrons in bismuth might be anomalously low—and indeed Landau himself hints at something like this in order to explain the fact that the steady diamagnetism of bismuth is anomalously large. As was already implicit in his theory, this would greatly increase the period of the oscillation so that field inhomogeneity would be less of a handicap. And secondly, even the great demands on homogeneity required to observe the

W J de Haas (1878–1960), P M van Alphen (1906–1967) and (below) L V Shubnikov (1901–1945).

short period characteristic of a metal more typical than bismuth are by no means impossible to meet. Many years later I asked Landau why he had made such a categorical remark about practical limitations. He replied that since he knew nothing about practical matters, he had asked Kapitza, whom he was visiting in Cambridge at the time, and Kapitza had told him that the required homogeneity was impracticable. Fortunately, what may have been true in 1930 was no longer a serious limitation for the technology of 30 years later. It is now quite a routine matter to observe the very short period oscillations that Landau had envisaged but dismissed.

In 1933 Peierls calculated the form of the de Haas–van Alphen (DHVA)

R E Peierls (b. 1907) and L D Landau (1908–1968).

oscillations for an electron gas in which the effective mass, m, and number density, n, were adjustable parameters but the surfaces of constant energy were assumed spherical. By assuming m to be only about 0.02 of an electron mass and n to be only of the order of 10^{-6} per atom he was able to obtain quite a reasonable fit to the de Haas and van Alphen data, though of course the dependence on field direction (see figure 5.2(a)) could not be explained by a calculation based on spherical symmetry. Some progress towards explaining the anisotropy was made in 1938 by Peierls' graduate student Blackman who, after laborious computations, somewhat improved the agreement with experiment by assuming that the energy surfaces were ellipsoids in sets of three (three to take account of the trigonal symmetry). It must be emphasised that in the calculations of both Peierls and Blackman the expression for the magnetisation was essentially a fairly complicated integral of a summation involving the adjustable parameters m and n as well as H, T and any assumed anisotropy. Since the integration and summation had to be carried out numerically on a desk calculator for each combination of variables, obtaining any sort of fit to experimental data was a challenging task.

At this point some personal history is appropriate. In 1932 I had been introduced to the peculiarities of bismuth by Kapitza who was my PhD supervisor at the Royal Society Mond Laboratory in Cambridge. Somewhat later I had, together with Zaki Uddin, an Indian graduate student, made some measurements of the de Haas–van Alphen effect (I think I invented this nomenclature), particularly as regards its great sensitivity to the addition of other metals to the bismuth. Peierls who was the 'house theoretician' in the Mond Laboratory was at once able to interpret this sensitivity in the light of his theory. The effect of alloying was rather like introducing or withdrawing a small volume of liquid into or from a bottle in which the level was in a very narrow neck. Thus there could be quite a big change of level (i.e. of the effective number of electrons) even though the volume change was only a tiny fraction of the total (i.e. of all five valence electrons per atom).

In this work we had followed de Haas and van Alphen in using the Faraday method, which measures the magnetisation by the force exerted on the sample in an inhomogeneous field. An essential limitation of this method is that the field varies slightly over the sample and if the variation is comparable with the period of an oscillation, the oscillations will begin to be smeared out (essentially the effect that Landau had envisaged as making an oscillatory effect unobservable). During a visit by K S Krishnan to Cambridge he described his ingenious techniques of studying magnetic anisotropy by the torque exerted on the sample and this stimulated me to think of applying the torque method to study the DHVA effect. Since bismuth is strongly anisotropic there should be plenty of torque and since no field inhomogeneity is required to produce a torque the phase-smearing difficulty would be largely eliminated, thus making possible a more detailed study of the oscillations.

A convenient opportunity for carrying out this idea was provided by an invitation from Kapitza to work for a year in the Institute for Physical Problems in Moscow, which had just been built for him in 1936 after the Soviet authorities had refused him permission in 1934 to continue working in Cambridge. The torque experiment was just about simple enough to carry out in a year and in fact I was able to observe striking oscillations (figure 5.2(*b*)) within seven weeks of getting to Moscow. I can still remember my delight in seeing the spot of light which measured the torque move smoothly back and forth on the scale as the field was steadily changed. As expected, the torque method and the use of liquid helium rather than liquid hydrogen made it possible to follow the oscillations to much lower fields and to study in detail how the effect varied with crystal orientation, field and temperature.

Once again, I was lucky in having the guidance of a house theoretician. This time it was Landau and by what seems to have been a rather extraordinary coincidence he had just then developed a theory of the detailed form of the oscillations and showed me a formula which I could test against every aspect of my experimental results. His formula marked an advance in two important respects. First, he used an ingenious mathematical trick to evaluate in explicit form the expressions that Peierls and Blackman had had to calculate numerically and of course the existence of an explicit formula enormously simplified the problem of interpreting the experimental data. Secondly, although he still assumed the energy surfaces were ellipsoids, he went beyond

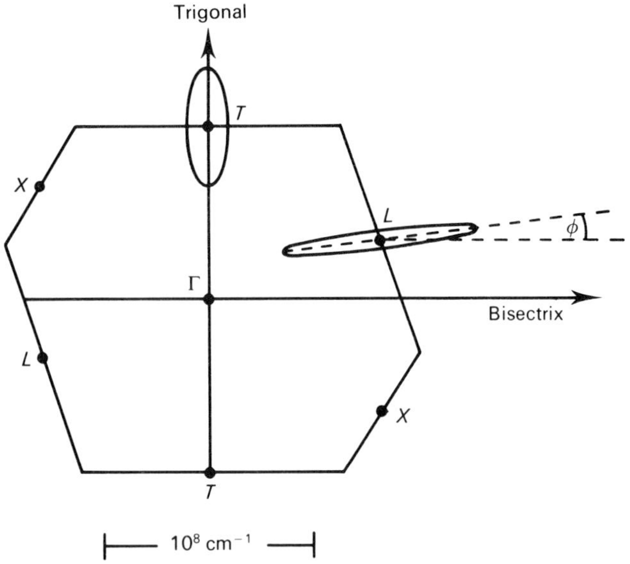

Figure 5.3 Section normal to the binary axis of the FS of bismuth in the Brillouin zone. Only one of the three electron ellipsoids (at L) is shown, the others are located at the two other points L obtained by rotations of $\pm 120°$ about the trigonal axis. The tilt angle ϕ is 6.4° and for clarity the dimensions of the ellipsoids are exaggerated by a factor of about 6 relative to those of the Brillouin zone. The hole ellipsoid at T has rotation symmetry about its long axis, that at L, however, has all 3 axes unequal; in the binary axis direction the ellipsoid is about 30% thinner than in the plane of the diagram.

Blackman in taking fuller account of the crystal symmetry of bismuth, by introducing a fourth effective mass parameter, essentially specifying a tilt of each of the ellipsoids.

In fact Landau's formula proved to fit nearly every detail of my experimental results and I was able to determine the values of all the parameters involved. With hindsight I could claim that this was the first experimental determination of a Fermi surface† (figure 5.3). However, in those days one did not think in such terms. If the surfaces of constant energy were ellipsoids and the band parabolic, as was usually assumed, specification of the mass tensor and the Fermi energy was all that was needed to describe the electronic structure. It was only 15 or 20 years later that the concept of a Fermi surface was developed when, as we shall see below, all sorts of weird and wonderful shapes were needed to explain the behaviour of metals other

† It is difficult to track down the first use of this nomenclature. It was certainly in use soon after the war (e.g. Onsager 1952) but rarely before, though one mention of it does occur in my 1935 PhD thesis.

than bismuth. Although the three-ellipsoid electron FS of bismuth has needed only slight adjustments in the light of later experiments, it is not the whole story. Some twenty years later, Brandt in Moscow showed that there was also a single ellipsoid of holes which had not shown up in my experiments because of the higher effective mass of the holes. There were also two discrepancies of detail between Landau's formula and my data, which later proved to be important. One was an extra amplitude damping factor. This is usually named after Dingle, who in 1952 interpreted it as caused by electron scattering. The other discrepancy concerned the phase and this was later interpreted in terms of electron spin.

The publication of my results was not without drama. Early in 1938, Landau was arrested in the Stalinist purges and was released only a year later, following a courageous intervention by Kapitza. Since at that time an 'enemy of the people' (as anyone arrested was called) inevitably became a 'non-person', I was forced to delete all mention of Landau's vital contribution in the Russian version of my paper. However, the appropriate acknowledgement did appear in the English version, together with an appendix in which Peierls, who was familiar with Landau's way of thinking, reconstructed the derivation of Landau's formula on the basis of some rough notes I had taken. Ironically, Landau was released shortly before the English version (Shoenberg 1939) had appeared in print.

5.3 POST-WAR EFFORT AND THE RISE OF FERMIOLOGY

In spite of an explicit comment by Peierls that the DHVA effect should occur in all metals, most experimenters had tacitly assumed that it was only in bismuth that it could be observed in practice. This complacent attitude was shaken in 1947 when Jules Marcus at Yale observed quite strong oscillations in zinc. This quickly started a search by the torque method for the effect in other metals, both in Kharkov by Verkin, Lazarev and Rudenko and in Cambridge by myself. It soon became evident that the oscillations could be observed in nearly every polyvalent metal of which good single crystals were available. It was not too surprising that nothing was observed in monovalent metals because order of magnitude estimates suggested that the amplitude would have been too small in the fields available, even if the homogeneity and steadiness of the field had allowed resolution of the very closely spaced oscillations.

The various polyvalent metals showed a bewildering variety of behaviour, both as regards the multiplicity of periodicities and the orientation dependence of the periods. It was evident that Landau's ellipsoidal model, which had served so well for bismuth, was quite inadequate to explain these features, though the observed temperature dependence of amplitude and the regular periodicity in reciprocal field were entirely consistent with his formula. Fortunately there was a major theoretical breakthrough at just about this time. Together with new experimental advances this eventually led to a much better understanding of what was going on.

Once again the theoretical breakthrough was associated with two people—L Onsager and I M Lifshitz. At various American conferences I had heard Onsager comment on the DHVA effect and its relation to the FS in a way which I found intriguing but cryptic (such as 'what if the FS is a hyperboloid?'). So, when in 1951 he came from Yale for a sabbatical year in Cambridge I kept urging him to put his ideas in writing so that I could achieve some understanding. Eventually, at the very end of his visit, he did produce a three page paper which has since become a classic. Quite independently, I M Lifshitz had presented essentially the same idea in an unpublished lecture to the Ukrainian Academy of Sciences in 1950. However, it was only later that Lifshitz's contribution became known in the West.

The new idea related the periodicity of the DHVA oscillations in a very elegant way to the geometry of the FS. From a simple semiclassical treatment of how the electrons move in a magnetic field, using the Bohr–Sommerfeld condition to quantise the motion, Onsager and Lifshitz obtained the formula†

†This is usually called the Onsager relation. By 1955 Lifshitz and Kosevich had developed the idea into a detailed theory for the amplitude and phase of the oscillations as well as the frequency; their formula is usually named after them (LK formula for short). Until about 1962 the periodicity was usually specified by the period P in units of $1/H$, but following a suggestion by B R Watts (unpublished) it has become customary to use the more convenient DHVA frequency, F $(=1/P)$.

L Onsager (1903–1976) and I M Lifshitz (1917–1982).

which relates the DHVA frequency, F, to A, the extremal area of cross section of the FS normal to the field direction

$$F = (c\hbar/2\pi e)A. \tag{5.1}$$

The importance of the formula is, of course, that it provides a powerful tool for quantitative determination of size and shape of the FS, but it took several years before this new tool could be properly exploited. Experimental techniques had to be improved to provide sufficient data for constructing a complete picture. Also, band structure calculations were not at first refined enough to give a sufficiently reliable guide to the interpretation of the observed results. It was only gradually realised that the FS was often made up of a multitude of sheets of complicated shape and that comparatively minor changes in a band structure calculation could introduce qualitative changes in the whole pattern.

On the experimental side, even though the torque method was refined in various ways, high frequency oscillations characteristic of relatively large sheets of FS could not be detected without going to fields higher than those of an ordinary electromagnet†. It occurred to me that a pulsed magnetic field in the style of Kapitza might be the solution, with detection of the oscillatory magnetisation by the corresponding oscillatory EMF induced in a pick-up coil as the field rose and fell. By discharging a large capacitor through a liquid-nitrogen-cooled coil, fields of the order of 10^5 G were achieved, rising to a peak and falling off again in a time of the order of 10 ms. Such fields were some five or ten times higher than those used before.

After a good deal of exploratory work, in the course of which higher frequency oscillations were indeed observed both in metals already studied and many new ones, the method was first systematically used in 1958 by A V Gold in Cambridge to determine the FS of lead from a study of the orientation dependence of the rather complicated frequency spectrum. His results led him to an interpretation which has proved to be an important landmark in the understanding of polyvalent metals. This interpretation was in terms of the FS of a nearly-free-electron model—essentially Bloch's weak binding limit—a model which at first sight seemed particularly implausible for a metal of high atomic number. If the free-electron Fermi sphere corresponding to the four valence electrons per atom of lead is dissected and reconstituted in the basic BZ, the FS would consist of the various sheets shown in figure 5.4. These offer a wealth of extremal areas normal to any direction. By only slight adjustments of this idealised model Gold was able to account for nearly all the features of his frequency spectrum in terms of the extremal areas of the 'monster'. In later, more refined, experiments at Ames (Iowa), Anderson and Gold in 1965 were able to obtain even better agreement.

† However, with sufficiently careful stabilisation, Joseph and Thorsen were subsequently able to observe high frequency oscillations in a large electromagnet.

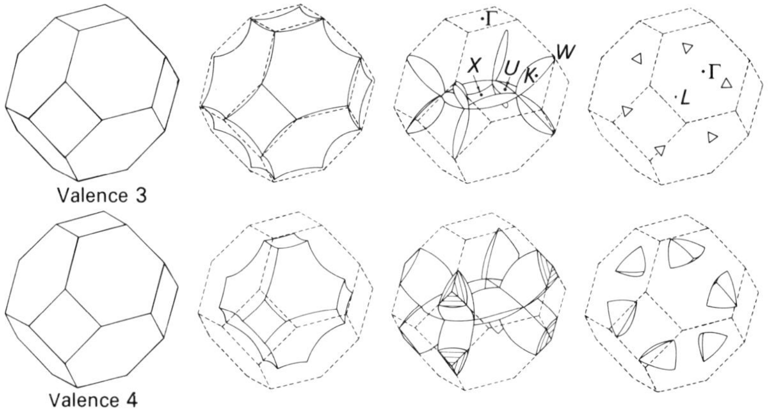

Figure 5.4 The various sheets of FS of face-centred cubic metals of valence 3 (e.g. aluminium) and 4 (e.g. lead) shown within the Brillouin zone. The surfaces are obtained by remapping into the first zone the various pieces of a free-electron sphere overlapping into higher zones. Note that the third band surfaces are displaced by half a reciprocal lattice vector in the direction ΓX to exhibit the form of the 'monster' more clearly and the fourth band surfaces are similarly displaced in the direction ΓL (Harrison 1960).

The success of the nearly-free-electron model was an important piece of evidence in the development of pseudo-potential theory in the late 1950s. Heine in 1957 had already found that a 'first principles' OPW band structure calculation for aluminium gave a nearly-free-electron-like FS even though the real periodic potential was much more than a feeble perturbation of the free-electron model. Thus the success of the nearly-free-electron model for lead suggested strongly that some new principle was involved. This principle was soon uncovered by Phillips and Kleinman, Cohen and Heine, and Harrison (for a review see Heine *et al* 1970) and led to the concept of a pseudo-potential. Roughly speaking, by some mathematical juggling the Schrödinger equation of the conduction electrons moving in the real and strong potential can be recast to look like the Schrödinger equation for a much feebler 'pseudo-potential'. The recast equation is in terms of a 'pseudo-wavefunction', but over the energy range of interest it has eigenvalues for the conduction electrons nearly identical to those of the original problem. In this way the very complicated real problem reduces to the much simpler one of solving the Schrödinger equation for a nearly-free-electron model with a weak perturbing periodic potential. As we shall see below, this approach subsequently found wide and successful application to many polyvalent metals.

Although the impulsive field technique was successful in showing up the oscillations associated with major sections of the FS of polyvalent metals and thus paving the way for a much better understanding of their band structures, several years elapsed before the DHVA effect was discovered in any monovalent

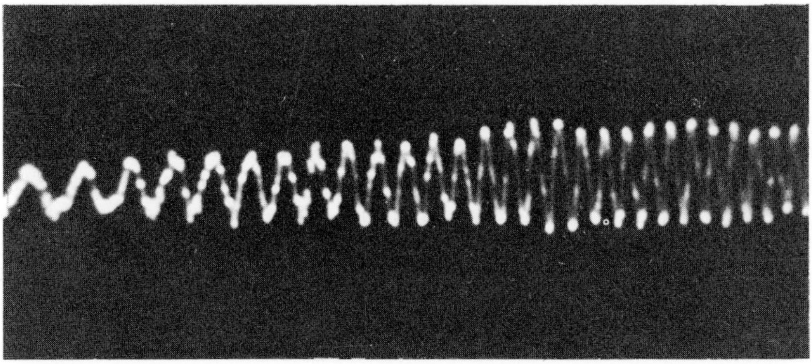

Figure 5.5 First observation of oscillations in copper (Shoenberg 1959). The sample was a (111) whisker at 1.1 K. During the 0.6 ms sweep time shown, the field drops by about 250 G from 72 kG at the left and the presence of 28 oscillations over this range implies that $F = 5.8 \times 10^8$ G. The crowding of the oscillations as the field falls is because the field falls more and more rapidly away from its maximum.

metal. With hindsight we can see the reasons for the negative results obtained in many preliminary attempts. Effort had been concentrated on the least promising metals and poor quality samples had been used. The very first attempts were made with sodium in the belief that it was the simplest metal, though in fact, because of its martensitic transformation at about 40 K, it is difficult to make a crystal sample which is not badly damaged by cooling to helium temperatures. Then there were many unsuccessful attempts with copper crystals, which probably failed mostly because of poor quality and unsuitable crystal orientation. It was not then appreciated that for the noble metals free-electron-like oscillations occur only over a fraction of all possible orientations. Nor was it appreciated that of the three noble metals copper had the weakest amplitude of DHVA oscillations. It was a frustrating situation, rather like looking for a black cat in a dark room, without even having complete assurance that there was a cat.

Eventually the breakthrough came by a combination of luck and slightly faulty reasoning. It occurred to me that whiskers might provide much more perfect crystals than ones grown from the melt, and during a summer visit to the General Electric Research Laboratories at Schenectady in 1958 I was able to get hold of some copper whiskers which had been prepared by Mrs E Fontanella. At last, on 19 December 1958, one of these whiskers produced oscillations (figure 5.5) which were unmistakably above noise level and of just the expected DHVA frequency. The slightly faulty reasoning lay in the argument that whiskers should be more perfect than crystals from the melt. In fact it is only extremely thin whiskers—too thin to be of much use—which are so perfect, while the whiskers I used were relatively thick (diameters of a few tenths of a millimetre). They were probably little more perfect than crystals grown from the melt. It was not long before much better crystals were available from the melt and the whiskers, having served their historical role, were no longer needed.

There were elements of both drama and farce in the further developments. The oscillations had been seen in the morning and both I and my assistant Harry Davies went off to lunch in a rather elated mood. I was so excited that I came back early and in order to make sure the effect was genuine I tried to work the equipment myself but alas, I was not quite familiar with the circuit. When I closed the switch to produce the pulse I heard a colossal bang instead of seeing pretty oscillations on the oscilloscope screen. I had in fact made a direct short circuit of the condenser bank, completely bypassing the magnet coil. Luckily the damage was not too serious, though I have been a little deaf ever since, and we were soon able to make a systematic study of the oscillations not only in copper but also in silver and gold. But before describing the results I must go back a bit and describe a quite different development—the anomalous skin effect—which had enabled Pippard to make an earlier determination of the FS of copper.

This anomalous skin effect had first been noted in 1940 by H London in his pioneer study of the absorption of very high frequency electromagnetic radiation by metals. London's main interest had been the absorption in the superconducting state but he noticed also that the absorption in the normal state was appreciably greater than it should have been according to standard skin effect theory. He suggested that this was because the electron mean free path was much longer than the classical skin depth at the low temperature of the experiment. Soon after the war, A B Pippard in Cambridge successfully applied the new microwave techniques that had been developed for radar and was able to take London's work much further, both as regards the superconducting properties and the anomalous skin effect in the normal state.

After he had reached a detailed understanding of the anomalous skin effect in an isotropic metal, Pippard went on to consider the theory for a single crystal. He showed that the surface impedance of the metal for any particular direction of the RF field could be related to the geometrical properties of the FS. Although the relation was a rather complicated one, involving the curvature of the FS, he suggested it might be used for determining the FS from systematic measurements of the anisotropy of the RF surface impedance in anomalous skin effect conditions. During a sabbatical visit to the University of Chicago in 1957 he succeeded in determining the FS of copper in this way with sufficient precision to suggest that the surface probably bulged out from the free-electron sphere enough to make contacts with the zone faces normal to (111), as in figure 5.6(*a*) and rather like Sommerfeld and Bethe's sketch of a constant energy surface in figure 5.1. This was an important pioneer effect and proved particularly useful in the guidance it provided for the later DHVA experiments. However, the relation between the geometry of the FS and the anomalous skin effect is so complicated that it is usually hardly possible to go uniquely from

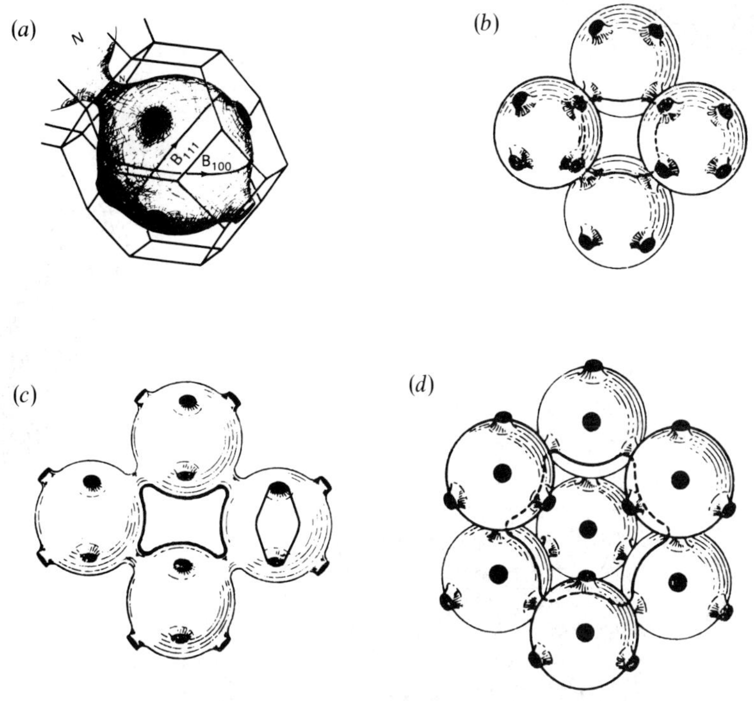

Figure 5.6 Schematic sketches of the FS of the noble metals. (*a*) the FS within a single zone joined by a neck (N) to the surface in a neighbouring zone. The belly orbits normal to (111) and (100) are also indicated. In (*b*), (*c*) and (*d*) (see Lee 1969), the FS in the extended zone scheme is viewed normal to (100), (110) and (111) respectively and shows the location of the four-cornered rosette in (*b*), the dog's-bone and lemon in (*c*) and the six-cornered rosette in (*d*). The hidden parts of the orbits in (*b*) and (*d*) are shown broken.

the experimental data to the FS, particularly if the FS consists of several sheets. Rather, it is the other way round, that once the FS has been determined, it is possible to calculate what the anomalous skin effect should be, thus providing a useful check on the determination.

Returning to my own experiments on the DHVA effect in copper, an element of farce came in when I tried to check whether Pippard's inference of contact between the FS and the (111) zone faces was really correct. The crucial test was to compare the behaviour of a (100) or (111) whisker with that of a (110) whisker. As can be seen from figure 5.6, if there is indeed contact, there should be no continuous 'belly'† orbit round a central plane normal to (110). in

† Note the curious, partly anatomical, jargon that rapidly came into use to describe the geometrical features of the FS.

contrast to the continuous belly orbits normal to (100) or (111). When, however, the test was made at the end of January 1959, the (110) whisker gave much the same high frequency belly oscillations as had been seen in December with a (111) whisker. This caused some consternation, but it soon turned out that it was all a mistake—it was simply that the whisker orientation had been wrongly determined (it was actually (100)). It was only some time later that the check could be properly made and it showed conclusively that Pippard was right.

In the meantime, good oscillations had been observed in both silver and gold crystals grown from the melt and it was soon established that their Fermi surfaces conformed to the Pippard model. The most direct evidence was the observation (first in gold but eventually in silver and copper too) of much lower frequency oscillations when the field was along (111). These correspond to the minimum area of what was immediately called a 'neck' and the corresponding frequencies directly measure the area of contact on the (111) zone face. Once the validity of the model was confirmed it was realised that there should be several other extremal areas. These are (see figure 5.6) the non-central dog's-bone normal to (110) and four-cornered rosette normal to (100), each about 0.4 times the belly area, and the central six-cornered rosette of nearly twice the belly area. The eventual observation of the expected corresponding frequencies provided striking confirmatory evidence for the Pippard model as well as additional data for a detailed specification of the FS.

The problem of how to specify a geometrical surface from knowledge of all its extremal areas of cross section is not a trivial one and indeed, as Lifshitz and Pogorelov showed in 1955, a unique answer can be given only if certain restrictive conditions apply. However, if, as here, there is cubic symmetry and the shape is qualitatively known, the equation of the surface can be expressed as an expansion in suitable functions with arbitrary coefficients and the values of the coefficients determined. This approach was successfully used by Roaf with a three-dimensional Fourier expansion—essentially an extension of the original tight binding formula given by Bloch but with coefficients devoid of physical meaning. This was only the beginning of a considerable further effort on the noble metals, but it will be convenient to defer further discussion until later.

In parallel with the start of systematic exploitation of the DHVA effect to determine Fermi surfaces there was a rapidly growing development of other methods. Pippard's use of the anomalous skin effect has already been mentioned and he was also instrumental in developing the magneto-acoustic method. This came about through his ingenious interpretation of the oscillations that Bömmel had observed in the attenuation of ultrasonic waves passing through a tin crystal when the magnetic field was varied. Pippard suggested that there should be increased attenuation whenever the classical electron orbit size could be fitted into an odd number of half-wavelengths of the waves and that the periodicity could therefore be related to a 'caliper' dimension of the FS. Such caliper measurements could ideally be used to determine the FS independently, but in any case could serve as a useful check on a determination based on any other technique.

Other important contributions in the late 1950s came from the theoretical group of I M Lifshitz at Kharkov. One such contribution, by Lifshitz, Azbel, Kaganov and Peschanski, showed how the orientation dependence of the magnetoresistance of a metal crystal could reveal the topology of the FS, i.e. could indicate whether or not sections of the FS by planes normal to the field direction contained any open orbits. This idea soon began to be exploited by Alekseevski's experimental group in Moscow and often provided valuable guidance in the unravelling of complicated Fermi surfaces. An early confirmation of the theory was pointed out by Priestley in 1960 who showed that the Russian magnetoresistance results were consistent with the FS of gold as deduced from my DHVA data.

Again, in 1956 Azbel and Kaner invented the idea of a new kind of cyclotron resonance in which the magnetic field is parallel to the metal surface and there should be increased RF surface resistance whenever the RF was a multiple of the cyclotron frequency (i.e. the frequency of rotation of the electrons). Measurement of the periodicity of the corresponding oscillations in RF surface resistance should give the cyclotron frequency and hence the cyclotron mass. The effect was soon discovered by Fawcett in Cambridge and later studied in detail in a number of metals, particularly by Kip and his group at Berkeley. The cyclotron mass can also be determined from the temperature dependence of the DHVA amplitude and it was reassuring that both methods gave consistent answers. The importance of the cyclotron mass is that it effectively measures the derivative dA/dE of the extremal area A with respect to the energy E of the surface of constant energy, taken at the Fermi energy. Thus, in principle, if dA/dE and A are known for all directions it is possible to construct the form of surfaces in the neighbourhood of the FS and hence to deduce how the electron velocity varies over the FS and also the value of the density of states at the Fermi energy.

Finally, we should mention that many other physical properties besides the magnetisation show oscillations as the field is varied, e.g. the temperature of the sample (magnetothermal effect), the physical dimensions (magnetostriction), the velocity of sound, the electrical properties (e.g. Shubnikov–de Haas effect) and the attenuation of sound. Some of these provide alternative techniques for determining the DHVA frequency and hence the FS, while others also provide additional information such as the stress dependence of the FS.

5.4 COOPERSTOWN 1960 AND AFTER: FERMIOLOGY COMES OF AGE

By 1960 a considerable activity had grown up in the investigation of the electronic structure of metals, both theoretically through band structure calculations and experimentally by study of the various phenomena mentioned above. This field of study gradually came to be called 'Fermiology', a convenient name for a subject which is not very easy to define, and it is fair to say that by 1960 it had become an established branch of solid state physics. This was marked by an important conference on Fermi surfaces at Cooperstown, New York, at which both experimental and theoretical progress was reviewed in a way which stimulated mutual interaction (Harrison and Webb 1960). A review of the many-body theory of electrons in metals at about this time (Falicov and Heine 1961) was able to give an unequivocal 'yes' in answer to the subtitle of the review; 'Has a metal really got a Fermi surface?'

D Shoenberg and A B Pippard holding a model of the Fermi surface of aluminium at the Cooperstown conference, 1960.

Progress after Cooperstown was surprisingly rapid on a number of fronts as Fermiology became increasingly popular and considerable improvements were made in both theoretical and experimental techniques. One of the hopes expressed at the conference was that it should soon be possible to apply many different techniques to a single metal and thus check that the FS was a concept which was independent of the particular phenomenon used to measure it. This hope was soon fulfilled for a number of metals, and indeed several new techniques such as Gantmakher's RF size effect and positron annihilation were developed which gave further confirmation. However, it gradually became clear that usually the most versatile and accurate method of determination of the Fermi surfaces was in fact the DHVA effect, partly because of the very high phase† of the oscillations which makes for high precision and partly because it is in principle such a simple technique.

Added confidence in the DHVA method was provided by a rather general argument presented by Luttinger at the conference which showed that the constant of proportionality between DHVA frequency, F, and FS area, A, is indeed as equation (5.1) implies, a universal constant which should be unaffected by many-body interactions. This was contrary to suggestions by Falicov and Stern who had argued that the electronic charge e in equation (5.1) should be modified to an effective electronic charge e^* which might be a few per cent different. There was indeed some experimental support for such a difference from my preliminary measurements on the noble metals, but this proved to be yet another hiccup in the orderly development of the subject and it was soon shown to be a spurious result, due to a small error in the calibration of the magnetic field.

A puzzling mystery mentioned at the Cooperstown conference was Priestley's observation that one of the DHVA frequencies he observed in magnesium was too high to make any obvious sense. It corresponded to a FS extremal area much larger than any which could be expected on the basis of plausible models. This mystery was soon solved in principle by Cohen and Falicov and worked out more fully by Blount. It turned out that the electron was experiencing 'magnetic breakdown', a kind of quantum tunnelling between states in different zones of equal energy but separated by a small potential barrier. Thus, effectively, the electron could switch tracks from one orbit to another and in this way describe quite new and sometimes much larger orbits. This opened up a whole new field of investigation which was not only of considerable interest in its own right but provided valuable supplementary information about subtle details of the band structure.

The main technical advance of the 1960s came from the introduction of superconducting magnets. Even though the available fields were rather less than those which could be reached impulsively, the fact that the field was steady could be exploited to improve greatly the signal to noise ratio and the precision of measurement. Instead of using the time variation of an impulsive field to induce an EMF, part of which was proportional to $\mathrm{d}M/\mathrm{d}H$, detection was achieved by superimposing a small alternating field on the steady field and observing how the pick-up signal amplitude varied as the steady field was slowly changed. One of my former students, J K Hulm, then at Westinghouse,

† The oscillations can be described by a periodic function of the phase $2\pi F/H$ (note that F/H is typically several hundreds) and the precision of determination of F increases with the magnitude of the phase since more oscillations appear over a given field range. In most other oscillatory effects the relevant phase is much lower than in the DHVA effect.

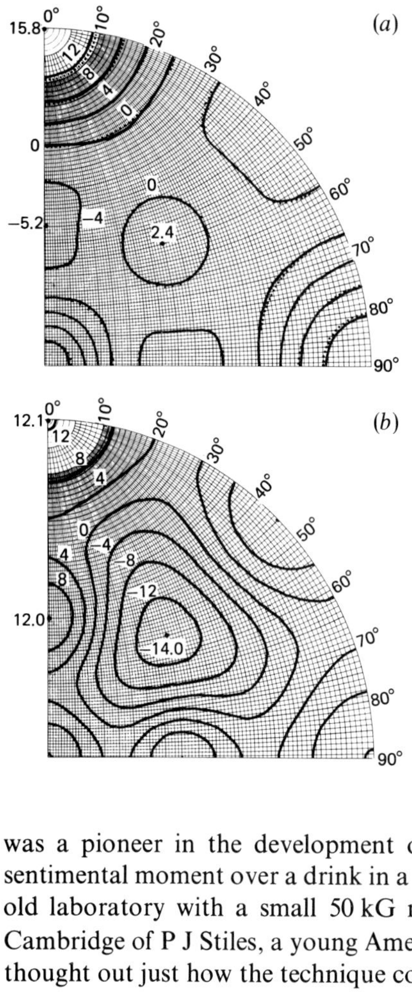

Figure 5.7 Stereographic projections illustrating departures of the potassium FS from an exact sphere (Templeton 1981). (*a*) Contours of $\Delta A/A_0$ where A is the area of cross section of the FS normal to a given direction, A_0 is the cross section for the free-electron sphere and ΔA is $(A - A_0)$. (*b*) Contours of $\Delta k/k_0$ where k is the radius of the FS in the given direction and k_0 the mean radius. The units in both (*a*) and (*b*) are parts in 10^4. The broken lines in (*a*) are the contours as originally determined by Shoenberg and Stiles (1964).

was a pioneer in the development of superconducting magnets and in a sentimental moment over a drink in a Pittsburgh bar he offered to present his old laboratory with a small 50 kG magnet. This, together with a visit to Cambridge of P J Stiles, a young American postdoctoral researcher who had thought out just how the technique could be made to work, enabled us to be the first to use a superconducting magnet to study the DHVA effect. It also made possible the use of a nice trick for precise determination of the slight departures of the nearly spherical FS of an alkali metal from a perfect sphere. The trick was to record the oscillations as the sample was rotated in an exactly constant field rather than as the field was varied at fixed orientation. Since the phase of the oscillations was of the order of $10^4 \times 2\pi$, a single oscillation implied an area change of only 1 in 10^4, so that the departures from sphericity could be mapped out with great precision (as in figure 5.7). The inversion from areas to radii was achieved in a very simple way by describing the surface in terms of cubic harmonics.

This kind of method was soon widely taken up and with the introduction of various ingenious new features, made possible by advances in electronic and computing techniques, it gradually became a precision tool for measuring

DHVA spectra. The Fermi surface has been described as the 'visiting card' of a metal (Lifshitz and Kaganov 1980) and since Cooperstown the Fermi surfaces of nearly all the metals have been determined, some with great precision. Thus the FS of copper is now known to a precision of the order of 0.1% and agrees with that based on reliable band calculations to well within the uncertainties of the latter (typically of the order of 1%).

Further studies of the noble metals have also revealed interesting new features which have given even more confidence in the validity of the FS concept. Thus in 1969 Morley Halse demonstrated that the frequencies of subsidiary oscillations observed over certain orientation ranges by Joseph and Thorsen in 1964 were indeed exactly those to be expected from non-central extremal areas of the FS, the form of which had been deduced from the central extrema alone. Halse also observed oscillations corresponding exactly to his prediction of a lemon-shaped non-central extremum normal to the (110) plane (see figure 5.6(*c*)). Studies of the cyclotron masses by Halse and more recently by Lengeler and his group in 1977, analysed along the lines indicated earlier, have led to a detailed picture of how the electron velocity varies over the FS and also to values of the density of states in almost exact agreement with those deduced from the electronic specific heats. The cyclotron masses are enhanced over the values indicated by band structure calculations because of electron–phonon interaction and the precise data of Lengeler's group combined with band calculations by Martin Lee showed that the enhancement factor $(1+\lambda)$ is appreciably anisotropic (varying by 20% or so in copper).

An interesting aspect of the noble metal studies is that the FS determined from the DHVA effect can be specified not only by the purely empirical Fourier expansions of the Roaf type, but by a somewhat more meaningful parametrised band structure calculation. This approach was pioneered by Lee (1969) who, with the use of only four fitting parameters—three phase shifts and the Fermi energy—carried through a band structure calculation for copper which gave an FS in excellent agreement with the DHVA data. Further calculations on these lines for all three noble metals have been carried out by Shaw, Ketterson and Windmiller in 1972 and Holzwarth in 1975.

The studies of bismuth, the noble metals and the alkali metals have been described in some detail partly because of their historical interest and partly because they are relatively simple. We must now turn to a few other case histories of Fermi surfaces which, though rather less simple, illustrate particular points or are of some aesthetic appeal.

The first attempt to determine the FS of aluminium was made in 1957 by Gunnersen who tried to analyse a rather complicated pattern of low frequency oscillations by the Lifshitz–Pogorelov approach and came up with three small cushion-shaped surfaces which could be plausibly associated with features of a 'first principles' OPW band structure calculation by Heine. This, however, proved to be yet another hiccup! Thus Harrison in 1959 pointed out that Gunnersen's data could be equally plausibly fitted to a very slight distortion of

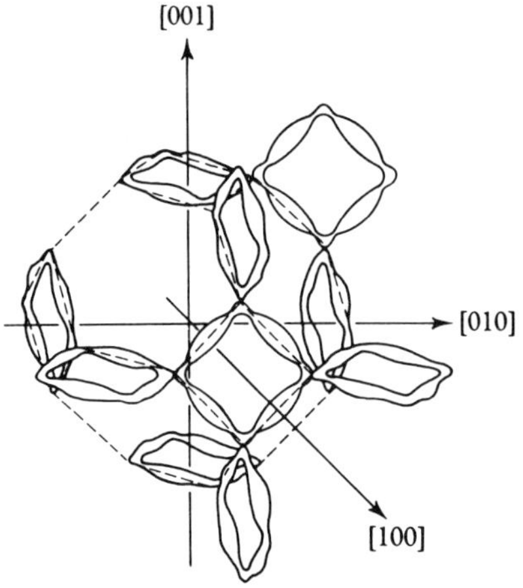

Figure 5.8 The third zone FS of aluminium (Volski 1964). The multiply connected 'monster' of figure 5.4 breaks up into separated 'rings of four' after Ashcroft's pseudo-potential has been applied. This diagram is shifted from that of figure 5.4 by half a reciprocal lattice vector in the direction $X\Gamma$. Several of the rings of four shown are outside the basic zone.

the multiconnected sheet of the exactly free electron surface in the third zone shown in figure 5.4. Not long afterwards Ashcroft (in Cambridge) did a systematic pseudo-potential calculation and was able to show that the low frequency data could only be fitted in detail by a particular combination of the pseudo-potential parameters V_{200} and $|V_{111}|$ which broke up the 'monster' of figure 5.4 into the 'rings of four' of figure 5.8. By a curious coincidence Volski (in Moscow) came to hear of Ashcroft's calculations only after he had completed a more thorough study of the DHVA effect and the reliability of both theory and expreiment was strikingly demonstrated by the excellent agreement he found between his frequency data and the calculated values which Ashcroft sent. A few years later a still more detailed experimental study by Larson and Gordon provided further confirmation for Ashcroft's model. The much higher frequencies corresponding to the second zone nearly-free-electron surface of figure 5.4 were observed by Priestley in 1962 and though the modifications to the exactly free electron surface by the small pseudo-potential were only slight, it was satisfying that Priestley's data did confirm them. In a way it is paradoxical that the low frequencies associated with the fine detail of the FS provided the key to the accurate theoretical determination of the whole surface—rather like the tail wagging the dog!

Arsenic has the same crystal structure as bismuth, but the overlap of the FS and the Brillouin zone is much more considerable and the FS is consequently rather more complicated. From a careful analysis of the rather complicated frequency spectrum (figure 5.9(*a*)) Priestley *et al* (1967) concluded that as in bismuth there are three tilted electron surfaces which are, however, only roughly ellipsoidal and considerably larger. The hole surface is however dramatically different: instead of the single ellipsoid of bismuth there are six sections multiply connected by thin necks. These results agreed well with a pseudo-potential calculation by Lin and Falicov (1966) according to which the hole surface has the rather curious carrot-like form shown in figure 5.9(*b*).

(*a*)

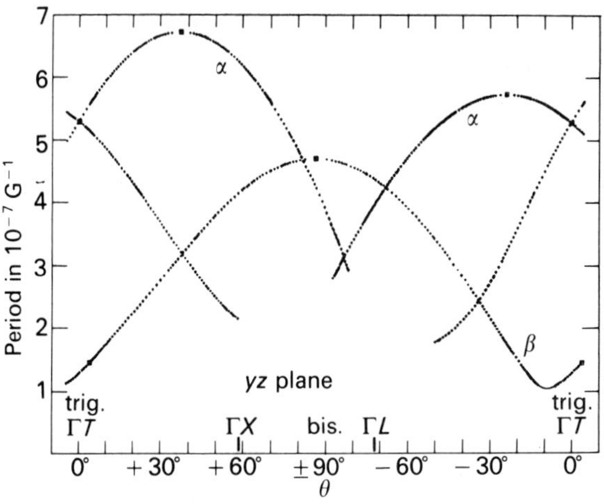

(*b*)

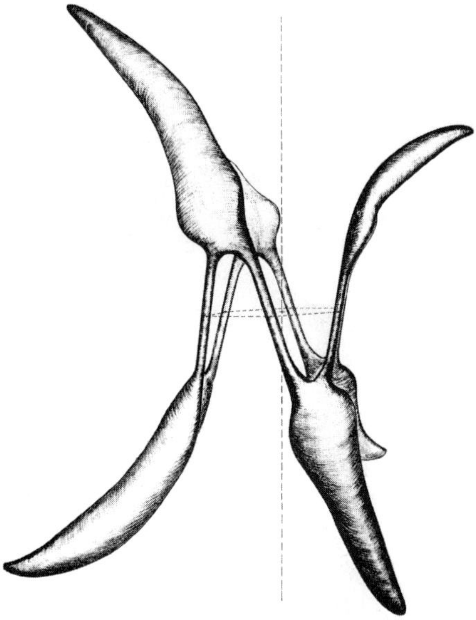

Figure 5.9 (*a*) Variation of DHVA periods of arsenic in the trigonal-bisectrix plane (Priestley *et al* 1967). The α-oscillations correspond to major sections of the hole FS shown in (*b*); the β-oscillations correspond to three, approximately ellipsoidal, electron surfaces. (*b*) A perspective sketch of the multiply connected holes FS (Lin and Falicov 1966); the narrow necks connecting the 'carrots' give rise to periods (not shown in (*a*)) of the order of 50 times greater than those shown in (*a*).

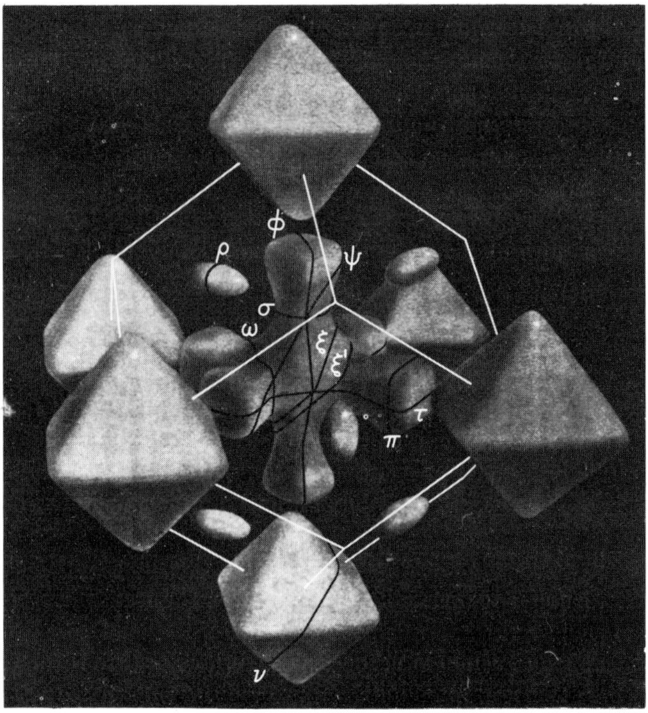

Figure 5.10 Perspective sketch of the FS of tungsten (Girvan *et al* 1968). The various orbits of extremal area are indicated by the Greek letters and give rise to the complicated frequency spectrum actually observed.

This picture is an artist's impression by Falicov's wife Martha, who confessed it was one of the most difficult drawings she had ever made. The theoretical FS in fact has extremal areas corresponding very closely to the experimentally observed frequency spectrum and indeed the considerable interaction between the theorists and experimenters at the same Institute contributed greatly to the successful outcome of the whole project. This was the Institute for the Study of Metals, now the James Franck Institute, at the University of Chicago; it was here too that Pippard made his classical study of the anomalous skin effect in copper.

Our last example of an interesting FS is that of the transition metal tungsten shown in figure 5.10—again an artist's impression. The FS is based on the complicated frequency spectrum measured by Girvan, Gold and Phillips (1968) with guidance from a theoretical band structure calculation. The calculated areas of the model agree well with the observed spectrum. The calculation of the band structures of transition metals is complicated by the dominant role of the d-electrons close to the FS, and in the early days Conyers

Herring once compared such calculations to the mixing of a cocktail for which each theorist has his own ingredients. Nowadays the recipes have settled down and there are generally accepted methods which give reliable results (see review by Mackintosh and Andersen 1980). An interesting feature of the FS of the non-ferromagnetic transition metals pointed out by Lomer is that a qualitative picture can be derived for all of them from a common band structure by merely altering the Fermi level appropriately.

Although the emphasis of this chapter has been on Fermi surfaces and how they are determined, the information provided by the experimental techniques goes a lot further and new applications are still emerging. We have already mentioned that cyclotron masses can be determined from the temperature dependence of the DHVA oscillation amplitude or from cyclotron resonance and such determinations give information on the electron–phonon interaction which plays an important role in the theory of superconductivity. The damping of the oscillations as the field is reduced is partly controlled by electron scattering and partly by sample imperfections (the kind of phase-smearing mentioned earlier). Measurements of this damping (the Dingle factor) in suitable samples provide important information for the theory of dilute alloys in which impurity scattering is dominant, or about dislocations in pure but slightly deformed samples.

Again, there are characteristic features of the oscillation amplitude due to interference between the spin-up and spin-down contributions. By suitable measurements of the amplitude, the effective g-factor of the spins can be determined and related to spin–orbit and many-body electron–electron interactions†. If the electron–electron exchange interaction is strong enough, the metal may become ferromagnetic and the FS of the spin-up and spin-down electrons become well separated. The observation of the DHVA effect in a ferromagnetic metal (iron) by Anderson and Gold in 1963 opened up a whole new field of study. This, supplemented by neutron diffraction, is throwing new light on the basic theory of ferromagnetic and nearly ferromagnetic metals (Lonzarich 1984, Lonzarich and Taillefer 1985). Another important development, made possible by improvements in metallurgical techniques, has been the study of stoichiometric metallic compounds, thus enormously enlarging the scope of the field. The striking progress since the first observation of DHVA oscillations in BiIn by Thorsen and Berlincourt in 1961 and the beginning of systematic studies on several other compounds soon after by J-P Jan's group at NRC, Ottawa, has been reviewed by Sellmyer (1978). More recently, various weakly ferromagnetic (e.g. Ni_3Al, Sigfusson *et al* 1984) or nearly ferromagnetic (e.g. Ni_3Ga, Hayden *et al* 1986) compounds have extended the range over which magnetic theory can be tested; a rather pleasing looking part of the FS of Ni_3Ga is shown in figure 5.11. Another

† The g-factor, which is two for a free electron, specifies the ratio of the magnetic moment to the angular momentum of the electron spin in units of $(e/2mc)$. It was introduced by Landé in 1923 in his analysis of the anomalous Zeeman effect.

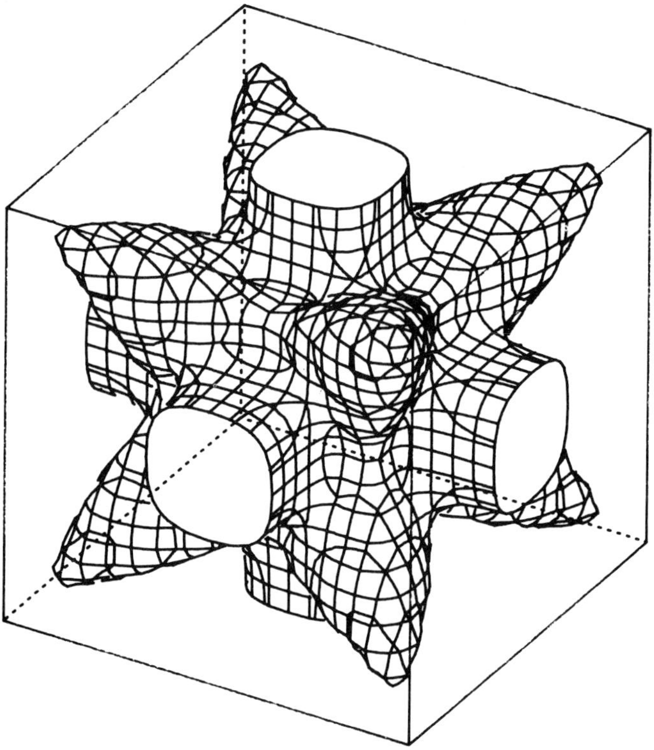

Figure 5.11 Part of the hole FS of Ni_3Ga shown in the simple cubic Brillouin zone (Hayden *et al* 1985). The necks give rise to prominent DHVA oscillations of frequency about 3×10^7 G, in good agreement with the band structure prediction.

application is in essentially two-dimensional systems such as silicon inversion layers (Fang and Stiles 1983, Störmer *et al* 1983) or suitable intercalated graphite samples (Iye *et al* 1980, Markiewicz *et al* 1985).

Finally, we return briefly to an important feature of Fermiology which has so far been mentioned only in passing. This is the rather remarkable stroke of luck that the simple-minded independent particle picture should have proved so successful, both in theoretical calculations of band structure and in the theory of electronic phenomena such as the DHVA effect. This is a good illustration of how sometimes physics can develop quite happily on the basis of only partially understood or even implausible assumptions. In fact it was only in the 1950s and 1960s that the justification for the independent particle model began to be properly understood in terms of 'quasi-particles' on the basis of Landau's Fermi liquid theory (for a recent review see Lifshitz and Kaganov (1980)). It turns out that for a system of strongly interacting electrons, phonons and magnons, for states close to the Fermi energy, the system is equivalent to that of non-interacting quasi-particles which behave in almost the simple-minded way assumed all along. The main difference is that the effective mass and spin-moment of the quasi-particles may be appreciably enhanced over the values suggested by band theory. The magnitudes of these enhancements as revealed by a combination of experiment and theory are providing important data to improve understanding of many-body theory.

FURTHER READING

Many important topics have been mentioned in this chapter with little or no detailed explanation and for more information a standard textbook, such as Ashcroft and Mermin (1976) should be consulted. Other important texts which illustrate the state of the subject at various earlier times are those of Mott and Jones (1936), Seitz (1940) and Wilson (1953). Collections of reviews on various relevant topics may be found in the 'Mottschrift' (Ziman 1969) and in Springford (1980); a few other reviews are mentioned in the text. My recent book (Shoenberg 1984) gives a systematic account of the DHVA and allied effects and contains a detailed bibliography. Interesting historical material is given in Chapter 1 of Wilson (1953), in a collection of reminiscent articles in *Proc. R. Soc.* A **371** (1980) and in Hoch (1983). Except for a few recent papers and papers from which diagrams have been copied, references to original papers have in general been omitted; such references may be found in the various reviews and books mentioned above and in other review articles quoted in the text.

ACKNOWLEDGMENTS

Portraits of W J de Haas, P M van Alphen, L V Shubnikov, R E Peierls, L D Landau, L Onsager and I M Lifshitz and many of the figures were taken from *Magnetic Oscillations in Metals* (Cambridge University Press 1984) by D Shoenberg with the kind permission of Cambridge University Press. Dirac's portrait was kindly provided by Professor R H Dalitz; it appears in *Tributes to Paul Dirac* (Adam Hilger, 1987). Original drawings for figures 5.9 and 5.11 were provided by kind permission of Dr M G Priestley and Dr S M Hayden respectively.

REFERENCES

Ashcroft N W and Mermin N D 1976 *Solid State Physics* (New York: Holt Rinehart and Winston)

Falicov L M and Heine V 1961 *Adv. Phys.* **10** 57

Fang F and Stiles P J 1983 *Phys. Rev.* B **28** 6992
Girvan R F, Gold A V and Phillips R A 1968 *J. Phys. Chem. Solids* **29** 1485
de Haas W J and van Alphen P M 1930 *Proc. Netherlands Royal Acad. Sci.* **33** 1106
Harrison W A 1960 *Phys. Rev.* **118** 1190
Harrison W A and Webb M B (ed.) 1960 The Fermi Surface *Proc. Int. Conf. Cooperstown (New York) 1960* (New York: Wiley)
Hayden S M, Lonzarich G G and Skriver H L 1986 *Phys. Rev.* B **33** 4977
Heine V, Cohen M L and Weaire D 1970 *Solid State Phys.* **24** 1, 38 and 250
Hoch P 1983 *Contemp. Phys.* **24** 3
Iye Y, Takahashi O and Tanuma S 1980 *Solid State Commun.* **33** 1071
Lee M J G 1969 *Phys. Rev.* **187** 901
Lifshitz I M and Kaganov M I 1980 in *Electrons at the Fermi Surface* ed. M Springford (Cambridge: Cambridge University Press) p 3
Lin P J and Falicov L M 1966 *Phys. Rev.* **142** 441
Lonzarich G G 1984 *J. Magn. and Magn. Mater.* **45** 43
Lonzarich G G and Taillefer L 1985 *J. Phys. C: Solid State Phys.* **18** 4339
Mackintosh A R and Andersen O K 1980 in *Electrons at the Fermi Surface* ed. M Springford (Cambridge: Cambridge University Press) p 149
Markiewicz R S, Meskoob M and Zahopoulos C 1985 *Phys. Rev. Lett.* **54** 1436
Mott N F and Jones H 1936 *The Theory of the Properties of Metals and Alloys* (Cambridge: Cambridge University Press)
Onsager L 1952 *Phil. Mag.* **43** 1006
Peierls R 1972 *Aspects of Quantum Theory* ed. A Salam and E P Wigner (Cambridge: Cambridge University Press) p 117
Priestley M G, Windmiller L R, Ketterson J B and Eckstein Y 1967 *Phys. Rev.* **154** 671
Seitz F 1940 *The Modern Theory of Solids* (New York: McGraw-Hill)
Sellmyer D J 1978 *Solid State Phys.* **33** 83
Shoenberg D 1939 *Proc. R. Soc.* A **170** 341
—— 1959 *Nature* **183** 171
—— 1984 *Magnetic Oscillations in Metals* (Cambridge: Cambridge University Press)
Shoenberg D and Stiles P J 1964 *Proc. R. Soc.* A **281** 62
Sigfusson T I, Bernhoeft N R and Lonzarich G G 1984 *J. Phys. F: Met. Phys.* **14** 2141
Sommerfeld A and Bethe H 1933 *Handbuch der Physik* **24** 333
Springford M (ed.) 1980 *Electrons at the Fermi Surface* (Cambridge: Cambridge University Press)
Störmer H L, Haavasoja T, Narayanamurti V, Gossard A C and Wiegmann W 1983 *J. Vac. Sci. Technol.* B **1** 423
Templeton I M 1981 *J. Low Temp. Phys.* **43** 293
Volski E P 1964 *Zh. Eksp. Teor. Fiz.* **46** 123 (*Sov. Phys.–JETP* **19** 89)
Wilson A H 1953 *The Theory of Metals* (Cambridge: Cambridge University Press)
—— 1980 *Proc. R. Soc.* A **371** 42
Ziman J M (ed.) 1969 *The Physics of Metals 1: Electrons* (Cambridge: Cambridge University Press)

C F Braun (1850–1918)

6 Semiconductors

D Weaire

6.1 INTRODUCTION

Whenever we look back at the history of technology, it is natural to seek to identify crucial events, the points at which new materials, methods or machines first appeared.

In the case of the Semiconductor Age, in which we may be said to be living at present, 23 December 1947 is often identified in such terms. That was the day on which the point contact transistor was born. It was not a very pretty baby—a small slab of germanium with various contacts crudely attached (figure 6.1). Its descendants, organised in the form of large computers, are taking over the running of many aspects of our lives.

The device invented by Bardeen, Brattain and Shockley (figure 6.2) also had ancestors which we can trace back well into the nineteenth century. In this chapter we shall sketch some of that prehistory and offer a few comments on the extraordinary progress since 1947, while M Kelly will attempt elsewhere (Chapter 11) to extrapolate into the future.

6.2 THE NINETEENTH CENTURY

Many of the key properties of semiconductors were observed in the nineteenth century but they could not take their place within a coherent classification and interpretation of materials since none existed.

Michael Faraday, in addition to his pioneering work on the basic laws of electricity and magnetism, took great interest in material properties. In particular, he noted that various materials, such as silver sulphide, were capable of conducting electricity but showed an anomalous variation of conductivity with temperature. As he put it 'the conducting power increases with the heat'. His observation contrasted with the typical behaviour of metals, in which the conductivity decreases with increasing temperature. This was in 1833—it required a hundred years before the puzzle was resolved.

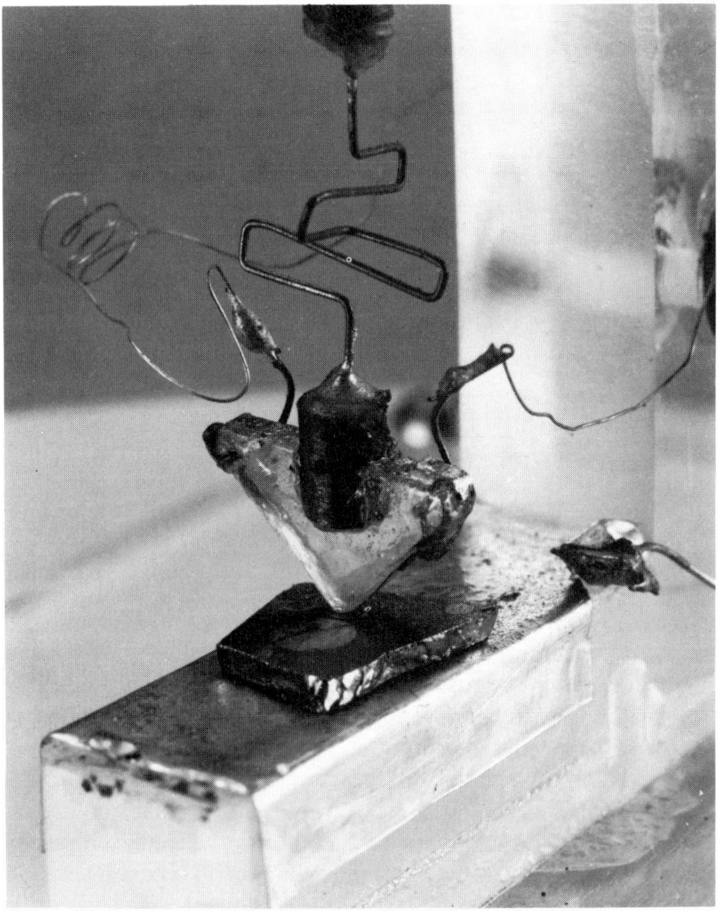

Figure 6.1 The first point contact transistor.

Many other relevant discoveries followed, including the photovoltaic and photoconductive effects and rectifying contacts (for lead sulphide, by Braun in 1874). All of these offered promising opportunities for applications, but only compounded the mystery of such materials, lying between the two familiar extremes of insulator and good conductor.

With the benefit of hindsight we can see that most of the important general properties of semiconductors had already been established for special cases by 1900. These are as follows:

(i) sensitivity to temperature ($d\sigma/dT$ large and positive);
(ii) sensitivity to light (photoconductivity);
(iii) rectifying junctions.

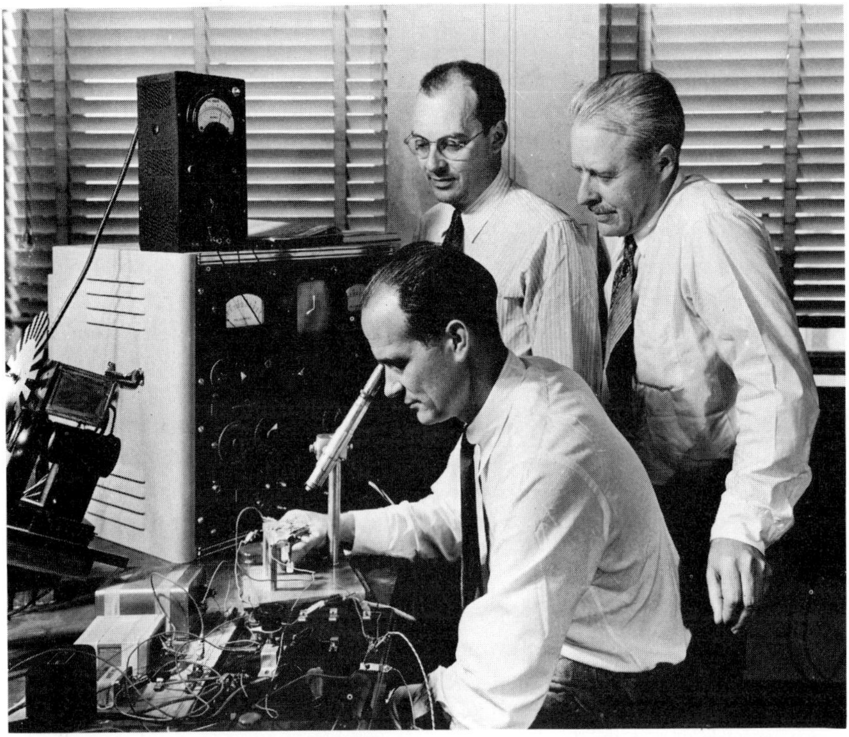

Figure 6.2 Bardeen, Brattain and Shockley (seated), inventors of the transistor. (Photograph courtesy of the Bell Telephone Laboratories.)

Together with these we should include a fourth:

(iv) sensitivity to impurities.

Much nineteenth century research was carried out on impure natural minerals. Indeed, it was not until around 1940 that techniques of crystal growth and the removal or introduction of impurities were adequate to the task of understanding the sensitivity of semiconductors to impurities.

In retrospect, Braun's discovery was particularly significant. The 'cat's whisker' for wireless telegraphy, which developed from his invention, was certainly an early and important semiconductor device. Braun received the Nobel Prize in 1909 for his work on wireless, but this was more a recognition of his later designs of circuits and antennae than that early and seminal discovery. By 1909 the crystal set was already on the way out as a practical method of wireless reception, although it remained of interest for another 20 years and still lingers on as a toy. A popular book on wireless, dated 1931 (Stranger 1931), lists 32 suitable crystals. As for the origin of the effect, even by then it was 'still a mystery'.

6.3 THE AGE OF THE VACUUM VALVE

Braun's rectifying contact enjoyed only a brief heyday in wireless, in the form of the Dunwoody carborundum detector, in the early years of this century. It was, together with other primitive receiving devices, rapidly overtaken by the invention of the vacuum valve. Compared with this the metal–semiconductor device was awkward and unreliable.

Beginning with the invention of the vacuum diode by Fleming in 1904 and the vacuum triode by de Forest in 1906, the new technology swept all before it. From time to time some visionary would suggest the hidden potential of the solid state for electronics. A curious example is that of Lilienfeld, who in the 1920s patented device designs for solid state triodes very similar to those of modern transistors. But such voices were crying in the wilderness. The rapid progress of solid state theory in the 1930s, which we shall shortly review, did not inspire the electronics industry. Rather, it was the metallurgical engineers that were excited. It was the Iron and Steel Foundation which gave a generous grant for the development of a solid state physics laboratory at Bristol University when Nevill Mott went there. Yet Mott at that time produced, with others, the first satisfactory explanation of rectifying junctions, in largely classical terms.

Of course, there were, throughout the period in question, various specialised commercial devices based on semiconductors—light detectors and meters, thermometers, robust devices and so on. One early application of great importance in the history of physics was the use of zinc sulphide in the 'Spinthariscope' or scintillation detector of radiation, invented by Crookes and used by Rutherford in such experiments as α-particle back-scattering. But 'electronics' was synonymous with the vacuum valve. It reached its zenith in the mid-1940s, when such giant computers as ENIAC were assembled. They used tens of thousands of valves and consumed kilowatts of power while computing on the astonishing time-scale of milliseconds. They seemed to herald the dawn of an era of yet more successful applications of the valve. As de Forest said (in 1950!) 'by and large the electron *in vacuo* will continue to rule the amplifier roost'. In fact ENIAC and its cousins were dinosaurs on the brink of extinction.

6.4 THEORETICAL BACKGROUND

As we have noted, the pre-war period was one of rapid advance in the theory of solids, having mostly to do with the role of the electron in determining solid state properties. At the turn of the century the electron had emerged, apparently as the last vital ingredient of the classical theory of matter. In fact the failure of classical theory to account for the properties of the electron precipitated quantum theory. Some of the early successes of quantum theory were specifically related to solid state properties, particularly Einstein's theory of the photoelectric effect and his (and Debye's) theory of the specific heat of solids. Generally speaking, the properties of metals were soon elucidated by the new ideas, in terms of the free-electron model and eventually the Bloch model, while some properties of insulators came to be understood in the opposite terms, as due to localised electrons bound at specific defects. But semiconductor properties involve both Bloch electrons, whose band structure is rather subtle, and bound electrons at impurities. This was a challenge which solid state theory could not meet in the 1930s and theoretical ideas regarding these materials remained rather nebulous and/or classical in nature.

6.5 EXPERIMENTAL BACKGROUND

In addition to the inadequacy of available theory there were not enough adequate experiments for comparison in the 1930s, although many vital components of experimental technique were already in place. One highly significant and quite venerable measurement was that of the Hall effect, discovered in 1879. This identifies the carriers of electrical current as positive or negative in charge and gives their density. Early Hall measurements were important in classifying various compounds as semiconductors and later work in the 1930s clearly distinguished 'excess' and 'defect' semiconductors. In modern terms, these are n-type (with electrons as current carriers) and p-type (with holes, or missing electrons, as current carriers).

Perhaps the greatest deficiency lay in the area of sample preparation. Semiconducting properties are, as we have noted, strongly modified by impurities and other defects. Modern semiconductor physics depends critically on preparation of samples (and hence, to a large extent, on the semiconductor industry which was its own creation).

It is sad to have to recall that the source of the major investment of effort and finance which rapidly improved the situation was World War II. The development of radar made demands upon valve technology which it could not readily meet—in the extension of existing radio techniques to higher frequencies. It was recognised that metal–silicon rectifiers did not suffer from the same limitations at high frequencies. Out of this realisation, together with requirements for infrared detectors, came the intense wartime effort to improve the characteristics of semiconductor devices. But the really crucial steps were taken just after the war, particularly at the Bell Telephone Laboratories.

6.6 THE BELL LABORATORIES

It must be the dream of every research manager to make plans for the future as farsighted and fruitful as those which the Bell Telephone Laboratories made in the mid-1940s. Solid state physics was 'given its head' in anticipation of progress in solid state telecommunication devices. The pay-off of this policy was not long in coming and it established the laboratories as the 'Mecca' of the subject, which they have remained until today. The first few years leading up to 1950 saw the invention of the point contact transistor, various techniques for growing and doping semiconductors, the p–n junction and the junction transistor, together with theoretical interpretations which still hold water today. Bell Laboratories contributed much to the detailed theory of semiconductors which we now have at our disposal. (See Chapter 5 for band structure.)

While it did not quite come first, the p–n junction is logically the prototypical modern semiconducting device. By doping two adjacent regions, p- and n-type, a rectifying junction is set up within the material itself. Not only does this provide the first step to the junction transistor (two such junctions back-to-back and close together), but it also suggests the essential spirit of later semiconductor technology—to build circuits *within* a solid block of material. We are still far from realising the full potential of this idea, as Kelly indicates in Chapter 11. The original point contact transistor, on the other hand, was a comparatively crude device, a direct descendant of Braun's 'cat's whisker', in which the cat has two whiskers! The public reaction (that this should be good for hearing aids) was not entirely unjustified in its moderate tone.

In the years since 1950 the junction (or bipolar) transistor, together with its cousin the field effect transistor, have been absorbed with other circuit elements into integrated circuits of ever greater complexity on single chips. The latest computers, based on these, approach the *nanosecond* time-scale and are composed of individual devices on the *micrometre* length scale. Meanwhile, the older applications of semiconductors, as light detectors and meters and luminescent devices, have benefited from the new techniques in many ways. Particularly significant at present is the semiconductor laser, used in optical communications. A spirited debate surrounds another related possibility—the use of semiconductors as components in all-optical circuits. If this comes about it will be a startling change of direction in what is already an extraordinary story. Another recent directional change has been towards the use of amorphous semiconductors for some purposes.

6.7 AMORPHOUS SEMICONDUCTORS

In the prehistory of solid state physics, which Smith has described in Chapter 2, the distinction between *crystals* and *glasses* was widely recognised and, indeed, of some commercial importance in such matters as the making of porcelain and glassware. It was realised that glasses occur commonly 'where

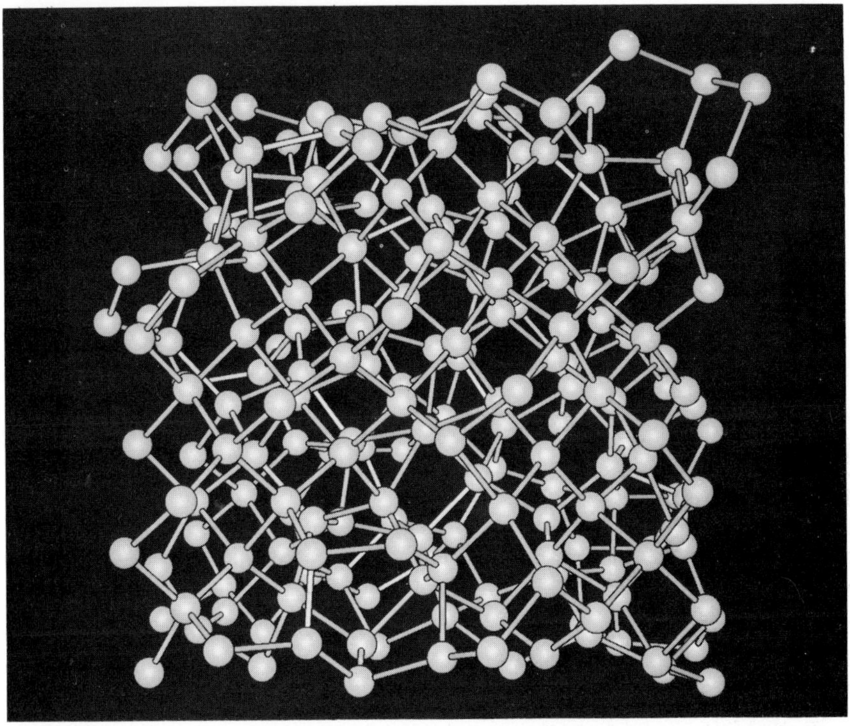

Figure 6.3 Computer-generated model for amorphous silicon. The model was developed by F Wooten, D Weaire and K Winer with graphics software designed by N Max. (Reproduced by permission of the above and the Lawrence Livermore National Laboratory.)

Figure 6.4 Flat screen television using amorphous silicon, developed by Sanyo in 1985.

their transition from the fluid to the solid state were not effected too hastily' (Keir 1776). The basic *microscopic* distinction (order/disorder) was also appreciated—the glass structure was an 'inextricably entangled skein', 'noise' as opposed to 'music' (Tyndall 1868).

Today, amorphous materials, which have the disordered structure of a glass but are not necessarily made by rapid cooling of a liquid, are of growing importance in solid state physics. They may be cheaply fabricated in thin films of large area; hence their use in xerography (amorphous Se), now several decades old. Amorphous semiconductors, especially hydrogenated amorphous silicon (figure 6.3), are of particular interest today. To a first approximation, this form of amorphous silicon has all the familiar properties of its crystalline counterpart, so that p–n junctions, transistors and other electronic devices can be made from it. Where large areas are desired (displays, solar cells) the advantage is obvious and will be exploited in the coming years (see, for example, figure 6.4).

Such materials have posed intriguing problems for the theorist. How do the properties of a disordered material differ from those of a crystalline one? How may we calculate them? Nevill Mott has pioneered this field, together with Philip Anderson in the specialised area of conductivity; they received the Nobel Prize for their work in 1977.

6.8 SEMICONDUCTOR PHYSICS

The intense industrial interest in the development of semiconductors has fed back into the supporting science of semiconductor physics. Invention and application have marched in step with fundamental theory and laboratory experiment in a manner which is the very model of the interdependence of pure and applied science.

At the most basic level, the band structure of the common semiconductors has been established by experiment and calculation with enough accuracy for most practical purposes. Chapter 5 reviews some of the early post-war theoretical work on this problem, while Chapter 10 mentions the sophisticated optical techniques which have served as experimental probes.

In parallel with this effort to establish the electronic band structure of the pure semiconductor, other experimental techniques have explored the levels associated with electrons or holes bound at impurities. Here the early theory of Wannier has served as a guide only to *weakly* bound states due to impurities not too grossly different from the host material. Only today is the problem of deep impurity levels being adequately analysed. This is because in such cases there are large local distortions around the impurity, difficult to calculate and almost impossible to measure. We are nevertheless on the brink of extending our understanding to such awkward (and important) impurities. Once that has been achieved, the classic problems of semiconductor physics may be said to have been solved. But new challenges emerge all the time—the electron–hole plasma in semiconductors, non-linear optical effects, amorphous semiconductors . . . . The materials which Faraday found so interesting in 1833 are here to stay, for both the engineer and the physicist.

REFERENCES AND FURTHER READING

Atherton W A 1984 *From Compass to Computer* (San Francisco: San Francisco Press)

Braun E and MacDonald S 1982 *Revolution in Miniature*, 2nd edn (Cambridge: Cambridge University Press)

Keir J 1776 *Phil. Trans. R. Soc.* **66** 530

Smith R A 1978 *Semiconductors*, 2nd edn (Cambridge: Cambridge University Press)

Stranger R 1931 *The Outline of Wireless for the Man in the Street* (London: G Newnes)

Tyndall J 1868 *Heat—A Mode of Motion* (London: Longmans Green)

N F Mott (1905–)

7 Magnetism

E P Wohlfarth

The science of magnetism is one of the oldest known to man and is now the basis of an extensive endeavour called magnetic information technology (MINT). Such a long period of research and development has given this science an enormous stability from which predictions of the future may be made with some degree of certainty. In spite of its unique scientific and technological importance, the image of magnetism and the intensity of its elementary and advanced study has varied considerably in time and place. A historical survey of the subject in isolation cannot explain this phenomenon, which must be ascribed to locally varying perceived needs and the presence or otherwise of sufficiently enthusiastic practitioners. Such enthusiasts were around as early as 1269 when Petrus Peregrinus wrote his famous Letter about the lodestone, a subject which gained its most elegant early scientific study at the hands of Gilbert in 1600. (See figure 7.1 for an armed lodestone of more recent date.) Not much research on the lodestone problem was carried out until the present century. For example, compass needles were increasingly produced from steel magnets and this led to one of many fields of research in applied magnetism which is more active than ever, as will be described.

Apart from the development of such useful magnetic materials, the question of the origin of magnetism has arisen many times in the past, as well as the present and presumably the future. The earliest theories, from pre-Christian days, involved effluvia or invisible emanations (a not infrequent phenomenon even now!). Many centuries later Ewing proposed ideas involving tiny bar magnets to simulate magnetic molecules. This was at the end of the nineteenth century, in which French and other European scientists had made many classical contributions to the theory and practice of magnetism. The present century began with two major events in this subject; Ewing's visit to Tokyo some years earlier provoked an unprecedented Japanese effort in magnetism at the hands of Honda, and Weiss proposed the two key concepts of the *molecular field* and *domains* as basic to any understanding of the subject. These enthusiastic practitioners laid the foundations of modern magnetism.

The molecular field concept of Weiss explained quite well many of the

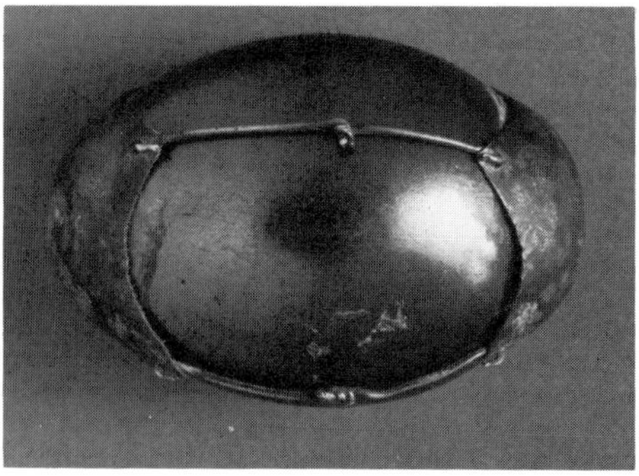

Figure 7.1 An armed lodestone. (Courtesy of the late Professor M Blackman.)

observed characteristics of real magnets, such as the temperature dependence of the magnetisation M. The molecular field coefficient, the ratio of this field to M, was, however, so large ($\sim 10^4$ for iron) that its origins could not be magnetostatic. It took the introduction of quantum mechanics for Heisenberg to propose exchange, i.e. an electrostatic origin for this field. This Heisenberg model, as well as the equally influential, though classical, Ising model, has been used in discussions of the magnetic properties of a vast range of materials for which the magnetic carriers are localised electrons, such as magnetite (which is equivalent to lodestone), rare earth chalcogenides such as EuO and spin glasses. The models have been more widely applied in discussing critical phenomena.

Within the framework of the Heisenberg model a central outcome of applying statistical mechanics was the establishment by Bloch that the elementary excitations determining the temperature dependence of the low-temperature magnetisation, $M(T)$, are the magnetic analogues of excitons called spin waves. The resulting law, $M(T) = M(0)(1 - \alpha T^{3/2})$, where α is determined by the exchange energy, is the most famous law in magnetism. It agrees with experiment much more closely than the corresponding law derived directly from the Weiss molecular field model using not quantum, but classical, statistics.

In metals the magnetic carriers are not localised but itinerant electrons. That is, the electrons are relatively free to move through the lattice and are described as being distributed in energy bands. For the 3d transition in metals such as Fe these bands are usually several electron-volts wide; for the 4f electrons in the rare earths, such as Gd, the bands are so narrow that the localised description is regained and for the 5f electrons in the actinides, such as U, the bands are again of finite width, typically 1 eV. Apart from having to know all about the structure of these energy bands it is again essential to apply statistical mechanics. The spin wave excitations continue to be described by Bose statistics but the additional single particle excitations peculiar to the itinerant electron or Stoner model are most simply described by Fermi statistics. In this way the $T^{3/2}$ law is supplemented by other laws characteristic of the metallic state. The model has more recently received growing attention, with international conferences and much public acrimony. The relative importance of single-particle excitations and spin fluctuations in different temperature regions and the best ways of estimating exchange and correlation effects have been discussed. In addition, the complications of the metallurgy of the materials being investigated (metals, alloys, compounds, thin films, amorphous substances, etc) are clearly of great relevance and can lead to important magnetic effects when discussing such matters as the field dependence of the magnetisation. Fortunately, many modern techniques have been applied in these studies, including neutron scattering, solid state spectroscopy, Mössbauer effect, high magnetic field and high pressure studies. The theoretical interpretations cannot always keep pace with the resulting data.

The scientific study of magnetism has another aspect, analogous to the analysis of microstructure in metallurgy, which leads to the understanding of the desirable features of applied and technological materials. The nomenclature of 'secondary effects' is somewhat misleading as there is considerable overlap with the primary effects discussed so far. For example, the domain structure in bulk and thin film materials is dependent on a variety of energy contributions such as the exchange energy, magnetostatic contributions, magnetocrystalline anisotropy (a manifestation of spin–orbit coupling) and magnetostrictive strain energies. The problems of computing stable domain configurations and, often more important, of describing domain wall dynamics (e.g. in bubble domain materials) are mathematically difficult and form a part of micromagnetics. Domain walls (early examples of solitons) can be observed experimentally by a number of methods, such as by powder pattern, electron microscopy, Kerr and Faraday magneto-optic effect, x-ray and neutron methods.

A particularly important branch of secondary magnetism concerns the behaviour of fine particle magnetic materials. These particles, if energetically appropriate as regards size, shape etc, are single domains and their properties are thus more easily understood. The resulting rotations of the magnetisation are determined by the anisotropy of the particles and may be so difficult that the materials are magnetically hard, i.e. have a high coercive force. In this way the properties of a broad family of materials, spin glasses, rocks, ferrofluids, some permanent magnets, some living organisms (e.g. magnetotactic bacteria) and, in particular, magnetic recording media, may be understood, leading to an unusually active field of research with both scientific and technological aspects. In Great Britain this research is in the hands of the Northern Fine Particle Magnetism Association, a loosely knit organisation of theorists and experimentalists. Figures 7.2 and 7.3 show examples of fine particles.

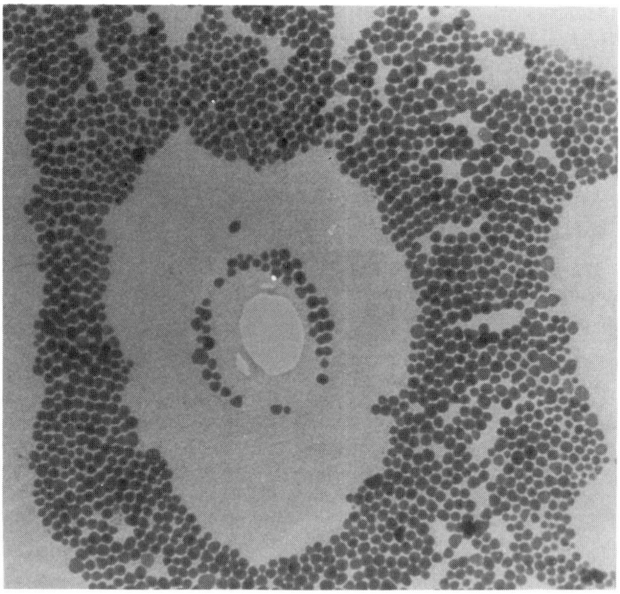

Figure 7.2 Cobalt particles for ferrofluids. (Courtesy of Dr Kevin O'Grady.)

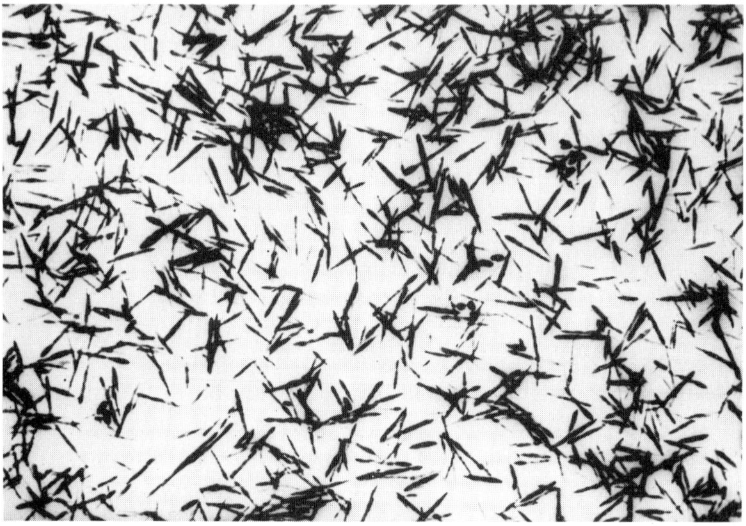

Figure 7.3 Iron oxide magnetic recording media. (Courtesy of Dr Ariello Corradi.)

To further illustrate the applicability of the subject, let us run through some specific examples of interest.

Rapidly quenched amorphous magnetic materials

Here we have a new class of materials with low losses and high magnetisation for applications such as power transformers, magnetic heads and shields to rival the classical iron–nickel and iron–silicon based crystalline alloys. The materials have zero overall magnetocrystalline anisotropy and thus high permeability and low coercive force. They are produced by rapid quenching, as described by Cahn in Chapter 4, and have a typical composition Fe–Ni–X, where X is a combination of the metalloids B, P, C, Si etc. The problem of stresses is still serious and is partially solved by an annealing treatment.

High Si content iron–silicon alloys

It was known for many decades that sheet steel with a low Si content, produced by classical rolling techniques, has a Si content limited by ductility considerations to about 3%. Crystalline 6% Si alloys have now been produced by Tsuya using the same rapid quenching method as above. Again, an annealing treatment must be applied subsequently and the process brings about columnar grain growth and alignment and very low coercive forces, typically down to 60 mOe. The losses are typically half of those of conventional, best, 3% Si steel.

Crystalline and amorphous Invar alloys

Since Guillaume received the Nobel Prize in 1920 for inventing Invar (Fe–Ni) alloys, pure and applied research has revealed a whole range of other related materials with interesting magnetoelastic properties. Applications are possible to devices such as glass–metal seals, bimetals, wave-guide tubes and mechanical watch springs (not completely excluded by quartz; the IWC Engineer costs more in its mechanical than electronic form). The low thermal expansion (Invar effect) arises from a balance between positive phonon and negative magnetic contributions. Another manifestation is a large volume magnetostriction. Crystalline alloys not containing iron (e.g. Ni–Pt) and amorphous alloys such as Fe–B are among materials of this type with a great potential for future device applications.

Magnetostrictive rare earth compounds

Transducers based on magnetism require a large magnetomechanical coupling effect and the materials must thus have a large magnetostriction. The polycrystalline compound $TbFe_2$ has a saturation magnetostriction coefficient of almost 2×10^{-3} at room temperature, which is some 50 times larger than that of nickel. The effects are due to the large anisotropic 4f charge density of the Tb^{3+} ions. The major development was carried out by the American Navy in line with applications in underwater detection.

Amorphous rare earth–iron alloys for magneto-optic recording
Using this technique Japanese firms have claimed an eight inch disc holds 20 000 pages of A4! The technique uses a number of concepts in magnetism, including compensation point writing (the magnetisation of ferrimagnets vanishes due to magnetic sublattice compensation), the temperature dependence of the coercive force and the Kerr magneto-optic effect which allows a written bit to be read. The materials used in this important development include amorphous Gd–Co, Tb–Fe and Tb–Fe–Co alloys, produced by sputtering or evaporation.

Particulate magnetic recording media
This is the most widely used application of fine particle magnetism, involving many billions of dollars in the magnetic information technology (MINT) field. The Japanese lead is so overwhelming that the American effort will have the greatest difficulty in competing; the European research and development effort may never compete, in spite of the European academic excellence in this field. Video and audio tapes and rigid and floppy discs must have the following properties; large magnetisation and coercive force, a narrow switching field distribution, low noise (i.e. small particles) and a smooth surface. Optimum properties have been developed using elongated γFe_2O_3 particles (see figure 7.3), including those with cobalt adsorption to add magnetocrystalline anisotropy to shape anisotropy. CrO_2 particles, with their excellent surface condition, are also used. The problem of dipole interactions in practical tapes and discs also makes this a prime example of academic research.

Co–Cr films for perpendicular recording
Since the ordinary longitudinal recording mode at high bit density produces a non-uniform magnetisation with unwanted perpendicular components, a major (mainly Japanese) effort on thin films with perpendicular anisotropy has been carried out leading to a perpendicular recording mode. The 80:20 Co–Cr films may be produced by sputtering which seems to lead to hexagonal Co-rich crystals perpendicular to the film. Scientific questions of great interest concern the precise magnetic character of these materials. Are they best described as domain or as fine-particle materials? (See figure 7.4 for a fine-particle model.) Answers to such questions will lead to improvements in material quality.

Ferrofluids
These are fine ferromagnetic particles of cobalt (see figure 7.2), magnetite etc, covered with a surfactant and dispersed in a liquid such as oil. Applications include rotary hermetic seals of shafts in some machines and ink-jet printing. (See figure 7.5 for an example of ink-jet printing.) Scientific problems in fine-particle magnetism again concern the dipole interactions which can be assessed by computer simulations or by magnetic measurements.

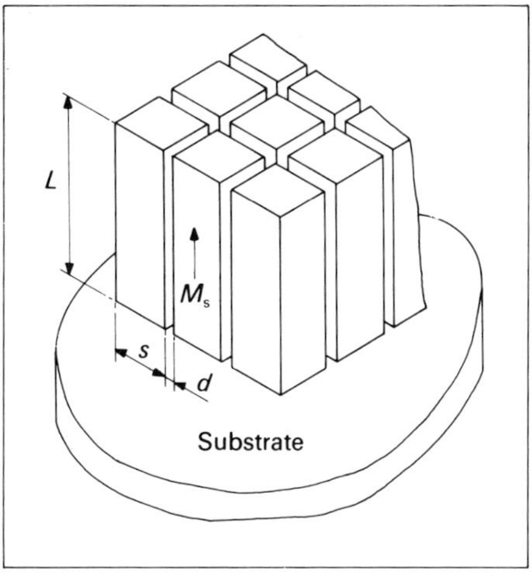

Figure 7.4 A columnar fine-particle model for Co–Cr films. (Thesis of T Wielinga 1983 Technical University, Twente.)

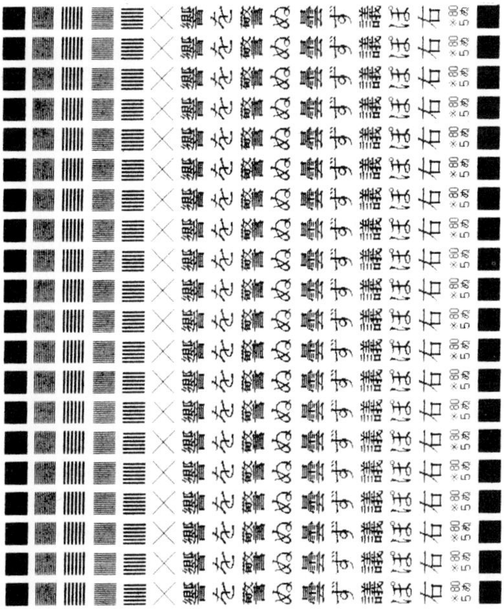

Figure 7.5 An example of ink-jet printing. (Courtesy of Matsushita Company.)

Fe–Nd–B permanent magnets

For many years $SmCo_5$ permanent magnets have led in their figure of merit, the maximum energy product (giving the energy in the magnet gap). Figures of over 3×10^7 gauss-oersted were obtained, leading to many novel applications, e.g. in medicine. In 1983 General Motors produced Fe–Nd–B magnets by rapid quenching leading to rather high values of this product. However, Sumitomo Special Metals produced magnets by more normal metallurgical techniques with product values of over 4×10^7 gauss-oersted. This extreme hardness arises as a result of the tetragonal crystal structure, leading to a very high magnetic anisotropy which is instrumental in determining the pinning of domain walls and/or rotational processes. The resulting uses of these non-cobalt-containing magnets have not yet been fully evaluated.

This brief review is intended to highlight the fact that although magnetism is an ancient subject it has kept pace with the needs of society where those who organise matters have allowed this to occur. The Japanese pre-eminence is manifest and arises historically from the influence of Ewing falling on extremely fertile ground. It is the interplay between theory and practice in this environment which has had the required effect. Can this not also occur in the West, in the United States and particularly in Europe where there is such a long tradition in academic as well as practical research? Or is the image of the 'old' subject too destructive when compared with that of other more modern-sounding disciplines such as semiconductor physics? The British record is not hopeful; in the 1970s, soon after all the talk of the white heat of technology, EMI stopped research on magnetic recording and the Permanent Magnet Association was disbanded! Can anything be done in the 1980s and beyond? Can the information technology revolution not encompass MINT, or does *hardware* sound too old-fashioned compared with *software*?

Since the future technological developments of magnetism look so uncertain as far as Europe is concerned, and since they may move even further towards Japan, it is pleasant to project a much more optimistic outlook for the purely scientific research effort in this subject. The computer revolution will enable ever more sophisticated calculations to be carried out. In metallic magnetism energy bands will be calculated to greater and greater accuracy. Calculations of equal quality will be made of the dynamic susceptibility and the resulting improvements in the discussion of elementary excitations will enable a number of outstanding problems to be solved, e.g. the question of the existence of spin waves at higher temperatures. The influence of orbital effects will be considered. Sophisticated measurements using neutron and electron scattering will help towards the same end. Extreme conditions of pressure and magnetic field will become more and more routine. Fundamental researches on oxides, alloys, compounds, rare earths, actinides, amorphous materials and disordered substances like spin glasses will reveal a rich variety of phenomena and a return to fashion of lateral thinking will lead to the recognition of a universality in the description of a variety of properties of different materials. Not all of these substances will remain forever academic and now and again a technological breakthrough will emerge from this improved understanding. Perhaps this will, after all, occur in Europe!

FURTHER READING

Blackman M 1983 The lodestone: A survey of the history and the physics *Contemp. Phys.* **24** 319–31

Chikazumi S 1982 Evolution of research in magnetism in Japan *J. Appl. Phys.* **53** 7631–6

White R W 1985 Research opportunities in magnetic materials *J. Appl. Phys.* **57** 2996–9

Wohlfarth E P and K H J Buschow (ed.) *Ferromagnetic Materials*, vols. 1, 2 (1980), 3 (1982), 4, 5 (in preparation) (Amsterdam: North-Holland)

8 Superconductivity

T D Clark

8.1 INTRODUCTION

Just over three-quarters of a century ago Kamerlingh Onnes and his co-workers at the University of Leiden succeeded in liquefying helium, for those days a real technological *tour de force*. There existed, of course, a strong scientific drive behind this development. It was appreciated that many materials possess intriguing properties at low temperatures. Onnes, for example, was particularly interested in the problem of low-temperature residual resistance in metals and its dependence on impurity content. With the newly liquefied helium opening up the temperature range from 4.2 K downwards, Onnes set out to measure this resistance in the purest metal then available to him—mercury. To his undoubted astonishment the resistance of his mercury samples did not tend towards some constant value as the temperature was reduced below 4.2 K, as would have been the case had impurity scattering of electrons been the process principally responsible for the resistance. Instead, at a precise temperature of 4.15 K, all DC resistance in these samples vanished. Onnes had discovered superconductivity. His first report on this behaviour in mercury was published in 1911, just three years after the first liquefaction of helium (Onnes). Superconductivity can perhaps lay claim to being the first child of low-temperature physics, even though an adequate explanation of its physical origins had to wait for almost another half-century. Even so, Onnes clearly recognised superconductivity as something quite remarkable. As he wrote in 1913, 'No doubt was left of the existence of a new state of mercury in which its resistance has practically vanished .... Mercury has passed into a new state which on account of its extraordinary electrical properties may be called the superconductive state' (Onnes 1913, Shoenberg 1965). It is certainly a curious twist of fate that a phenomenon which is such a striking manifestation of quantum mechanical behaviour on the macroscopic scale should have been discovered before the development of the quantum theory. With the benefit of hindsight it is easy to see that it is just this macroscopic quantum coherent behaviour which made the observation of superconductivity so simple. Superconductivity is, in general, not critically dependent on impurity level. Compared with semiconductors, relatively extremely dirty materials can display superconductivity. Local impurities, provided they are not magnetic, do not scatter the highly correlated electrons responsible for the superconducting state.

What perspective could Onnes have had on his discovery? He would have expected important applications in the generation of magnetic fields and, at the time, he could have had no obvious reason to suspect that superconductivity was restricted to low temperatures, given, of course, that mercury was not the only superconductor in nature. He was not to know in 1911 that the majority of superconductors can only maintain their superconducting state in rather low magnetic fields. Really high-field superconducting materials, suitable for engineering applications, only became available in the 1950s. The notion that superconductors might have applications in electronics surely could not have occurred to Onnes. Electronics was in its infancy; after all, de Forest did not invent the vacuum triode until 1906. Again, it was not until the late 1950s that serious thought was given to the development of practical superconducting electronics. The steady evolution of high-field superconductors and the varying fortunes of superconducting electronics over the last 25 years form the two principal themes of this chapter.

It could be argued that the 74 years that have passed since Onnes' discovery represent a very long time on the scale on which technologies evolve. In that time the science of semiconductivity has been developed to a quite extraordinary degree and, as is quite evident, currently provides the basis of the dominant world industry—electronics. By comparison, superconductivity has not fared very well. As its protagonists would claim, superconductivity has many possible and varied applications. However, even they would admit that, as yet, no major industry based on superconducting materials or electronic devices has been established. Such a major industrial base almost came into being with the development of Josephson junction computing systems. Unfortunately, but not unexpectedly, semiconducting electronics has proved a most formidable and effective adversary. For example, it is sadly apparent that superconducting (Josephson junction) electronics will play no role in the next generation of mainframe computers. In the longer term this triumph of semiconductivity over superconductivity may turn out to be incomplete. It now appears that circuits which must be described in purely quantum mechanical terms at the macroscopic level can be constructed from superconducting materials. Clearly such macroscopic quantum systems constitute a very radical departure from present day classical electrodynamic circuits and may well have revolutionary potential. It is not yet clear, at a fundamental level, how electronic engineering based on macroscopic quantum circuits can be realised. However, given that this can be achieved, there seems little doubt that superconductivity will have an important part to play in the future development of electronics. This would complement other, already established, areas of application where superconducting technology cannot realistically be displaced—for example, in high magnetic field electrical engineering, high-power and ultra-high quality factor microwave resonators and ultra-sensitive magnetic flux sensors.

8.2 THE SUPERCONDUCTING STATE

The most striking and well known property of superconductors is their lack of electrical resistance. Such behaviour, bizarre though it may be, does not offer much insight into the nature of the superconducting state. The first real indicator came in Berlin in 1933 when Meissner and Ochsenfeld found, most surprisingly, that when a pure metal is cooled through its superconducting transition temperature (T_c) in the presence of an applied magnetic field, all magnetic flux is expelled from within its bulk (Meissner and Ochsenfeld 1933). Furthermore, when Meissner and Ochsenfeld removed the field with the metal sample superconducting they could detect no residual magnetic dipole moment. As they were well aware, if a superconductor merely acts as a perfect electrical conductor, such a trapped dipole moment should have been observed in the sample. Thus they were confronted with superconductivity being both a zero resistance and zero inductance (i.e. perfectly diamagnetic) state. Static magnetic fields could, it seemed, sustain persistent screening currents. If so, a new relation had to exist connecting magnetic field with electric current in a superconductor.

The spontaneous exclusion of magnetic flux from a superconductor could not be quite complete of course, even in principle. Currents in metals arise from the flow of electrons and no physical current sheet can be made infinitely thin due to the repulsive forces between these electrons. Similarly, screening supercurrents generated in response to an externally applied magnetic field cannot flow in an infinitely thin layer on the surface of a superconductor. The supercurrent sheet must have a finite thickness which, since the current is sustained by a magnetic field, requires that the field be able to penetrate an equal distance, invariably referred to as the superconducting penetration depth (λ).

The connection between field ($\boldsymbol{B}$) and supercurrent density ($\boldsymbol{J}_s$) was established within two years of Meissner's and Ochsenfeld's discovery, again by two German physicists, albeit then living as refugees in England. The Londons (Fritz and Heinz) demonstrated that for the Meissner effect (i.e. $\boldsymbol{B}$ finite just outside and $\boldsymbol{B}=0$ deep within a superconductor) to be accommodated theoretically a necessary modification of Maxwell's equations took the form

$$\boldsymbol{B} = -\text{curl}(\Lambda_s \boldsymbol{J}_s) \tag{8.1a}$$

where, for an electron mass and charge of m and e respectively, and a superconducting electron density of n_s, $\Lambda_s = m/n_s e^2$ (London and London 1935).

Equation (8.1a) can, of course, be written down in the alternative (vector potential $\boldsymbol{A}$) form

$$\boldsymbol{A} = -\Lambda_s \boldsymbol{J}_s \tag{8.1b}$$

which, for a superconductor, replaces Ohm's law relating current density to electric field in a normal metal. Combining (8.1a) with the Maxwell equation, curl $\boldsymbol{B} = \mu_0 \boldsymbol{J}$, yields

$$\nabla^2 \boldsymbol{B} = \frac{\boldsymbol{B}}{\lambda^2} \tag{8.2}$$

which for a plane (superconductor/vacuum) interface in a parallel field has the solution required for the Meissner effect, i.e.

$$\boldsymbol{B} = \boldsymbol{B}_0 \exp(-z/\lambda) \tag{8.3}$$

where $\boldsymbol{B}_0$ is the value of the field at the interface, z is the distance into the superconductor from this interface and $\lambda = (m/\mu_0 n_s e^2)^{1/2}$, the superconducting penetration (or screening) length.

The equations (8.1a) and (8.1b), although phenomenological, have a deep significance at a quantum mechanical level and, as such, have been influential in the development of more fundamental theories of superconductivity. Thus, following the reasoning of Fritz London (1961), we can write the electronic momentum in the presence of a field as

$$\boldsymbol{P} = m\boldsymbol{v}_s + e\boldsymbol{A} \tag{8.4}$$

where $\boldsymbol{v}_s$ is the velocity of the superconducting electrons, and given, as Bloch has argued (London 1961), that with no field applied the ground state should have zero net momentum then the local velocity in the presence of a field is given by

$$\langle \boldsymbol{v}_s \rangle = \frac{-e\boldsymbol{A}}{m}. \tag{8.5}$$

From this, the expression (8.1b) for $\boldsymbol{J}_s = n_s e \langle \boldsymbol{v}_s \rangle$ can be retrieved by inspection and the equation for perfect conductivity of the superconducting electrons (i.e. that these electrons are accelerated steadily by an applied electric field $\boldsymbol{E}$) is given by

$$\boldsymbol{E} = \Lambda_s \dot{\boldsymbol{J}}_s \tag{8.6}$$

and can be obtained by taking time derivatives of both sides of (8.5) with the appropriate choice of gauge, div $\boldsymbol{A} = 0$ (the London gauge).

We can see, therefore, that the implication built into the Londons' equation (8.1) is that superconductivity arises because of a condensation of conduction electrons into a $\langle \boldsymbol{p} \rangle = 0$ ground state. Fritz London's prescience in this respect is remarkable. In 1950 he wrote (London 1961):

> The long-range order of the momentum would be a specific quantum effect .... It would arise from wave packets of wide extension in space assigning the same local mean momentum to the entire superconductor, and from a peculiar stability enabling these wave packets to persist essentially undisturbed in a not too strong magnetic field. It might well be that the wide extension of the wave packets is correlated, by the exigencies of the uncertainty relation, to some cause forcing the momentum statistics to be unusually sharp, perhaps to something like a condensation in momentum space.

We can also see, as no doubt could London, that such a condensation will only come about if the conduction electrons in a metal can undergo a boson-like condensation into the superconducting state. This is, of course, incompatible with electrons acting as Fermi particles. London might have argued (though there is apparently no record that he did) that although single electrons are fermions, suitably paired electrons could behave as bosons. In these circumstances a superconducting condensate, described by a single particle-like wavefunction, could form below some critical temperature dependent on the strength of the pairing interaction between the electrons. The wavefunction of such a single quantum state can be written as

$$\Psi = |\Psi| \, e^{i\phi} \tag{8.7}$$

where the phase ϕ may be a function of position and $n_s = |\Psi|^2$ is the density of paired superconducting electrons of effective charge q, where $q = 2e$, and mass m_p, where $m_p = 2m$.

The macroscopic quantum mechanical nature of such a condensate state becomes obvious when we consider the behaviour of a superconducting ring, thick compared with the penetration depth λ, in an applied magnetic field with the possibility that magnetic flux Φ can be trapped within the ring as it is cooled through the superconducting transition temperature. If the condensate can be described by the wavefunction given in equation (8.7), then it is apparent that in order to keep this wavefunction single-valued at every point within the ring the phase of the wavefunction must change by zero or an integer value $(n) \times 2\pi$ along a path around the ring inside the bulk superconductor. If, then, we assume an effective mass and charge of $2m$ and $2e$ respectively, for the paired state with the corresponding quantum mechanical probability current

$$\boldsymbol{J}_p = \frac{-i\hbar}{4m} \left[\Psi^* \left(\nabla - \frac{2ie}{\hbar} \boldsymbol{A} \right) \Psi - \Psi \left(\nabla + \frac{2ie}{\hbar} \boldsymbol{A} \right) \Psi^* \right]. \tag{8.8}$$

The above phase quantisation condition in the ring can be shown to require that any flux threading the ring must be quantised as

$$\Phi = n\Phi_0 \qquad n = 0, \pm 1, \pm 2 \ldots \text{etc} \tag{8.9}$$

where $\Phi_0 = h/2e$ is the unit of flux associated with a $q = 2e$ condensate (Grassie 1975).

The integer flux states given in equation (8.9), which can have no classical analogue, will exist in a quantum mechanical sense as macroscopically different states of the ring and are a consequence of the assumed macroscopic quantum mechanical nature of the condensate wavefunction of equation (8.7). Each state can, in principle, be accessed by cooling the ring through its superconducting transition temperature in the presence of an appropriate externally applied magnetic field. If the external magnetic flux applied to the superconducting ring is Φ_x, and we accept from equation (8.9) that the total included flux in the ring must be quantised as $n\Phi_0$, then

$$n\Phi_0 = \Phi_x + \Lambda I_s \tag{8.10}$$

where I_s is the diamagnetic screening supercurrent flowing in the ring of geometric inducatance Λ. Since the magnetic energy stored in the ring is $W = (\Lambda I_s^2)/2$ this becomes, for the nth quantised flux state,

$$W_n = \frac{(n\Phi_0 - \Phi_x)^2}{2\Lambda}. \tag{8.11}$$

As we assumed that the thick superconducting ring can be described by a macroscopic single-particle wavefunction, we can now write a time-independent Schrödinger equation for the ring

$$W_n D_n = E_n(\Phi_x) D_n \tag{8.12}$$

where D_n is the quantum mechanical amplitude for the ring to exist in the $(n\Phi_0)$th flux state. The solutions $E_n(\Phi_x)$ of equation (8.12), which are shown in figure 8.1, consist of a set of intersecting, but non-connecting, parabolas. As is apparent, there exists no mechanism for coupling different n states together while the ring remains superconducting.

Within seven years of Fritz London's prophetic remarks on superconductivity being intimately related to a condensation of conduction electrons in momentum space, a successful microscopic theory of superconductivity had been developed. As is common knowledge today, this theory, which originated with John Bardeen, Leon Cooper and J Robert Schrieffer (Bardeen *et al* 1957), shows that superconductivity can arise because it is possible for conduction electrons close to the Fermi surface in a metal to interact attractively via the metal lattice. In quantum mechanical language this attraction develops through the exchange of virtual phonons between electrons. In the simplest and least sophisticated treatment these exchange phonons can be assumed to have energies close to $\hbar\omega_D$, where ω_D is

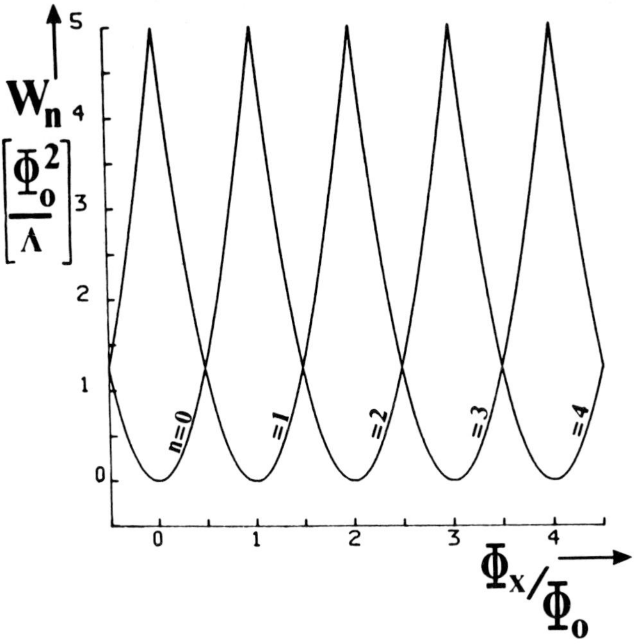

Figure 8.1 External magnetic flux-dependent (Φ_x) energy levels for a thick superconducting ring with trapped flux, 0, Φ_0, $2\Phi_0$, . . ., $n\Phi_0$.

the Debye frequency of the metal. The importance of the lattice in mediating the interaction between electrons became obvious with the discovery in 1950 that for many superconducting elements the transition temperature is a function of isotopic mass (Maxwell 1950, Reynolds *et al* 1950). The connection between the electron–lattice interaction and superconductivity had already been suggested by Fröhlich (1950) prior to the discovery of the isotope effect. Following this discovery the stage was set for the Bardeen–Cooper–Schrieffer (BCS) theory by Cooper in 1956 who demonstrated that, provided there exists a net attractive interaction between electrons, no matter how weak, the Fermi sea is always unstable against the creation of at least one bound pair (Cooper 1956). In the BCS theory these pairs are composed of electrons, coupled together via the electron–phonon–electron interaction, with zero orbital angular momentum and opposed spins (s-type pairing). These pairs condense to form the BCS ground state. With the possible exception of some recently discovered exotic systems, superconductivity always appears to involve s-type pairing.

The energy difference between the normal and superconducting phases of a metal is very small ($\simeq 10^{-8}$ eV per atom) compared with, for example, average binding energies in solids (a few eV per atom). It is, therefore, not surprising that the BCS theory has provided remarkable qualitative insights into the nature of the superconducting state but has been of limited value in calculating critical temperatures and maximum sustainable (critical) magnetic fields from first principles, i.e. from a detailed knowledge of the band structure of the metal or material in question. In the BCS theory, the condensation of pairs leads to the formation of an energy gap, Δ, in the electronic excitation spectrum of the superconductor. Thus, by definition, if one electron state in a pair is occupied, then so is the other, this means that at least 2Δ of energy is required to break a pair. Since the BCS model requires that coupling energy for pair creation is constant over an energy width $\hbar\omega_D = k_B\theta_D$ (θ_D being the Debye temperature) about the Fermi surface, the energy gap at zero temperature can be written simply as

$$2\Delta \simeq 4k_B\theta_D \exp(-1/N(0)V) \tag{8.13}$$

where $N(0)$ is the density of single electron states of one spin in the normal state at the Fermi surface and V is the matrix element for the pairing interaction which is assumed constant over the energy range $k_B\theta_D$. Expression (8.13) is valid when $N(0)V \ll 1$ which, in practice, means in the limit of weak coupling between the pairing electrons. In this limit of weak coupling the critical temperature is given by

$$2\Delta \simeq 3.52k_B T_C. \tag{8.14}$$

So, from equation (8.13),

$$T_C \simeq \theta_D \exp(-1/N(0)V) \tag{8.15}$$

where, typically, θ_D is in the range 100 to 1000 K and, for most known superconductors, $N(0)V \lesssim 1/3$ to $1/4$. If, for example, we take $\theta_D = 500$ K and $N(0)V = 1/4$ then we find from equation (8.15) that $T_C \simeq 10$ K, which is close to the middle of the range of critical temperatures found in real superconducting materials.

The low critical temperatures found for superconductors reflect, of course, the very weak forces of attraction leading to pair formation which means that, on average, paired electrons are a large distance apart, roughly 10^{-5} cm in a typical superconductor. Each Cooper pair, of diameter $\simeq 10^{-5}$ cm, is therefore required to coexist in its own volume with a large number ($\sim 10^6$) of other pairs (London 1961, p30). Thus there exists a large overlap of the pair wavefunctions in this and every other pair volume in a superconductor. This then requires that the phases of these wavefunctions should be locked together over macroscopic length scales. This macroscopic phase coherence, which forces different points in a superconductor to have built-in phase relations, is the fundamental feature of the superconducting state. It follows then that, firstly, a superconductor which is, in principle, a many-bodied system, can be represented by the 'single particle' wavefunction (8.7), i.e. a single macroscopic wavefunction for a very large number of particles, and secondly, it is extraordinarily difficult for such a macroscopic quantum state to be

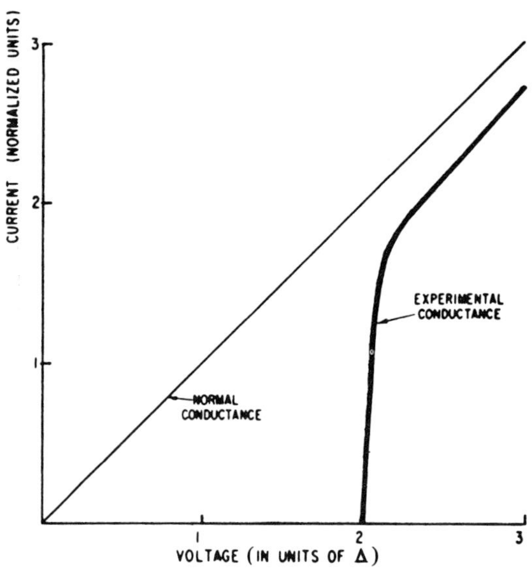

Figure 8.2 Normal and superconducting current against voltage ($I-V$) characteristics of a tin/tin oxide/tin thin film tunnel junction (Giaever 1962).

scattered by the microscope impurities and defects which lead to resistance in normal metals.

The predictions of the BCS theory were subsequently checked very thoroughly, although, perhaps surprisingly, direct experimental evidence for the existence of an energy gap in the electronic excitation spectrum of superconductors came before the observation of flux quantisation in superconducting rings. Following the earlier work of Glover and Tinkham (1956, 1957) on infrared absorption edge studies of superconducting films, Giaever (1960) was able to show that the then new technique of electron tunnelling between metal films separated by a very thin oxide barrier provided an ideal way of investigating the superconducting gap. Figure 8.2 shows some of Giaever's early, and very elegant, results for a tin/tin oxide/tin tunnel junction (Giaever 1962). With the aluminium films in their normal state (at 1.3 K) there is no energy gap in the electronic excitation spectrum so tunnelling currents can flow even in the limit of negligibly small voltages applied to the junction. At $T=0.3$ K, however, where the energy gap in superconducting aluminium is almost fully developed, no current can flow until the applied voltage is sufficient to break up electron pairs, i.e. at $V \geqslant 2\Delta/e$. This tunnelling technique also provided an unexpected bonus which directly verified the BCS interpretation of superconductivity in terms of a phonon-mediated interaction between electrons. Close inspection of the voltage against current (V–I) characteristics of these tunnel junctions revealed

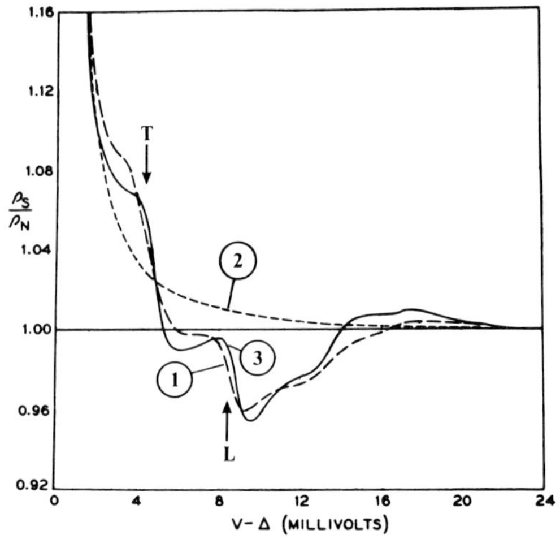

Figure 8.3 Dynamical conductance (ρ_S) plots. (1) Normal aluminium/superconducting lead tunnel junction normalised to the normal state junction conductance (ρ_N) showing the transverse (T) and longitudinal (L) phonon peak structures. (2) The BCS density of states. (3) The density of states calculated using the simplified phonon spectrum of lead. (Rowell and Kopf 1965.)

structure, measured relative to the energy gap, at the energies of the transverse and longitudinal phonon peaks of the superconducting electrodes. This structure was observed to be most pronounced in relatively strongly coupled superconductors such as lead. Even so, more sensitive methods which involved plotting the dynamic conductance (dI/dV), or its derivative (d^2I/dV^2), against V were required to reveal weak structures. Of particular interest at that time was the dynamical conductance plot dI/dV against V for a normal/tunnel–insulator/superconductor junction, since this reflects directly the density of excited allowed states in the superconductor as a function of energy. Small deviations from the BCS density of states (an inverse square root singularity, centred on the energy gap) are always seen, arising from the weighted effect of the phonon spectrum of the superconductor on the electron pairing interaction. These derivative techniques were first introduced by Giaever (1962) but reached their fully developed form through the work of Rowell and his colleagues at the Bell Telephone Laboratories. In figure 8.3 we show a typical dynamical conductance plot for an aluminium/aluminium oxide/lead junction with only the lead superconducting, taken from the data of Rowell and Kopf (1965). The deviations from the simple BCS density of states function can be seen quite clearly. The structures denoted T and L correspond to the transverse and longitudinal peaks in the phonon spectrum of lead. Subsequent refinements of these derivative tunnelling experiments provided much detailed information on the phonon spectra and frequency-dependent pairing interactions of a variety of superconductors (McMillan and

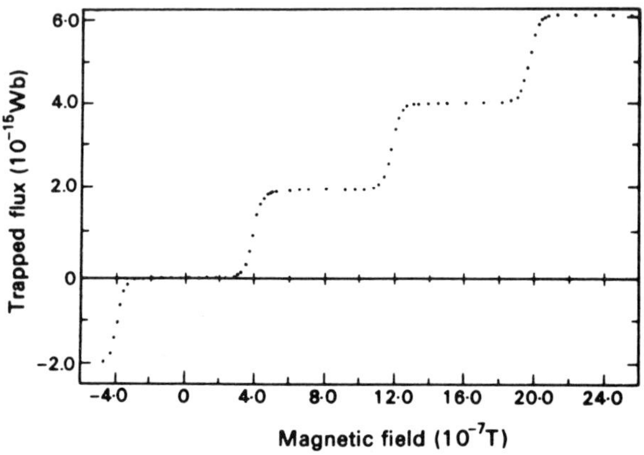

Figure 8.4 Plot of the magnetic flux trapped in a superconducting tin cylinder 24.0×10^{-1} cm long, approximately 5.0×10^{-5} cm thick and 5.6×10^{-3} cm inner diameter, as a function of the ambient magnetic field in which the cylinder is cooled through its superconducting transition temperature. (Goodman *et al* 1971.)

Rowell 1965). Today, electron tunnelling in its various guises is used on a routine basis as one of the principal diagnostic techniques for the investigation of new superconducting materials.

Although Giaever single-electron tunnelling clearly demonstrated that superconductivity arises because of a phonon-mediated interaction between electrons, with the implication that pairing takes place, it did not provide direct evidence for the $q=2e$ effective charge in superconductors. This came in the year following Giaever's first observations of the superconducting energy gap in tunnel junction characteristics. Thus, Deaver and Fairbank (1961) and Doll and Näbauer (1961) reported independently on measurements of the magnetic flux Φ trapped within superconducting cylinders made by vacuum depositing or electroplating the superconducting metal onto the outside of very fine filaments. Although these early results displayed considerable scatter, they did suggest very strongly that flux confined within a superconducting ring is quantised in units of the flux quantum $\Phi_0 = h/2e$ ($=2.07 \times 10^{-15}$ Wb), i.e. the effective charge associated with the superconducting state is $2e$, not e. Later investigations showed that this flux is confined precisely in integer units of $h/2e$, as the data of Goodman *et al* (1971) demonstrate very convincingly in figure 8.4. It still seems extraordinary that, given the outstanding importance of these flux quantisation experiments to the theoretical understanding of superconductivity, they should have been delayed until such a late date.

Flux quantisation played a key role in an extension to a description of the superconducting state which may be thought of as complementary to the BCS theory. As a natural development of the Landau theory of second-order phase transitions (Landau and Lifshitz 1970), Ginzburg and Landau, well before BCS, suggested that an order parameter characterising the condensed superconducting electrons could be expressed as a complex pseudo-wavefunction ψ with the condition that $n_s = |\psi|^2$, the local density of superconducting electrons (Ginzburg and Landau 1950). No consideration was given to any excitations in the system. Even though, at the time, their approach was purely phenomenological, they clearly recognised the macroscopic quantum mechanical nature of the superconducting state. As with London (1961), they did not know the effective mass and charge for superconducting electrons. However, they could reasonably assume that ψ was a function of position so, following the Landau theory, the free energy density of the ordered, superconducting phase could be written as that of the disordered, normal phase plus an expansion in powers of ψ and $\nabla\psi$. This led them to derive two equations, now known as the time-independent Ginzburg–Landau equations, the first being of the form

$$\frac{1}{2m^*}(i\hbar\nabla - e^*\boldsymbol{A})^2\psi + \beta|\psi|^2\psi = -\alpha(T)\psi \qquad (8.16)$$

where m^* and e^* are, respectively, the effective mass and charge of the superconducting electrons and α and β are the coefficients of the first two expansion terms in $|\psi|^2$ in the superconducting free energy density. The second equation is simply an expression for the superconducting current density $\boldsymbol{J}_s = e^*\boldsymbol{J}_p$, where $\boldsymbol{J}_p$ is the quantum mechanical probability current density for particles of mass m^* and charge e^* (see equation (8.8)). The emergence of equation (8.16), so similar to the Schrödinger equation for a free particle, but with an additional non-linear term, together with the quantum mechanical form of the expression for the supercurrent density, underlines the macroscopic quantum mechanical view implicit in the Ginzburg–Landau approach. It is therefore of great interest that Gorkov (1959) was able to demonstrate that close to the ordering temperature (T_c), where Landau theory holds, Ginzburg–Landau theory is a limiting form of the microscopic theory of BCS with ψ being directly proportional to the energy gap parameter, Δ. Thus, by this connection, ψ takes on the role of the wavefunction of the centre of mass motion of the condensed Cooper pairs.

Unlike the earlier electrodynamics of the Londons (1935), the Ginzburg–Landau theory is adequate to deal with both the effect of magnetic fields strong enough to change n_s and with spatial variation of n_s. In order to deal with spatial variations Ginzburg and Landau introduced a minimum distance below which it is energetically unfavourable for ψ to vary, i.e. below which the superconducting phase is unstable with respect to the normal phase. Physically, this distance corresponds roughly to the size of a Cooper pair and is known generally as the Ginzburg–Landau coherence length $\xi(T)$. The other characteristic length, which determines the distance over which the magnetic field varies in a superconductor, is the penetration depth λ. The ratio of these lengths sets the Ginzburg–Landau parameter

$$\kappa=\frac{\lambda}{\xi}. \tag{8.17}$$

In typical, pure superconductors $\lambda \simeq 5\times 10^{-6}$ cm and $\xi \simeq$ a few $\times\, 10^{-5}$ cm, so $\kappa \ll 1$. Since these were historically the first kind of superconductors to be investigated, they are referred to as type I in their behaviour. Abrikosov (1957), in the year that the BCS theory was published, demonstrated that another kind of superconductor should exist—type II—where $\xi < \lambda$. Naively, we might expect the change over from type I to type II to occur at $\xi=\lambda$. Exact calculation shows that the critical value of the Ginzburg–Landau parameter is $\kappa = 1/\sqrt{2}$. For type I behaviour ($\kappa < 1/\sqrt{2}$) there is a single critical magnetic field (H_c) above which the superconducting state cannot be sustained. These critical fields are always low ($\leqslant 0.1$ T). For $\kappa > 1/\sqrt{2}$ (type II behaviour) there is a first and second critical field, H_{c1} and H_{c2} respectively. Abrikosov proved that at H_{c1} magnetic flux begins to penetrate the superconductor in distinct and separate units of $h/2e$, the superconducting flux quantum. Each $h/2e$ tube of flux can be thought of as having two diameters. The first (where the order parameter is close to zero) is approximately equal to 2ξ. The second, and larger (where the flux penetrates), is approximately equal to 2λ. Since each flux tube with a 'normal' core acts as a multiply connected region in the superconductor, circulating screening currents flow around the core region out to distances of approximately λ.

The quantised flux tubes (or bundles) will naturally form a minimum energy configuration. Abrikosov predicted, incorrectly, that they arrange themselves into a square lattice; in fact a triangular lattice array has a lower energy. Such lattice arrays have been observed by electron microscopy using a magnetic powder enhancement technique. In figure 8.5 we show a particuarly beautiful example of a triangular array from the original work of Essmann and Träuble (1967). In the field range between H_{c1} and H_{c2} a type II superconductor is said to be in the mixed state. Physically, it can be seen that if flux bundles enter the superconductor at H_{c1} their surface density will grow continuously as the field is increased beyond this value. However, at some higher field it is obvious that the core regions of the bundles will start to overlap. This corresponds roughly to the upper critical field H_{c2} of a type II superconductor beyond which the superconducting state cannot be sustained.

Second critical fields can be very high indeed (several tens of tesla) in type II superconductors so we might expect such materials to be ideal candidates for the construction of high-field magnets. Of course, to make use of the current-carrying capacity of mixed-state superconductors it is essential that power is not dissipated, i.e. the flux bundles must not move when a supercurrent is flowing. In clean type II superconductors it is actually rather easy for these flux bundles to move, even collectively, as has been shown in the very elegant

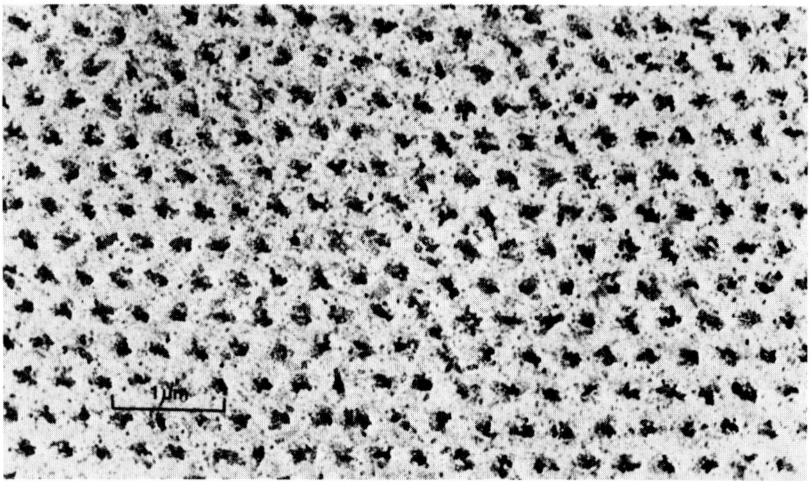

Figure 8.5 Electron microscope image of a mixed-state flux bundle array in the type II superconductor $Pb_{96}In_4$, using a magnetic decoration technique (Essmann and Träuble 1967).

radiofrequency synchronisation experiment of Martinoli and his co-workers (Martinoli 1975, Fiory 1971, Clark 1973). Fortunately it is simple to 'pin' flux bundles in the mixed state to defects (e.g. dislocations) in the superconducting materials, thus allowing the combination of high supercurrent-carrying capability, high field operation and zero power dissipation required for magnet construction (see, for example, Saint-James *et al* (1969)).

8.3 SUPERCONDUCTING MATERIALS

If we lived in an ideal world we would be able to predict the superconducting properties (if any) of a material directly from detailed considerations of band structure, electron–electron interactions, electron–phonon interactions etc. Unfortunately, as we have seen already, the interaction leading to the superconducting state is very weak ($\sim 10^{-8}$ eV per atom) which makes this direct approach extremely difficult in practice. In general, the microscopic theory of superconductivity (BCS) has been of limited value in calculating critical temperatures and critical magnetic fields from first principles. Also, it has to be said that the most important superconductors technologically (i.e. those with the highest T_c and H_c values) are the most complicated structurally. Consequently, all the significant advances as far as superconducting materials are concerned have been and still are being made by experimentalists, with theoreticians following a long way behind. For example, Matthias's semi-empirical rule (Hulm *et al* 1981)—that the T_c of a superconductor is

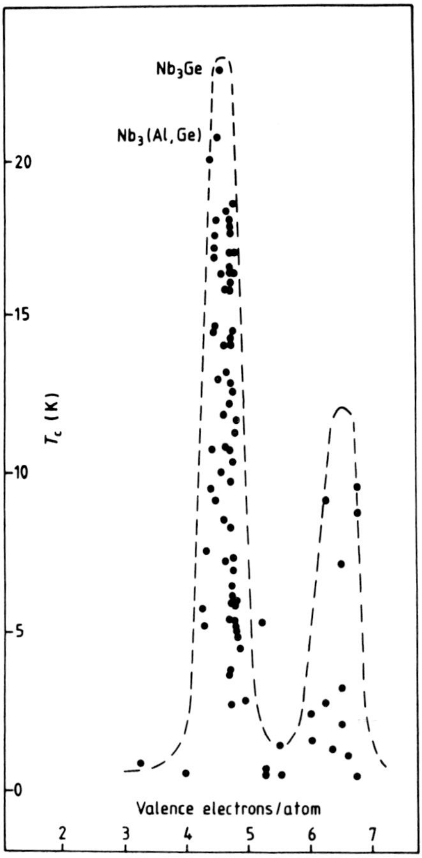

Figure 8.6 Plot of critical temperature, T_c, against ratio of valence electrons, N_e, to atoms, N_a, for the A15 superconducting compounds.

determined by its ratio of valence electrons, N_e, to atoms, N_a—is still the best available to us for making predictions about critical temperatures. This applies for both compounds and solid solutions and, so far, all the technologically useful superconductors have been discovered by applying the rule. As an illustration, figure 8.6 shows a plot of T_c against the number of valence electrons per atom (which is equivalent to N_e/N_a) for the technologically extremely important class of superconductors known as the A15 compounds. As can be seen, there is a very strong maximum for $4.5 < N_e/N_a < 5$ and a weaker maximum for $6 < N_e/N_a < 7$. All such plots for known superconductors show similar maxima in T_c within these ranges of N_e/N_a.

Matthias was always a very outspoken man but on his record of achievement he no doubt had a right to be. In 1969 he wrote:

> Today ... after many excellent theoretical treatises one may ask: What are the experimental results with regard to the criterion for the occurrence of superconductivity or the increase in the superconducting transition

> temperature? Without exaggeration one must say, in effect, not a single result was achieved. While extremely successful in print, these papers were unable to predict any real superconductivity ever and at the same time failed entirely to raise any of the existing superconducting transition temperatures. Until this day, it has been impossible to verify a single prediction along these lines, and yet we are told time and time again that the problem has been solved and is fully understood. We are also told that it is extremely difficult to predict transition temperatures or even superconductivity (Matthais 1969).

These are cautionary words for aspiring theoreticians. How does the picture look today? Its not much different. Figure 8.7 shows a plot of highest attained T_c against year, from 1911 when Onnes discovered superconductivity in mercury. The last two data points for $Nb_{3.8}Al_{0.75}Ge_{0.25}$ ($T_c = 20.98$ K) and Nb_3Ge ($T_c = 23.2$ K) are due to Matthias and his co-workers. From this curve it seems that if the state of superconductivity as a subject is to be judged by improvement in critical temperatures, no advances have been made since the early 1970s. This is not completely fair. In the twelve years since the 1970s discovery of Nb_3Ge, in the A15 (β-tungsten) class of compounds many interesting new superconducting systems have been investigated. Even so the record still remains firmly set with Nb_3Ge. What then are the chances of developing or discovering high-temperature superconductors? The simple answer is, nobody knows. Without a predictive theory which can guarantee success we are really guessing. It may be that a new class of superconductors with very high transition temperatures (say $\geqslant 100$ K) will suddenly appear (Ginzburg 1976, Ginzburg and Kirzhnitz 1982). On the other hand it is equally likely that high-temperature superconductivity is impossible in principle (Philips 1972, Cohen and Anderson 1972).

An ideal practical superconductor would combine a very large critical field and high critical current density at room temperature (300 K). We might

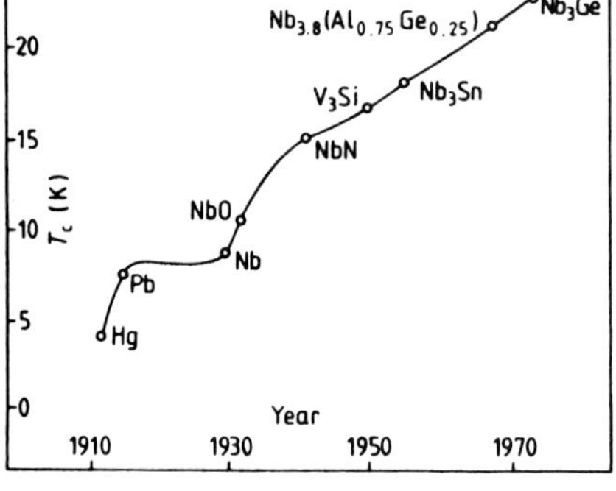

Figure 8.7 Plot of the highest attained critical temperature against year, from 1911.

expect critical fields of a few hundred tesla and critical current densities in thin samples greater than or equal to 10^8 A cm^{-2}. How far can the only definitely established mechanism for superconductivity—the phonon-mediated electron–electron interaction—take us in this regard? For superconducting metals the best we can expect is $\theta_D \simeq 500$ K and $1/N(0)V \simeq 1/3$, so from equation (8.15) the maximum attainable critical temperature (in the BCS approximation) is 25 K, very close to that measured in Nb_3Ge. More detailed arguments (Ginzburg and Kirzhnitz 1982) put the upper boundary for phonon-mediated superconductivity at no greater than 40 K. It does not seem unreasonable, therefore, to anticipate phonon-mediated superconductors with transition temperatures close to 25 K, thus making a superconducting technology based on cheap liquid hydrogen cooling feasible. This may come through new alloys and compounds or by better preparative techniques. What does not seem reasonable on our present understanding is to expect phonon-mediated superconductors to yield transition temperatures $\geqslant 100$ K, i.e. suitable for a technology based on liquid nitrogen.

One material which, as yet, does not exist on earth, but may well display phonon-mediated superconductivity with a critical temperature of approximately 100 K, is metallic hydrogen. Best estimates put metallic and molecular hydrogen in thermodynamic equilibrium at a pressure of somewhere between 1 and 2.5 Mbar. With a predicted Debye temperature of about 3000 K, and assuming that $N(0)V = 1/3$, the expression (8.15) yields $T_c = 150$ K (Ginzburg 1985). Apparently, it has been demonstrated theoretically (Brovman *et al* 1971, 1972, 1977, Saltpeter 1972 and Chapline 1972) that metallic hydrogen, although metastable with respect to the molecular form, has a finite lifetime under zero pressure. It is not clear, however, whether this lifetime will be sufficiently long to make metallic hydrogen of technological importance. If the lifetime of metallic hydrogen, metallic deuterium or alloys of these with heavier elements turns out to be long at low pressures, there can be little doubt that the study of these materials will assume great importance with, it should be hoped, the study of high-temperature superconductors, compatible with a liquid-nitrogen-based technology, also benefiting.

At present, any discussion of metallic hydrogen—normal or superconducting—is academic, since we still require considerable improvements to be made in the creation of ultra-high pressure environments. However, metallic hydrogen almost certainly sets the upper boundary on the transition temperature for phonon-mediated superconductivity. What of real superconductors and superconducting systems actually being studied in the laboratory today? We can make the following qualitative assessments, bearing in mind figure 8.8 which shows a plot of critical current density, together with critical field, against temperature for various superconducting materials.

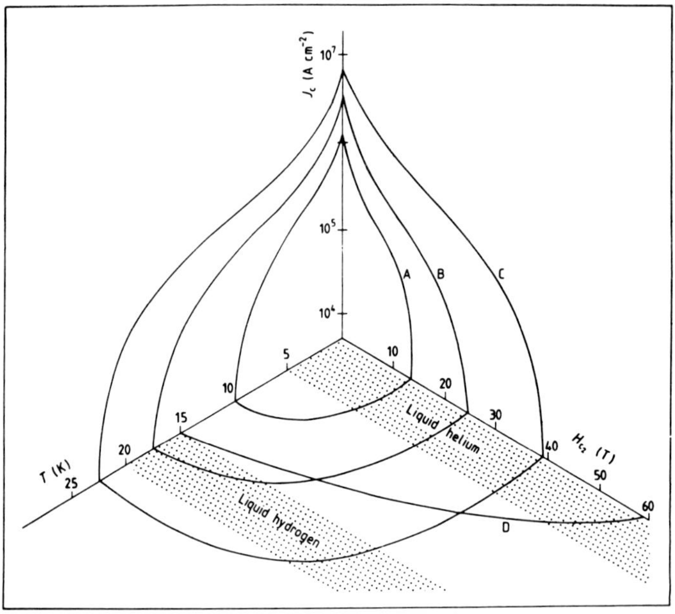

Figure 8.8 Nested plots of critical current density with critical field as a function of temperature for a variety of superconducting materials. (A) NbTi. (B) Nb_3Sn. (C) Nb_3Ge. (D) V_3Ga (projected). (Hulm *et al* 1981.)

Well established superconductors

The A15 family. As we have mentioned previously, all the superconductors with the highest known transition temperatures are members of this family. These superconductors are intermetallic compounds with the A15 structure. Their ideal composition is V_3X and Nb_3X, where X is a non-transition metal which is positioned on a body-centred cubic lattice with the V or Nb atoms arranged as non-intersecting chains closely spaced on each cubic face. It is known that single crystals of V_3Si and Nb_3Sn undergo weak structural transitions (cubic to tetragonal) at temperatures just above their transition temperatures (17 K and 18 K respectively). This is usually the sign of a rather 'soft' lattice, i.e. the electron–phonon interaction, as characterised by $\alpha^2(\omega)F(\omega)$, the product of the square of the electron–phonon matrix element and the phonon density of states averaged over all phonon modes of energy $\hbar\omega$, increases at low frequencies.

The precise reason for the remarkably high transition temperatures of the A15 compounds has still not been settled but these have been ascribed to peaks in the electronic density of states at the Fermi energy and incipient metastability of the A15 structure arising from a strengthening of the weighted

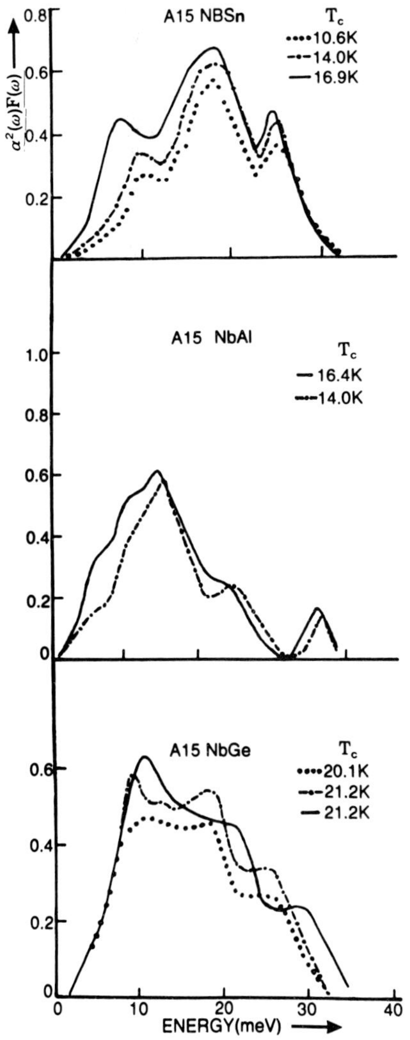

Figure 8.9 Dependence of $\alpha^2(\omega)F(\omega)$ against energy of composition for the metastable A15 compounds NbSn, NbAl and NbGe, where increased T_c always correlates with softening of $\alpha^2(w)F(\omega)$. (Beasley and Geballe 1984.)

electron–phonon interaction (through $\alpha^2(\omega)F(\omega)$) at low frequencies. It is interesting to note that all known superconductors with transition temperatures greater than 18.5 K are metastable A15 compounds, which certainly upholds the view that high T_c values and metastability of structure are closely linked. Systematic tunnelling derivative spectroscopy studies (following Rowell and Kopf (1965) and McMillan and Rowell (1965)) provide reasonable experimental evidence for a connection between the softening of the spectral density function $\alpha^2(\omega)F(\omega)$ and the increase in T_c. In figure 8.9 we show $\alpha^2(\omega)F(\omega)$ against energy for the metastable A15 compounds, NbSn, NbAl and NbGe, as compiled by Beasley and Geballe (1984). In all cases an increase in X from the stoichiometric Nb_3X leads to an overall softening of $\alpha^2(\omega)F(\omega)$, and greater metastability, accompanied by an increase in T_c.

The B1 family. These cubic (sodium chloride) structure compounds could be taken as a good example of Matthias's judgment on anticorrelations in superconductivity (Matthias 1969). Many superconductors have this B1 structure, with the highest transition temperatures (up to 18 K) being recorded for compounds based on niobium, for example NbN. Curiously, compounds with the highest known melting points also form part of this family. This argues against the phonon softening mechanism presented previously to explain the very high transition temperatures of some A15 compounds. Some calculations of the transition temperature of B1 type compounds based on band structure have been made. For example, it is predicted, but not yet observed by experiment, that the metastable B1 compound MoN should have a particularly high transition temperature (Klein and Pickett 1982, see also Liuken *et al* 1984). Another metastable B1 compound of obvious interest, given our previous discussion of metallic hydrogen, is the hydride PdH which can be synthesised by low-temperature hydrogen implantation. When alloyed with silver the PdH system is observed to have a rather high transition temperature, around 17 K. No theoretical understanding of PdH seems to have emerged and it is worth noting that this compound is one of the few hydrides which displays superconductivity.

Ternary compounds. It is no surprise that, with three components to consider, there exists such a large number of ternary compound superconductors. One well known example is the Chevrel system where it is possible for two kinds of long-range order—superconductivity and magnetism—to compete with one another in the same compounds. In this system it appears that the magnetic component originates with the magnetic moment of the Y cation while the electrons giving rise to superconductivity are restricted principally to the Mo_6Y_8 grouping. Even now, it is apparent that suitable manipulation of these competing ordering processes will lead to superconductors with very novel properties (Shirane 1983).

Divergent superconductors

In recent years several new systems have emerged where superconductivity may not arise by the standard BCS phonon-mediated s-type pairing interaction. For want of a better description we shall refer to these as divergent superconductors. There is, apparently, no physically unifying connection between these various systems. This appears to demonstrate the subtlety and widespread nature of superconductivity as a possible ordering process in condensed matter. With no heed to possible, or even conceivable, future importance, we list the following.

Organic superconductors. The idea of organic-based materials with high superconducting transition temperatures was first put forward by Little in the 1960s (Little 1964, 1966, 1967 and 1969). The discovery of quasi-one-dimensional metallic behaviour in the organic charge transfer salt (Coleman *et al* 1973) tetrathiofulvalinium tetracyanoquinodimethane (TTF-TCNQ) certainly encouraged the proponents of Little's ideas. This salt displays a high conductivity maximum at reduced temperatures but does not undergo a superconducting transition. Such compounds have a tendency to instability at low temperatures; indeed it was thought at one time, as an example, that the Peierls metal–insulator transition (see, for example, Patton and Sham (1973)) would suppress any incipient superconductivity in quasi-one-dimensional organic conductors. However, superconductivity has been observed in several organic salts, although the transition temperatures are rather low (less than or approximately equal to a few K). An example of such a salt is tetramethyl tetraselenafulvalene in combination with organic cations, i.e. the so-called Bechgaard salts, $(TMTSF)_2C$, where C is ClO_4, $FSO_3 . PF_6$ or ReO_4. Organic superconductivity has also been observed in a salt of ReO_4 incorporating a sulphur-containing ring (Jerome and Shulz 1982). It is intriguing to note that, according to some authors (for example, Dunne *et al* (1983)), the behaviour of the very large diameter (10^{-6} cm) organic ring compound bacteriorhodopsin, in which the π-orbital electrons are delocalised around the ring, may have strong parallels with the superconducting state. Bacteriorhodopsin plays an essential part in the transport of ions across biological cellular membranes.

Inorganic low-dimensional superconductors. Several examples of low-dimensional inorganic superconducting systems exist. Probably the most novel is the polymeric sulphur nutride $(SN)_x$ which displays metallic conductivity with a superconducting transition temperature at about 0.3 K. This must encourage the search for other inorganic systems with conductivity comparable with metals but containing no metal atoms. Graphite, though not a metal, can display metallic conductivity and undergo a transition to the superconducting state when suitably intercalated with alkali metal cations. Another fairly recently discovered inorganic superconductor, the trichalcogenide $TaSe_3$, has a transition temperature of about 2.3 K. What makes this so interesting is that its 'companion' compound $NbSe_3$ can carry current in the presence of a moderate electric field by means of depinned, sliding, incommensurate charge density waves. Both $NbSe_3$ and $TaSe_3$ have a chain-like structure in which the Nb or Ta atoms have essentially one-dimensional interactions.

Heavy-fermion superconductors. In the last few years considerable attention has been focused on a collection of metallic compounds in which the electron effective mass (and hence the electron density of states) is extremely high—two orders of magnitude, or more, greater than the mass of the free electron. These

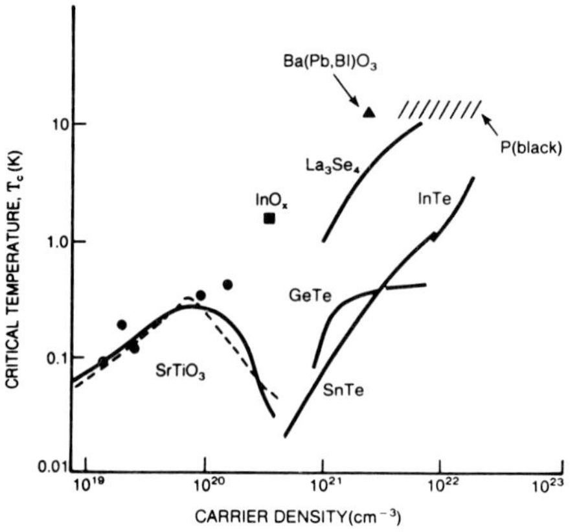

Figure 8.10 Plot of the variation in T_c with electron density for a variety of superconducting materials. (Beasley and Geballe 1984).

are known as heavy Fermi liquid systems. Superconductivity in heavy-fermion compounds was found first in $CeCu_2Si_2$ at a temperature of 0.5 K (Steglich *et al* 1979). Subsequently, superconductivity was discovered in two more heavy fermion systems, the actinide intermetallic compounds UBe_{13} and UPt_3. It may well be that the interaction leading to superconductivity in heavy-fermion systems involves p-wave pairing, i.e. where the pairs of electrons have both spin and orbital momenta. If this turns out to be the case, heavy-fermion superconductivity may not require a phonon-mediated interaction; other mechanisms become possible, for example, localised spin fluctuations.

Reduced electron density superconductors. It is now well established that, for superconducting materials with electron densities varying between approximately 10^{19} cm^{-3} (degenerate semiconductors) and approximately 10^{22} cm^{-3} (semi-metals to metals), the transition temperature, broadly speaking, increases as a function of electron density. This trend can be seen quite clearly in figure 8.10 where the transition temperature for a variety of systems is plotted against electron density (Beasley and Geballe 1984). Although the general dependence of transition temperature on electron concentration is apparent, the sheer physical diversity of the various systems shown in figure 8.10 certainly presents any theoretician wishing to put superconductivity on a systematic and rigorous basis with the most formidable problems.

Excitonic superconductors. The excitonic superconductor is an idea waiting on a discovery. It has been suggested as a possible way of increasing superconducting transition temperatures dramatically. Electronic excitations known as excitons are a well established feature of condensed matter physics. For example, in a molecular crystal an exciton would be an excited state of a molecule which propagates from one molecule to the next along the crystal. In semiconductors an exciton can be a bound electron–hole pair. The energy required to excite an electronic exciton in a solid may vary from a few hundredths to several electron-volts. Now, the exchange of excitons can generate an attractive interaction between conduction electrons, just as can the exchange of phonons. If these exchange excitons could have relatively high energy (say, 1 eV$\triangleq 10^4$ K) and the interaction between the electrons could be maintained sufficiently strongly ($N(0)V \geqslant 1/5$ to 1/4) then the BCS expression (8.15) predicts a maximum superconducting transition temperature close to 200 K.

Exciton-mediated superconductivity is not an established experimental fact. Two obvious approaches to the realisation of excitonic superconductivity come to mind. These are to prepare metal/highly polarisable dielectric sandwiches or to create microscopic precipitates of metal in highly polarisable dielectrics. In each case the excitonic interaction would occur close to the metal–dielectric interface. Metal–dielectric sandwiches have been prepared in a variety of ways to look specifically for excitonic superconductivity but this has never been observed in these systems. However, it is just possible that nature may have conspired to produce excitonic superconductivity by the second route, through the creation of microscopic metallic precipitates in simple inorganic compounds subject to extreme temperature cycling. The three best known examples are CuCl, CdS and NbSi. It was reported in 1978 that after CuCl samples had undergone such temperature cycling (i.e. extremely rapid quenching from a high temperature) under pressure, very high diamagnetism (superdiamagnetism) was observed at temperatures in the 150 to 200 K range (Brandt *et al* 1978, Chu and Rusakov 1978). It is certainly possible that this diamagnetism arises because of excitonic superconductivity in micro-clusters of metallic copper created by the quenching process. Similar enhanced diamagnetism was observed first in CdS samples, quenched in the same way, and recently in very thin films of NbSi, also quenched very rapidly from high temperature. It has to be said that, for the moment, none of these observations appears to be very reproducible. Nevertheless, the possibility of high-temperature superconductivity, which may be excitonic in origin, exists.

8.4 LARGE-SCALE APPLICATIONS

The discovery of type II superconductors with their large critical fields and supercurrent densities has revolutionised the design of high-field electromagnets. From our previous discussion it is apparent that if no new class of high transition temperature materials is found, then most large-scale applications of superconductivity will be based in the near future on A15 superconductors, probably the best known of the type II superconductors. There are, of course, other superconductors of technological interest, one example of which is the solid solution NbTi. This is ductile and can be used very easily in wire form to wind magnets up to 5 T with current densities of 10^5 A cm^{-2}. The novel Chevrel-phase ternary molybdenum sulphides, which have already been mentioned, can have extremely high critical fields. Thus the compound $PbMo_5S_6$ has a transition temperature of 14.4 K and a critical field of 50 T at 4.2 K. For comparison, the A15 compound Nb_3Ge, formed by the co-sputtering of niobium and germanium, has a maximum recorded transition temperature of 23.2 K and a critical field of 25 T at 4.2 K.

Referring back to figure 8.8, which shows the competing requirements of critical current density, critical field and temperature of operation for various superconducting materials, we can see very clearly the engineering scope provided by those superconductors already known. Although it may be desirable to find higher transition temperature materials, there is no essential need. When high quality, ductile and stabilised Nb_3Ge tape is developed, as it surely will be, superconducting magnets cooled by liquid hydrogen will become feasible, as can be seen from figure 8.8. Such a development will, no doubt, have a profound impact on high-power electrical engineering, particularly in the fields of thermonuclear fusion, electrical power transmission, magneto-hydrodynamic power generation and energy storage. High-energy particle accelerators, free-electron lasers and magnetic separators should also benefit. It is worth noting that all major new particle accelerators will be commissioned with superconducting beam-bending magnets in place, although cooled by liquid helium, not liquid hydrogen. It is also worth remembering that a 'hydrogen economy' is being considered seriously on its own merits, without reference to superconducting machines.

It seems likely that these, and other large-scale applications, will lead eventually to the creation of a major industrial base for superconducting products with a turnover of at least a few billion pounds per annum. This may come through the need to improve the efficiency of electrical power generation and transmission systems. Using superconducting techniques, operating efficiency in both areas combined may improve by up to 50% which, in this era of rising fuel costs, may be sufficient to overcome the inertia of power engineers brought up on conventional room-temperature systems. A second, and ultimately more important, reason for introducing large-scale superconducting technology is the absolutely crucial role this will play in any successful power generation based on controlled thermonuclear fusion. With our present perspective on future power demands and available resources controlled thermonuclear fusion is an economic and political necessity. Plasma fusion will require large containment fields which can only be provided efficiently by superconducting magnets. Even if the race to sustained fusion is won by inertial confinement systems using relativistic electron or ion beams, or even high-velocity macroscopic pellets, superconducting technology will still have an essential part to play. The high-power RF cavities, beam-bending and focusing magnets and linear induction systems needed to generate these beams will inevitably require the use of superconducting technology. Since this is spreading so rapidly, and on such a large scale, in high-energy particle physics it seems natural to assume that this will be the base from which superconducting technology will diversify into many other areas of engineering.

The rate at which this diversification takes place will, of course, depend very much on the level of development funds available. In this respect the Strategic Defence Initiative (SDI) proposed by the United States may be of considerable, if unsolicited, assistance to the superconducting community. Particle-beam, free-electron laser and kinetic energy (high-velocity pellet) antimissile defence systems may rely very heavily on superconducting magnet technology. Even the energy required to run these systems may be stored in superconducting inductors. Just as nuclear technology was promoted very rapidly because of the need to develop atomic weapons, so may the SDI decisively affect the progress of large-scale superconducting technology well into the next century.

8.5 SUPERCONDUCTING WEAK LINKS AND MACROSCOPIC QUANTUM CIRCUITS

In 1962 Brian Josephson, then a graduate student at Cambridge, published a paper in which he considered the quantum mechanical transfer of Cooper pairs across a tunnel barrier. In doing so he made use of the already established many-bodied, single-electron tunnelling Hamiltonian formalism (Bardeen 1961, Cohen *et al* 1962) which had been applied so successfully to the theoretical interpretation of Giaever (Giaever 1960, Giaever *et al* 1962, Rowell and Kopf 1965, McMillan and Rowell 1965). Josephson tackled the transfer of pairs as a barrier penetration problem assuming implicitly that the tunnel junction could be treated as singly connected, without reference to any external circuit attached to it. By this route he did not need to consider the full quantum mechanics (with canonically conjugate variables) of the circuit in which the pair-tunnelling superconducting junction was embedded. Josephson's theoretical predictions (Josephson 1962), based on the calculation of processes of single electron pairs tunnelling across the junction, are embodied in two equations. The first relates a difference, $\Delta\phi$, in the pair condensate wavefunction phase (equation (8.7)) with a macroscopic supercurrent, I, flowing through the junction. By analogy with bulk superconductors, if the maximum supercurrent which can flow with the junction in the zero voltage state is I_c, then

$$I = I_c \sin \Delta\phi. \tag{8.18}$$

It is important to remember that Josephson derived this expression in the limit of weak tunnel-coupling across the junction. By definition, this means in the limit where only single-pair charge transfer processes ($\pm q, q = 2e$) are considered. It is also important to remember that equation (8.18) requires the pair tunnel junction to be attached to an external source of current. Josephson's second equation connects any energy difference for electron pairs on opposite sides of the tunnel barrier to a frequency of oscillation of the macroscopic supercurrent flowing through the junction. In Josephson's analysis, if a true voltage source V is connected across the junction, so that there is an energy difference for single pairs of 2 eV, the time derivative of the pair condensate phase difference is given by

$$\frac{\mathrm{d}(\Delta\phi)}{\mathrm{d}t}=\omega_V=\frac{2\,\mathrm{eV}}{\hbar}. \tag{8.19}$$

Subsequently, Josephson widened his theoretical model to include any situation where two bulk superconductors are connected by some intermediary which allows the quantum mechanical transfer of single pairs between one superconductor and the other (Josephson 1964, 1965). Non-tunnel junction intermediaries could then be included, for example, planar metal microbridges (Anderson and Dayen 1964, Kirschman *et al* 1970) and point contacts (for example, Zimmerman and Silver 1968). The generic term 'weak link' was introduced to describe both these and electron-pair tunnel junctions. When the potential of weak links as electronic devices became apparent a circuit model, based on the traditional two-fluid (superconducting plus normal electrons) description (Gorter and Casimir 1934a and b) of superconductors, was proposed to account for the dissipation (i.e. the average product of voltage × current) in the voltage-carrying state. This became known as the resistively shunted junction (RSJ) model (see, for example, McCumber (1968)). The circuit model comprises two current-carrying channels in parallel, one for the superconducting condensate, the other for currents leading to dissipation. The latter are formally identified with a shunt resistor with resistance R. In the simplest view this resistance is not dependent on frequency (or, through equation (8.19), voltage). For the case of a weak link, described by equations (8.18) and (8.19), in the phenomenological RSJ model it can be shown that when the weak link is driven by a direct current source I the time-averaged voltage is given by the hyperbolic form

$$\langle V\rangle=R(I^2-I_{\mathrm{c}}^2)^{1/2} \tag{8.20}$$

which yields a straight line $\langle V\rangle$ against I characteristic in the limit $I \gg I_{\mathrm{c}}$, i.e. when the current V/R flowing through the shunt resistance is taken to be very large compared with I_{c}.

Applying the RSJ model and the Josephson expressions (8.1) and (8.2) to the problem of a weak link enclosed by a thick superconducting ring (i.e. the so-called superconducting quantum interference (SQUID) ring) yields the equation of motion

$$C\,.\frac{\partial^2\Phi}{\partial t^2}+\frac{1}{R}\frac{\partial\Phi}{\partial t}=-\frac{\partial U}{\partial\Phi}. \tag{8.21}$$

For this equation the 'Josephson' potential for the SQUID ring is given in terms of the externally applied flux Φ_x and total included flux Φ (see equations (8.10) and (8.11)) by the expression

$$U(\Phi, \Phi_x) = \frac{(\Phi - \Phi_x)^2}{2\Lambda} - \frac{I_c \Phi_0}{2\pi} \cos(2\pi\Phi/\Phi_0). \tag{8.22}$$

Here, the critical current I_c is defined as the value of the circulating screening supercurrent at which magnetic flux first enters the ring. We note that this circulating current is generated through an external magnetic flux and not by means of an external current source. It is usually assumed that equations (8.20) and (8.21) are applicable to weak links in general, i.e. that the pair-tunnelling model of Josephson is equally valid for weak links of the metal bridge type (microbridges and point contacts) as for tunnel structures.

Although the expressions (8.20) and (8.21) are based on a microscopic quantum mechanical calculation of pair transfer through a superconducting tunnel junction, their form is essentially classical, even though the potential (8.22) appears to require the quantisation condition (8.10) and a $\Phi_0 = h/2e$ periodicity in the coupling energy across the weak link. The resolution of this problem appears to be to deal with weak links in superconducting circuits in a fully quantum mechanical way and not to create a classical superstructure over the original Josephson predictions. As we shall demonstrate, such a quantum mechanical description derives ultimately from the physical ideas of Fritz London (London and London 1935, London 1961), namely the macroscopic quantum mechanical nature of the superconducting state, the superconducting penetration depth and flux quantisation in a thick multiply connected superconducting circuit. Indeed, there seems no reason, in principle, why London could not have arrived at the following conclusions himself.

We start with a thick superconducting ring, i.e. thick compared with the penetration depth λ. We know that given a macroscopic wavefunction for the superconducting condensate of the form of equation (8.7) and a path of integration deep in the superconducting ring compared with λ, the flux quantisation condition (8.9) follows. This quantisation was observed subsequently by Deaver and his co-workers (Deaver and Fairbank 1961, Goodman *et al* 1971) and by Dole and Näbauer (1961). In the quantum mechanical sense equation (8.9) provides for a set of macroscopically different quantum states of the ring which can be solved in terms of an externally applied flux, Φ_x, through the Schrödinger equation (8.12). These solutions, which have already been presented in figure 8.1, are perfectly consistent with London's thoughts in 1950. If he had then considered the effect of perturbations on these quantum states, and had restricted his attention to ring geometries, he would surely have ended up developing a description of at least some weak link phenomena.

Given a thick superconducting ring, and knowing that any superconductor

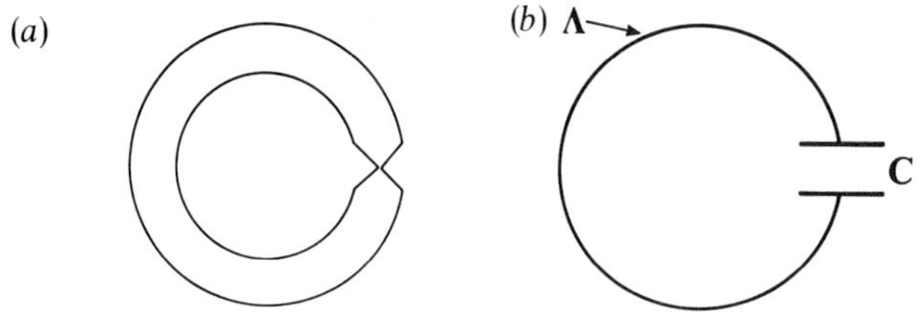

Figure 8.11 (*a*) Schematic weak link construction ring. (*b*) Canonical quantum mechanical circuit comprising a superconducting inductor of inductance Λ and a weak link constriction capacitor of capacitance *C*.

of a thickness comparable with λ will leak magnetic flux, the obvious way to perturb the solutions of figure 8.1 is to introduce a constriction in the ring with a cross sectional diameter at its narrowest point of roughly λ. If we assume that magnetic flux leaks across the constriction quantum mechanically, and we know from the mixed state that flux moving through a superconductor is always found in quantised units $n\Phi_0$ ($n=0, \pm 1, \pm 2, \ldots$ etc), then the introduction of the constriction will lead to a splitting of the energy parabolas of figure 8.1 at their crossing points. The strength of the splitting will depend on the rate at which flux can leak across the constriction quantum mechanically.

Our macroscopic quantum mechanical circuit, depicted in figure 8.11, consists of a superconducting inductor (the ring) of geometric inductance Λ and a constriction of local capacitance C. The constriction can obviously be identified with a point contact or microbridge weak link (Anderson and Dayem 1964, Kirschman *et al* 1970, Zimmerman and Silver 1968). The canonically conjugate quantum mechanical variables for this circuit, i.e. a SQUID ring (see, for example, Solymar (1972), Chapter 14), are the flux Φ threading the ring and $C\dot{\Phi}$ (which equals Q, the effective charge at the weak link capacitor). It follows that

$$[\Phi, Q] = \mathrm{i}\hbar \tag{8.23}$$

and

$$\Delta\Phi\,\Delta Q \geqslant \hbar/2. \tag{8.24}$$

Given that the minimum effective charge associated with the superconducting condensate is q, $q=2e$, and flux is quantised in a thick superconducting ring in units of Φ_0, we can identify from equation (8.23) two clear physical limits for a SQUID ring. These are;

(i) when $\Delta\Phi \gg \Phi_0$, $\Delta Q < q$ and the charge at the weak link is well defined about integer units of q, i.e. around Nq, where $N=0, \pm 1, \pm 2, \ldots$ etc and

(ii) when $\Delta Q \gg q$, $\Delta\Phi < \Phi_0$ and the flux threading the ring is well defined about integer units of Φ_0, i.e. around $n\Phi_0$, where $n=0, \pm 1, \pm 2, \ldots$ etc.

As a way of characterising the well defined variable for the two cases we refer to the first limit (i) as the charge mode, the second as the flux mode. In limit (i) the very large uncertainty in the flux is due to the ease with which flux can move into or out of the ring across the weak link. Thus large quantum mechanical flux currents can be sustained across the weak link but since the charge at the weak link is localised about Nq no macroscopic screening supercurrents can flow around the ring. Physically, this limit should be realisable when the weak link is very leaky to flux, i.e. when the weak link cross section is less than or equal to λ. In limit (ii) the very large uncertainty in charge is due to the ease with which charge (in units of q) can move through the weak link, in one or other direction. Hence large quantum mechanical charge currents can flow in the weak link and circulate around the ring as macroscopic diamagnetic screening supercurrents. On the other hand, the flux threading the ring will be confined to values close to $n\Phi_0$. This second limit should be realisable when the cross section is greater than or equal to a few multiples of λ, so that it becomes very difficult for flux to leak across but very easy for charge to flow through the weak link. In the extreme limit when the cross section is very much greater than λ (i.e. a uniformly thick superconducting ring) the flux is perfectly confined at the integer values $n\Phi_0$, which is the flux quantisation condition. With this physical description in mind we can place both limits of behaviour of a SQUID ring on the following, more formal, basis.

Charge mode behaviour in SQUID rings

By definition, in this situation charge is relatively well localised at the weak link about values Nq but quantum mechanical contact is still possible between different N states. Since $\Delta Q \ll q$ we can restrict ourselves, in the simplest case, to considering only nearest neighbour ($Nq \to (N \pm 1)q$) transitions between these charge states. This implies that the weak link constriction is small enough in cross section to allow all possible couplings between the flux states of the superconducting ring ($n\Phi_0 \to (n \pm 1)\Phi_0, (n \pm 2)\Phi_0, (n \pm 3)\Phi_0, \ldots$, etc). If Q is well defined about Nq states then the weak link may be described by a pair charge number amplitude A_N. We can write the total energy stored on the weak link capacitor as

$$W_N = \frac{Q_{\mathrm{tot}}^2}{2C} \tag{8.25}$$

where the charge Q_{tot} has two contributions, one of Nq from the localised charge state, and the other from an externally applied charge Q_x. Thus

$$Q_{\mathrm{tot}} = Nq + Q_x \tag{8.26}$$

and, for example, when the weak link is closed by a ring, Q_x may be applied by a Faraday law voltage, i.e. $Q_x = -C\dot{\Phi}_x$, where $\dot{\Phi}_x$ is a time-varying externally

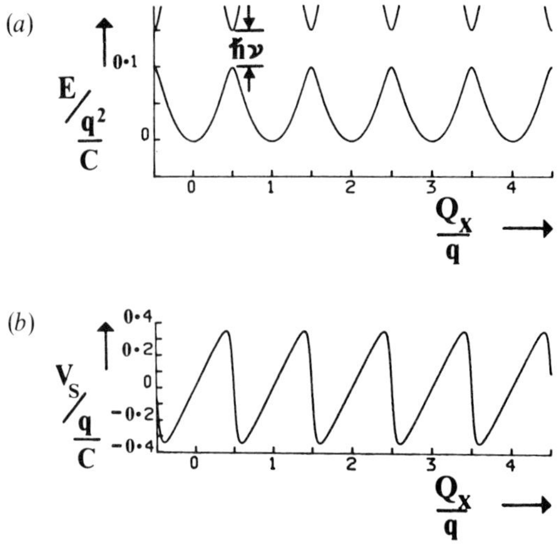

Figure 8.12 (*a*) E against Q_x ground state band in the charge mode limit for $\hbar\nu = 0.05q^2/C$. (*b*) Screening voltage pattern for the ground state charge band of (*a*).

applied magnetic flux. If $\nu/4\pi$ is the frequency at which single electron pair transfer $(N \rightarrow N \pm 1)$ takes place quantum mechanically through the weak link, then we can write down a time-independent Schrödinger equation for the link of the form

$$W_N A_N - (\hbar\nu/2)(A_{N+1} + A_{N-1}) = E(Q_x)A_N. \tag{8.27}$$

For relatively small values of $\hbar\nu$ (compared with the characteristic energy q^2/C of the weak link capacitor) equation (8.27) yields a ground state energy in external charge space (Q_x) which is a periodic function of q. In figure 8.12(*a*) we show the ground state solution for $\hbar\nu = 0.05q^2/C$. By analogy with energy bands in solid state physics we will refer to figure 8.12(*a*) as the ground state energy band of the weak link in external charge space.

Since flux and charge are taken to be canonically conjugate variables we can define a wavefunction in flux space in terms of the A_N values and an angular displacement, $\alpha = 2\pi\Phi/\Phi_0$, as

$$\psi(\alpha) = \sum_N A_N\, e^{iN\alpha}. \tag{8.28}$$

Equation (8.27) can then be rewritten in the form

$$(1/2C)\left(-qi\frac{\partial}{\partial\alpha} + Q_x\right)^2 \psi(\alpha) - \hbar\nu\cos[\alpha\psi(\alpha)] = E(Q_x)\psi(\alpha). \tag{8.29}$$

If we set $Q_x = 0$ we recover the quantum mechanical form of the familiar

Josephson pendulum equation (Josephson 1962, Mutton *et al* 1984, Widom *et al* 1984). We note that in equation (8.29) there is a matrix element $\hbar v/2$ for amplitude superpositions between N states varying by ± 1 but there is no macroscopic current term. This is as expected since the charge mode limit is concerned with charge localisation at the weak link.

It is possible to observe charge mode limit behaviour in superconducting weak link constriction devices by two experimental techniques, one dynamic the other quasi-static. The dynamical method makes use of the well known AC-biased SQUID magnetometer circuit configuration in which a SQUID ring is inductively coupled to a resonant parallel inductor–capacitor tank circuit (Zimmerman and Silver 1968, Solymar 1972, Long *et al* 1980). In this circuit, the basic layout of which is shown in figure 8.13, the AC (radiofrequency to microwave) voltage $\bar{V}_{OUT}$ is measured as a function of the AC input current, I_{IN}, with the phase of $\bar{V}_{OUT}$ preset with respect to I_{IN}. For a given AC operating frequency $\omega_R/2\pi$, and weak link capacitance C, the dynamical $\bar{V}_{OUT}$ against I_{IN} characteristic of the resonant tank circuit–SQUID ring system can be calculated quantum mechanically using the solution of equation (8.27) (see figure 8.12(*a*)). In order to do this we need to remember that amplitude superposition between adjacent N states (e.g. $N \rightleftharpoons N+1$) is only significant, for small v, close to the maxima in the graph of E against Q_x in figure 8.12(*a*). In between, the charge is screened by the quantum mechanical flux currents. We can thus define, for the weak link capacitor C, a screening voltage (by analogy with a screening current flowing around a superconducting ring)

$$\langle V_s(Q_x)\rangle = dE(Q_x)/dQ_x. \tag{8.30}$$

The screening voltage for the ground state charge band of figure 8.12(*a*) is shown in figure 8.12(*b*). For small v and for the regions of positive slope the weak link displays dielectric behaviour (i.e. it acts as a good capacitor) but, as can be seen, the link undergoes dielectric breakdown periodically along the Q_x axis when the charge state (N) of the capacitor changes. The screening voltage response on the weak link capacitor is obviously a sum of harmonics and can be written as

$$\langle V_s(Q_x)\rangle = \sum_{M=1}^{\infty} V_M \sin(2\pi M Q_x/q) \tag{8.31}$$

as is apparent from figure 8.12(*b*).

In the charge mode limit in which macroscopic charge currents do not flow through the weak link, the ring in figure 8.13 acts simply as an inductive element to couple a time-dependent magnetic flux across the link. For this purpose the ring can be either superconducting or normal since the above analysis of the charge mode limit does not require the existence of a phase-coherent (i.e. superconducting) external path between the two sides of the weak link. If the ring is normal (for example, copper), with a ring resistance R_{ring}, all that is required is that $R_{ring}/\Lambda > \omega_R/2\pi$. This prevents the AC flux from

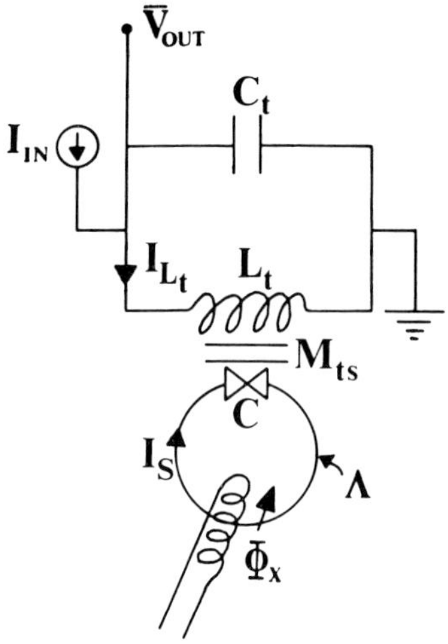

Figure 8.13 Basic circuit configuration for AC SQUID operation.

leaking out through the normal ring. Assuming that the ring is normal, a quantum mechanical calculation of the RF voltage $\bar{V}_{OUT}$ across the tank circuit at zero phase (in phase) with respect to the input RF drive current, I_{IN}, into the tank circuit yields the expression

$$\bar{V}_{OUT} = Q\omega_R L_t\left(I_{IN}\sin^2\delta - \frac{\mu}{\Lambda_r\omega_r\pi}\int_{-\pi}^{\pi} d\zeta \sin\zeta\langle V_s(Q_x^{tot}, \xi)\rangle\right) \quad (8.32)$$

where Q is the quality factor of the unloaded tank circuit, L_t, Λ_r and M_{ts} are, respectively, the tank circuit inductance, ring inductance and mutual inductance between the tank circuit coil and ring, with coupling coefficient $K_{ts} = (M_{ts}^2/L_t\Lambda)^{1/2}$, $\mu = M_{ts}/L_t$ and δ is the phase of the unrenormalised tank circuit impedance at frequency ω_r. The argument of the screening voltage $\langle V_s\rangle$ is $Q_x^{tot} = -C(\dot{\Phi}_x - \mu\dot{\Phi}_{xrf})$ where $\mu\Phi_{xrf}$ is the flux felt by the ring due to the radiofrequency current in the tank circuit. The interaction between the tank circuit and the ring renormalises the impedances of both, changing Λ to Λ_r where $\Lambda_r = \Lambda(1 - K_{ts}^2)$ and shifting the resonant frequency of the tank circuit from ω_R to ω_r where $\omega_r = \omega_R(1 - K_{ts}^2)^{-1/2}$. In the expression (8.32) for $\bar{V}_{OUT}$ the effect of coupling the quantum system (i.e. the weak link) to a dissipative external circuit oscillator mode (of the tank circuit) has been accounted for rigorously (Prance H *et al* 1985c, Feynman and Vernon 1963, Spiller *et al* 1985).

Figure 8.14(a) shows the zero temperature ($T = 0$), $Q_x = 0$ dynamical characteristic of a charge mode limit AC SQUID system calculated from equation (8.32) assuming, first, that the superconducting weak link operates in the ground state band with $\hbar\nu = 0.05q^2/C$ and, second, that typical values of SQUID circuit parameters can be used in the calculation (i.e. $Q = 40$, $K_{ts}^2 Q = 1.58$, $\Lambda = 5\times 10^{-10}$ H, $L_t = 2\times 10^{-7}$ H and $\omega_R/2\pi = 27$ MHz). As can be seen, the $Q_x = 0$, in-phase $\bar{V}_{OUT}$ against I_{IN} characteristic for ground state

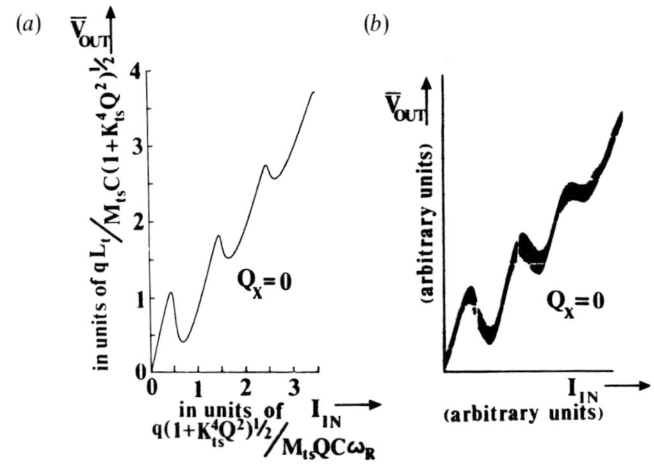

Figure 8.14 (*a*) Theoretical $Q_x=0$ plot of $\bar{V}_{OUT}$ (in phase) against I_{IN} for a charge-mode-limited SQUID ring operating in the ground state band with $\omega_R/2\pi=27$ MHz, $Q=40$, $K_{ts}^2Q=1.58$, $\Lambda=5\times10^{-10}$ H, $L_t=2\times10^{-7}$ H and $\hbar\nu=0.05q^2/C$. (*b*) Experimental $Q_x=0$ plot of $\bar{V}_{OUT}$ (in phase) against I_{IN} for a normal copper ring/superconducting niobium point contact SQUID system operating in the charge mode limit at $\omega_R/2\pi=27$ MHz and $T=1.5$ K.

band operation consists of a set of triangular features decreasing in amplitude with increasing number along the I_{IN} axis. If we fix I_{IN}, for example, at the maximum of the first triangular feature at charge Q_x we find, theoretically, that $\bar{V}_{OUT}$ is a periodic triangular function of Q_x with period q. In figure 8.14(*b*) we show an experimental $Q_x=0$ plot of the in-phase $\bar{V}_{OUT}$ against I_{IN} characteristic for a large dynamic range RF system (Prance R J *et al* 1985a, Prance H *et al* 1985a, b, c) operating at 27 MHz and driving a superconducting niobium point contact weak link enclosed by a standard Zimmerman SQUID block (Zimmerman and Silver 1968) machined from copper. The point contact is adjusted mechanically to have a very small cross section, roughly comparable with the superconducting penetration depth.

The experimental characteristic in figure 8.14(*b*) is not affected by an external DC flux (Φ_x) but the triangular features do split and invert when a sufficiently large Faraday law voltage ($V_x=-\dot{\Phi}_x$) is generated across the weak link (Prance R J *et al* 1985, Prance H *et al* 1985b). This sensitivity to $\dot{\Phi}_x$, together with the remarkable correspondence between the theoretical and experimental characteristics of figures 8.14(*a*) and 8.14(*b*), indicate that the weak link is operating in the charge mode limit. Figure 8.15 shows a plot of $\bar{V}_{OUT}$ (in phase), with I_{IN} fixed at the maximum of the first triangular feature in figure 8.14(*b*), as a function of a Faraday law voltage $\dot{\Phi}_x$ (which equals V_x) generated in a coil coupled inductively to the SQUID ring. As can be seen,

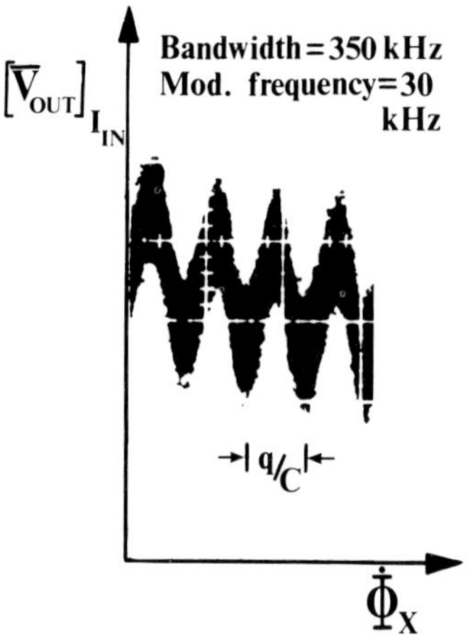

Figure 8.15 Experimental $(\bar{V}_{OUT})_{I_{IN}}$ against $\dot{\Phi}_x$ (equivalent to V_x) for a SQUID system RF current, biased at the maximum of the first triangular feature in figure 8.14(*b*).

$(\bar{V}_{OUT})_{I_{IN}}$ against $\dot{\Phi}_x$ (V_x) is a triangular pattern periodic along the $\dot{\Phi}_x$ (V_x) axis. If we assume that the effective charge is, as we have stated, $q=2e$, and we also assume that the SQUID ring is operating in the charge mode limit, then the period $\Delta V_x = q/C$. Knowing ΔV_x, we can calculate the effective capacitance, C, of the weak link. In the point contact devices looked at so far, C is in the range 10^{-16} to 10^{-15} F for charge-mode-limited behaviour. We note that essentially identical charge mode behaviour has been observed using superconducting SQUID rings (Prance R J *et al* 1985, Prance H *et al* 1985a) and that for normal SQUID rings the voltage V_x can be generated by passing a direct current through the normal region shunting the superconducting point contact (Prance H *et al* 1985b).

In the experimental dynamical characteristic of figure 8.14(*b*) we have, in effect, monitored the first derivative of the ground state charge band energy as a function of external charge (i.e. $\langle V_s \rangle = dE(Q_x)/dQ_x$, equation (8.30)). By taking the derivative again, i.e.

$$\chi^e(Q_x) = -C\,.\,(d\langle V_s(Q_x)\rangle/dQ_x) = -C\,.\,(d^2E(Q_x)/dQ_x^2) \qquad (8.33)$$

we determine the electric susceptibility of a superconducting weak link. This susceptibility χ^e can be measured as a function of Q_x (i.e. V_x) by the quasi-static technique of noise spectroscopy (Mutton *et al* 1984, Prance *et al* 1981, 1983, Clark 1986). This entails using a very-low-noise radio receiver to monitor the mean square noise voltage $\langle|\Delta V_t|^2\rangle$ across a tank circuit coupled to a charge mode limit SQUID ring (superconducting or normal) containing a superconducting weak link. The external voltage V_x can be varied either by means of a Faraday law generator coil or, for the case of the normal ring, a current passed through the shunt resistance external to the weak link. On a constant background, a plot of $\langle|\Delta V_t|^2\rangle$ against V_x provides direct information on χ^e. Figure 8.16 shows such a plot, taken at $\omega_R/2\pi = 27$ MHz

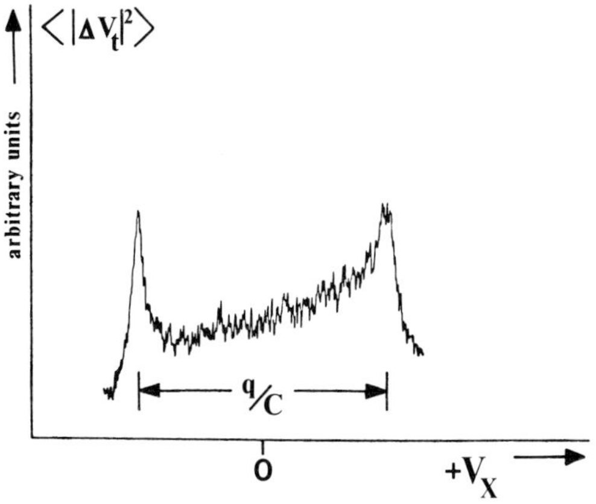

Figure 8.16 Plot of $\langle|\Delta V_t|^2\rangle$ against $\pm V_x$ for a hybrid copper ring/ superconducting niobium point contact SQUID at a temperature of 1.5 K.

and in a post-detection bandwidth of a few hertz, for a superconducting niobium point enclosed by a normal copper ring at 1.5 K. The voltage V_x is created by means of a slowly varying externally applied current. Now the second derivative $d^2E(Q_x)/dQ_x^2$ against Q_x of the ground state charge band shown in figure 8.12(*a*) consists of a set of cusp-like spikes, spaced at intervals of q along the Q_x axis, with a half-width determined by the matrix element $\hbar\nu/2$. These cusp-like spikes are quite apparent in figure 8.16. The linearly rising background between these spikes is caused by the breakthrough of the rather large external current drive used (± 1 A) into the 27 MHz detection electronics. A best theoretical fit for the noise pattern shown in figure 8.16 is found for $\hbar\nu = 0.05q^2/C$ (i.e. the ground state charge band of figure 8.12(*a*)). Given the measured half-width of the spikes in figure 8.16 and an estimate of the copper ring shunt resistance to calculate V_x, we find a charge mode limit pair transfer frequency $\nu/4\pi \simeq 4\times 10^{10}$ Hz at $T = 1.5$ K (Prance R J *et al* 1981, 1983).

Charge mode behaviour in SQUID rings as revealed, for example, in figures 8.14(*b*), 8.15 and 8.16, was discovered in 1985 (Prance R J *et al* 1985, Prance H *et al* 1985a, b, c). Given these observations, we are drawn to the following conclusion. If the Josephson effect is concerned with the coupling of charge states of a superconducting weak link by the quantum mechanical transfer of single electron pairs ($Nq \rightarrow (N\pm 1)q$), which it is clearly defined to be (Josephson 1962, Widom *et al* 1984, Ferrell and Prange 1963, de Gennes 1966, Widom *et al* 1982a), then we must identify what we term the charge mode with the true Josephson effect. In this mode flux is extremely ill defined, charge is well localised at the weak link around values Nq and macroscopic supercurrents do not flow. Exactly the opposite is implied in the traditional interpretation of the Josephson effects (Barone and Paterno 1982, Solymar

1972, Chapter 8 onwards) as characterised by equations (8.18) to (8.22). Here, flux is always considered to be well defined around integer values of Φ_0 (equation (8.22)). Even the second Josephson equation, (8.19), which can be rewritten as

$$\nu_V = \frac{\omega_V}{2\pi} = \frac{V}{\Phi_0} \tag{8.34}$$

is concerned with the motion of well defined (localised) Φ_0 flux bundles through the weak link (as has been pointed out by Anderson (1964)), while the charge states at the link are extremely ill defined since macroscopic supercurrents are flowing. The logic of this argument leads us to identify the traditional Josephson effects with limit (ii) on page 190, the flux mode.

Flux mode behaviour in SQUID rings

By definition, if a SQUID ring containing a superconducting weak link is to display manifestations of flux confinement (equation (8.12)), then phase coherence must be maintained around the ring, i.e. the ring has to be superconducting. With this in mind, we can simply transpose from charge mode to flux mode. Thus, we now require a relatively large cross section weak link (greater than or equivalent to a few multiples of λ) so that the flux in the ring is well localised about values $n\Phi_0$, while charge transfer processes $Nq \to (N\pm 1)q, (N\pm 2)q, (N\pm 3)q, \ldots$ etc up to very high order can take place through the link. Then the Schrödinger equation for the ground state band in external flux (Φ_x) space, periodic in Φ_0, can be written

$$W_n D_n - \frac{\hbar\Omega}{2}(D_{n+1} + D_{n-1}) = E(\Phi_x)D_n \tag{8.35}$$

with $W_n = (n\Phi_0 - \Phi_x)^2/2\Lambda$ (equation (8.11)) and where D_n is the amplitude for the ring to be in the ($n\Phi_0$)th flux state. Only nearest-neighbour transitions, $[n\Phi_0 \to (n\pm 1)\Phi_0]$, are considered and $\Omega/4\pi$ is the frequency at which these transitions take place quantum mechanically (Prance *et al* 1981, 1983). For small values of $\hbar\Omega$ (compared with the characteristic energy Φ_0^2/Λ of the superconducting weak link ring) equation (8.35) yields a ground state energy band in external flux space which is a periodic function of Φ_0. Such a ground state band for small $\hbar\Omega$ (which equals $0.05\Phi_0^2/\Lambda$) is shown in figure 8.17(*a*). The diamagnetic screening supercurrent response for the SQUID ring in the ground state band (cf equation (8.30))

$$\langle I_s(\Phi_x)\rangle = -\mathrm{d}E(\Phi_x)/\mathrm{d}\Phi_x \tag{8.36}$$

is shown in figure 8.17(*b*) where, as with the screening voltage (8.31), $I_s(\Phi_x)$ is necessarily the sum of harmonics (Prance R J *et al* 1981, 1983, 1985)

$$\langle I_s(\Phi_x)\rangle = \sum_{m=1}^{\infty} I_m \sin(2\pi m\Phi_x/\Phi_0). \tag{8.37}$$

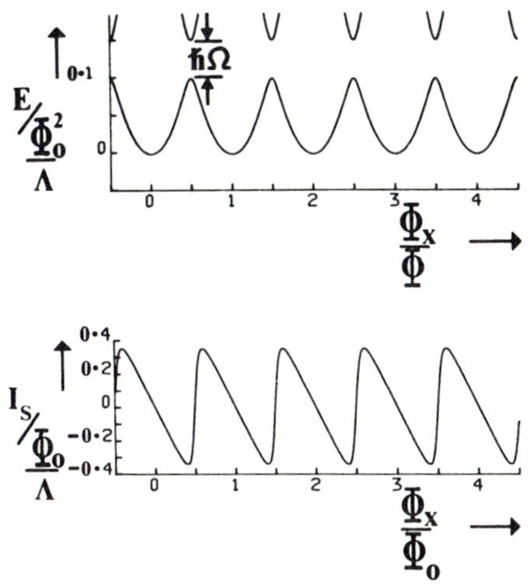

Figure 8.17 (*a*) E against Φ_x ground state band in the flux mode limit for $\hbar\Omega = 0.05\Phi_0^2/\Lambda$. (*b*) Screening current pattern for the ground state flux band of (*a*).

Again, we can define a wavefunction, this time in charge space, in terms of the D_n values and an angular displacement $\theta = 2\pi Q/q$, as (cf equation (8.28), see Widom *et al* 1982, 1984)

$$\psi(\theta) = \sum_n D_n \, \mathrm{e}^{-in\theta} \tag{8.38}$$

so that equation (8.35) can be rewritten in the form

$$(1/2\Lambda)(\mathrm{i}\Phi_0 \, \partial/\partial\theta - \Phi_x)^2 \psi(\theta) - \hbar\Omega \cos\theta \psi(\theta) = E(\Phi_x)\psi(\theta) \tag{8.39}$$

which is not the quantum mechanical form of the Josephson pendulum equation (Josephson 1962, Mutton *et al* 1984, Widom *et al* 1984).

We have described the ground state of a superconducting SQUID ring in the flux mode limit in terms of a winding number, n, (i.e. the number of 2π rotations of the pair condensate phase along a path around the ring, deep within the superconductor) and a quantum mechanical flux state transition frequency $\Omega/4\pi$. Since phase coherence is established around the ring, we can introduce excited states of the SQUID system in terms of quantised electromagnetic oscillations of the ring. If we define the quantised oscillation to have a frequency $\omega_c = (\Lambda C)^{-1/2}$, we can now describe the SQUID system in terms of nearest-neighbour $(n \to (n \pm 1))$ flux tunnelling transitions accompanied by photon number changes $(m \to m')$ at the same time as coupling $C_{mm'}$ which has a value given by

$$C_{mm'} = \sum_{r=0}^{\min\{m,m'\}} (m!m'!)^{1/2} [r!(m-r)!(m'-r)!]^{-1} \lambda_c^{(m+m'-2r)} \tag{8.40}$$

where λ_c is a dimensionless parameter and the combinatorial factors arise because photons are identical to bosons (Spiller *et al* 1985, Prance R J *et al* 1981, 1983, Prance H *et al* 1986).

The time-independent Schrödinger equation for the superconducting SQUID ring therefore changes to

$$W_{n,m}(\Phi_x)D_{n,m}-\frac{\hbar\Omega}{2}\sum_{m'}C_{mm'}(D_{n+1,m'}+D_{n-1,m'})=E_\kappa(\Phi_x)D_{n,m} \quad (8.41)$$

where $W_{n,m}(\Phi_x)$ is the unperturbed (i.e. zero flux tunnelling) Hamiltonian (Prance *et al* 1981, 1983, Solymar 1972, Chapter 8 onwards). This consists of the energy due to the superconducting screening current (I_s) which is given by $(1/2)\Lambda I_s^2$ plus the m photon energies (neglecting the zero point), and since $\Phi=n\Phi_0=\Lambda I_s+\Phi_x$ when no flux tunnelling occurs, it has the form

$$W_{n,m}(\Phi_x)=\frac{1}{2\Lambda}(n\Phi_0-\Phi_x)^2+m\hbar\omega_c. \quad (8.42)$$

The expectation value of the screening current for each band ($\kappa=1$ (ground state), $\kappa=2$, $\kappa=3$, ..., etc) then reads

$$\langle I_s(\Phi_x)\rangle_\kappa=-\frac{dE_\kappa(\Phi_x)}{d\Phi_x}. \quad (8.43)$$

When the superconducting SQUID ring is coupled to a resonant tank circuit (figure 8.13) we can calculate the $\bar{V}_{OUT}$ against I_{IN} AC dynamical characteristic fully quantum mechanically (with inclusion of loss in the tank circuit) by using path integral techniques, just as was done for the charge mode equation (8.32). The result in terms of the band screening current response $\langle I_s(\Phi_x)\rangle_\kappa$ is

$$\bar{V}_{OUT}\text{ (in phase)}=Q\omega_R L_t\left(I_{IN}\sin^2\delta+\frac{\mu}{\pi}\int_{-\pi}^{\pi}d\zeta\cos\zeta\langle I_s(\Phi_x^{tot}\zeta)\rangle_\kappa\right) \quad (8.44)$$

where Q, μ, ω_R, L_t and δ have the same meanings as in equation (8.32) and the argument of the screening current is the total applied flux felt by the ring, including a fraction of the AC flux ($\Phi_{xrf}(t)$) caused by $I_{IN}(t)$, so $\Phi_x^{tot}(\zeta)=\Phi_x+\mu\Phi_{xrf}(\zeta)$ (Spiller *et al* 1985, Prance R J *et al* 1981, 1983, Solymar 1972).

According to these quantum mechanical calculations, the AC voltage and current-scaling between the charge mode, (8.32), and flux mode, (8.44), expressions is in the ratio of $q/C:\omega_R\Phi_0$. In general, this means that charge mode features in $\bar{V}_{OUT}$ against I_{IN} plots are much larger (approximately 100 times) than the corresponding flux mode features. Consequently, it is relatively easy to observe in-band charge mode dynamics. However, it is an extremely difficult technical problem to design SQUID receiver electronics sufficiently sensitive to observe the conjugate in band flux mode dynamics. We

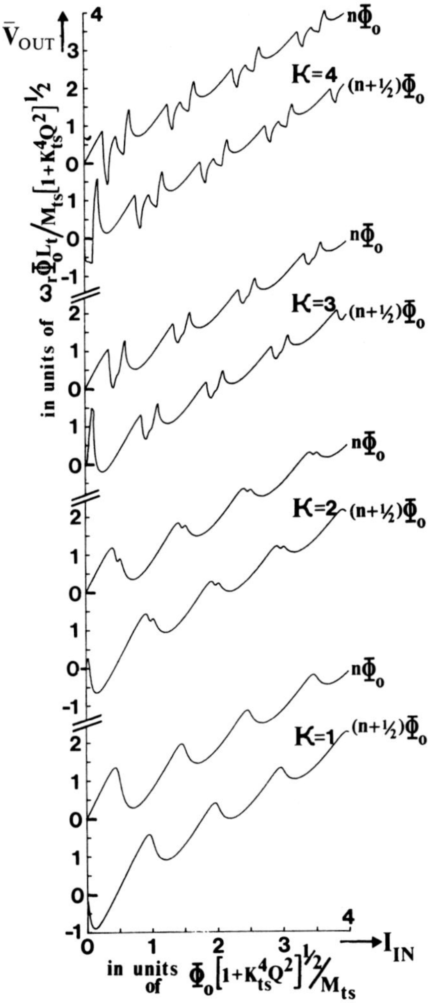

Figure 8.18 Theoretical $\bar{V}_{OUT}$ (in phase) against I_{IN} characteristics for ground state and first three excited states of flux band screening current patterns with $\hbar\Omega = 0.05\Phi_0^2/\Lambda$, $\hbar\omega_c = 0.078\Phi_0^2/\Lambda$, $K_{ts}^2 Q = 1.58$, $\Lambda = 5\times10^{-10}$ H, $L_t = 4\times10^{-8}$ H, $\lambda = 0.5$ and $\omega_R/2\pi = 430$ MHz.

have managed to achieve this using an ultra-low-noise UHF (430 MHz) SQUID electronics system, utilising cryogenically cooled GaAs field effect transistor (FET) pre-amplifiers, with a front-end receiver noise temperature of a few kelvin, as seen by the UHF tank circuit (Mutton *et al* 1984, Prance R J *et al* 1981, 1983, Prance H *et al* 1986).

Figure 8.18 shows a set of in-band dynamical flux mode $\bar{V}_{OUT}$ (in phase) against I_{IN} characteristics, calculated from equation (8.44), using typical UHF SQUID circuit parameters ($K_{ts}^2 Q = 1.58$, $Q = 40$, $\Lambda = 5\times10^{-10}$ H, $L_t = 4\times 10^{-8}$ H, $\hbar\Omega = 0.05\Phi_0^2/\Lambda$, $\hbar\omega_c = 0.078\Phi_0^2/\Lambda$, $\lambda_c = 0.5$ and $\omega_R/2\pi = 430$ MHz) for each of the first four bands ($\langle I_s(\Phi_x)\rangle_\kappa$, $\kappa = 1$ to 4) with the DC external flux Φ_x set at either $n\Phi_0$ or $(n+1/2)\Phi_0$, n being an integer. The $\Phi_x = n\Phi_0$ ground

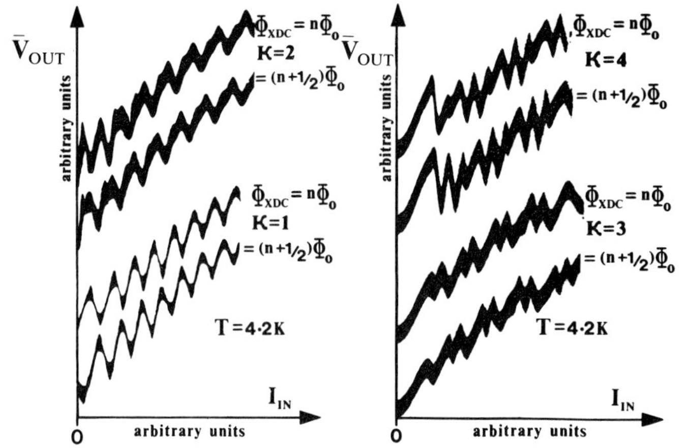

Figure 8.19 Set of experimental 430 MHz $\bar{V}_{OUT}$ (in phase) against I_{IN} characteristics for a flux mode, niobium point contact SQUID ring at $T=4.2$ K.

state band flux mode characteristic of figure 8.18 and the $Q_x=0$ ground state charge mode characteristic of figure 8.14(*a*) are identical, which is only to be expected since the charge and flux mode ground state bands of figures 8.12(*a*) and 8.17(*a*) are also identical.

Figure 8.19 shows a set of in-band experimental UHF ($\omega_R/2\pi=430$ MHz) flux mode $\bar{V}_{OUT}$ (in phase) against I_{IN} dynamical characteristics for the ultra-low-noise SQUID system described above, with all the characteristics being taken for the SQUID ring–tank circuit combination at 4.2 K (Solymar 1972, Barone and Paterno 1982). The superconducting niobium SQUID ring is of the Zimmerman two-hole point contact type (Zimmerman and Silver 1968, Long *et al* 1980). The niobium point contact is adjusted *in situ* in liquid helium to generate flux mode behaviour in the SQUID ring. We find experimentally that the SQUID system can either remain stably in the ground state band ($\kappa=1$, figure 8.19(*a*)) or, by adjusting the matching (and hence the noise level) of the first stage liquid-helium-cooled GaAs field effect transistor pre-amplifier, we can 'kick' the system in to one of the low-lying excited levels ($\kappa=2, 3, 4$). For each band the $\bar{V}_{OUT}$ against I_{IN} characteristics are plotted for the two DC applied flux states $\Phi_x=n\Phi_0$ and $(n+1/2)\Phi_0$. Large $\dot{\Phi}_x$ simply blur out the characteristics for each band. The correspondence between the characteristics in figures 8.18 and 8.19 is quite clear and provides strong evidence for the existence of energy bands in external flux space for flux mode superconducting weak link rings. Additional supporting evidence can be obtained through noise spectroscopy.

In the flux mode we expect macroscopic screening supercurrents to circulate around the ring while the confined flux is well defined around values of $n\Phi_0$. If the expectation value of this screening current is $\langle I_s(\Phi_x)\rangle_\kappa$, for a given energy band κ, then the ring magnetic susceptibility for this band is given by

$$\chi_\kappa(\Phi_x)=\Lambda \mathrm{d}\langle I_s(\Phi_x)\rangle_\kappa/\mathrm{d}\Phi_x \tag{8.45}$$

and if the tank circuit inductor L_t is coupled to the flux mode ring, with coupling coefficient K_{ts}, then the flux noise in the inductor is

$$\langle|\Delta\Phi_t|^2\rangle=k_B T L_t(1+K_{ts}^2\chi_\kappa(\Phi_x)) \tag{8.46}$$

and, with respect to frequency ω, the voltage noise is given by Spiller *et al* (in preparation)

$$\frac{\mathrm{d}\langle|\Delta V_t|^2\rangle}{\mathrm{d}\omega}=\frac{2k_B T L_t \omega_R^3\omega^2}{\pi Q}\frac{1}{(\omega^2-\bar{\omega}_r^2)+(\omega^2\omega_R^2/Q^2)} \tag{8.47}$$

where

$$\bar{\omega}_r^2=\frac{\omega_R^2}{1+K_{ts}^2\chi_\kappa(\Phi_x)}. \tag{8.48}$$

By using our ultra-low-noise 430 MHz SQUID receiver system (without an input drive current I_{IN}) we have been able to observe in-band ring susceptibilities for the ($\kappa=1$) ground state and first two ($\kappa=2$ and 3) weakly excited states. Examples of these band susceptibilities are shown in figure 8.20(*a*). As we have shown (Prance *et al* 1981, 1983), band susceptibility patterns almost identical to those of figure 8.20(*a*) can be generated from solutions of the matrix equation (8.41) using reasonable values of Ω and ω_c ($\hbar\Omega=0.05\Phi_0^2/\Lambda$, $\hbar\omega_c=0.078\Phi_0^2/\Lambda$). However, it is of considerable interest to calculate these susceptibility patterns through a variant of the quantum mechanical Josephson pendulum equation (8.29) (with $Q_x=0$), modified to describe the flux mode limit of behaviour of a superconducting weak link ring. The cosine term in equation (8.29), plus a magnetic energy storage term $(\Phi-\Phi_x)^2/2\Lambda$, make up the potential used in the quasi-classical description in equations (8.21) and (8.22) of a superconducting weak link ring. If we add this magnetic energy storage term to equation (8.29), and rewrite this explicitly in terms of flux Φ, again with $Q_x=0$ but with a DC flux Φ_x applied (i.e. the usual condition of operation of an AC-biased SQUID magnetometer), we arrive at the form

$$\left[-\frac{\hbar^2}{2C}\cdot\frac{\partial^2}{\partial\Phi^2}+\frac{(\Phi-\Phi_x)^2}{2\Lambda}-\hbar\nu\cos\left(2\pi\frac{\Phi}{\Phi_0}\right)\right]\psi(\Phi)=E(\Phi_x)\psi(\Phi). \tag{8.49}$$

There appears to be a contradiction in terms between this equation and equation (8.29), since the latter can only be arrived at if the charge mode limit of behaviour is invoked, i.e. where there exists an amplitude superposition of all possible flux states of the ring. However, flux mode descriptions (equation (8.21) and (8.22) quasi-classically or equations (8.35) and (8.39) quantum

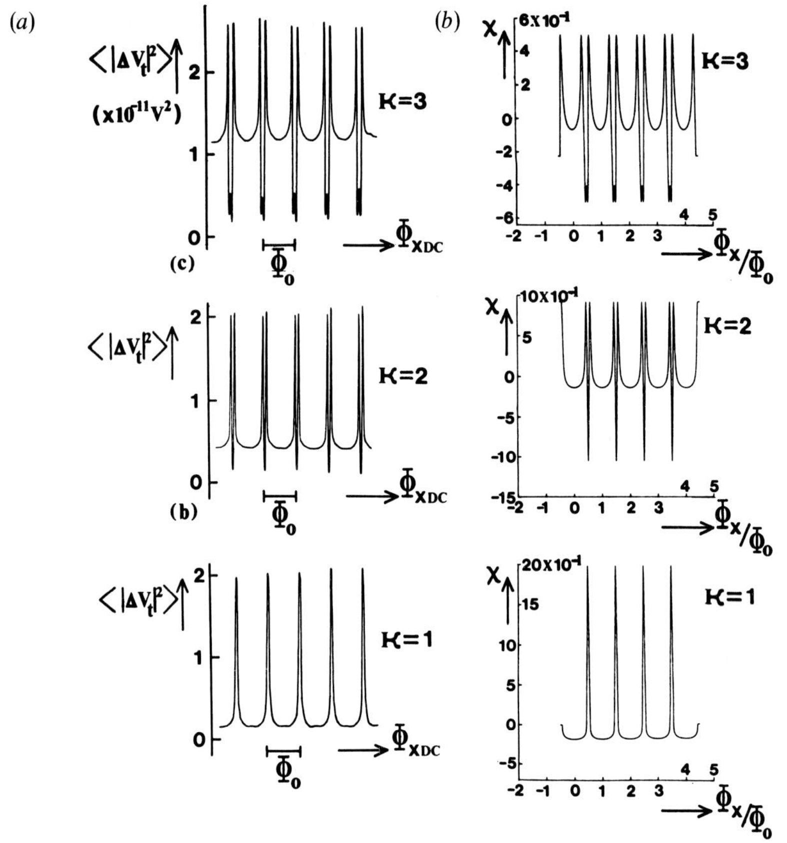

Figure 8.20 (*a*) Experimental in-band ($\kappa = 1, 2$ and 3) noise spectroscopy susceptibility patterns for a flux mode SQUID ring taken at $T = 4.2$ K using the ultra-low-noise UHF SQUID receiver system with $I_{IN} = 0$. (*b*) Theoretical $T = 0$ in-band ($\kappa = 1$, 2 and 3) susceptibility patterns computed using the flux mode tunnelling potential of equation (8.51) with $\Lambda = 5 \times 10^{-10}$ H, $r = 0.17$, $\hbar\nu = 0.062\Phi_0^2/\Lambda$ and $\hbar/(\Lambda C)^{1/2} = 0.03\Phi_0^2/\Lambda$.

mechanically) are concerned with the limit where the flux states of the ring are relatively well defined around states $n\Phi_0$ and macroscopic screening currents can flow around the ring. It is this macroscopic screening current which contributes a term $\Lambda I_s^2/2 = (\Phi - \Phi_x)^2/2\Lambda$ to the system Hamiltonian. This arises from processes where η pair charges ($\eta = 1, 2, 3, \ldots$ etc) transfer through the weak link quantum mechanically at some frequency $\nu_\eta/4\pi$ to give terms $-\hbar\nu_\eta \cos \eta\alpha$. Adding these to (8.49) yields our final equation for the flux mode limit,

$$\left[-\frac{\hbar^2}{2C}\frac{\partial^2}{\partial\Phi^2} + \frac{(\Phi - \Phi_x)^2}{2\Lambda} - \hbar\sum_\eta \nu_\eta \cos\left(\frac{2\pi\eta\Phi}{\Phi_0}\right)\right]\psi(\Phi) = E_\kappa(\Phi_x)\psi(\Phi). \quad (8.50)$$

As with equation (8.35), the solutions to equation (8.50), $E_\kappa(\Phi_x)$, are periodic bands in Φ_x space, with band number κ and period Φ_0. A physically reasonable form for the tunnelling potential takes the form

$$\sum_\eta v_\eta \cos \eta\alpha = v(\cos\alpha + r\cos 2\alpha + r^2\cos 3\alpha \ldots) = \frac{v(\cos\alpha - r)}{(1-2r\cos\alpha + r^2)} \tag{8.51}$$

where the parameter r $(0 \leqslant r < 1)$ governs the contributions from multiple pair charge transfer processes (Mutton *et al* 1985). Figure 8.20(*b*) shows the best-fit theoretical band susceptibilities. Compare these with the experimental patterns of figure 8.20(*a*), as calculated from solutions of equation (8.50) using $\Lambda = 5\times 10^{-10}$ H, $r=0.17$, $\hbar v = 0.062\Phi_0^2/\Lambda$ and $\hbar/(\Lambda C)^{1/2} = 0.03\Phi_0^2/\Lambda$. There is clearly good agreement between these results which we assume will improve with a better choice of the tunnelling potential (8.51). There has been considerable attention paid, particularly by Leggett and his co-workers (Leggett 1985), to the ground state, flux mode band in the limit of very small Ω, where it is hoped that quantum coherent oscillations between nearest-neighbour flux states of a superconducting weak link ring can be observed. Clearly, this is of interest, but technically difficult, given the ease with which large Ω weak link circuits, showing full macroscopic quantum behaviour, can be made. It also appears that, provided low-noise SQUID electronic systems are used to probe these weak link circuits, the external environment, which has been considered at great length by many workers (see, for example, Schön 1985), is not a problem. Macroscopic quantum mechanical behaviour is readily observable in such circuits, even at 4 K.

Clearly, for the quantum mechanical description of weak link ring circuits to be complete there must be a relationship between the two transfer frequencies v and Ω. At present this does not exist. However, it is reasonable to state that when one is small the other is large. We note from figures 8.12(*a*) and 8.17(*a*) that the smaller the transfer frequency the sharper the energy band. Sharp energy bands lead to easily recognisable dynamical and quasi-static behaviour, i.e. triangular features in SQUID magnetometer-type $\bar{V}_{OUT}$ against I_{IN} characteristics (figures 8.14(*a*) and 8.18) and spikes in noise spectroscopy susceptibility patterns (figures 8.16 and 8.20(*b*)). On the other hand, large transfer frequencies (in the conjugate space) generate very flat energy bands and, consequently, very weakly modulated dynamical and quasi-static behaviour.

The problem of external sources

Although we have been able to provide an adequate description of superconducting weak links in ring geometries, many experiments involve connecting weak links to external current sources. To illustrate the problem from the quantum mechanical viewpoint we can consider, for simplicity, the single cosine form of the superconducting SQUID ring potential

$$U(\Phi, \Phi_x) = \frac{(\Phi - \Phi_x)^2}{2\Lambda} - \hbar\nu \cos\left(\frac{2\pi\Phi}{\Phi_0}\right). \tag{8.52}$$

Breaking the ring and connecting the weak link to an external current source through normal leads effectively sets the circuit inductance Λ to infinity. The energy bands $E_\kappa(\Phi_x)$ disappear, since there no longer exists a parabolic term in the potential to provide the Φ_0-periodicity in external flux space. Connecting a current source I to the weak link leads to the familiar Anderson–Josephson washboard potential (Anderson 1968)

$$U(\Phi, I) = \hbar\nu \cos(2\pi\Phi/\Phi_0) - I\Phi \tag{8.53}$$

which, quasi-classically, is usually written in the form $(I_c\Phi_0/2\pi) \times \cos(2\pi\Phi/\Phi_0) - I\Phi$ (see equation (8.22)).

Provided quantum mechanical flux transfer frequencies are finite, potentials such as in (8.49) and (8.50) lead to the creation of Φ_0-periodic bands in Φ_x space. When the flux mode SQUID ring is effectively made singly connected by letting $\Lambda \to \infty$, these Φ_x-space bands disappear. However, with no parabolic part to the potential, bands can now form in Φ space. Thus, if the flux transfer frequency between ground state levels in adjacent wells of the potential $U(\Phi) = \hbar\nu \cos(2\pi\Phi/\Phi_0)$ is $\Omega/4\pi$, a ground state band of width $\hbar\Omega$ will form in Φ space. Excited levels in these adjacent wells will lead to the formation of higher bands in Φ space when $\Lambda \to \infty$. For weakly excited levels, in the simple harmonic oscillator approximation, these bands will be spaced in energy by $\hbar\omega_c$, centre to centre. Let us consider, for the moment, just the ground state band in Φ space. When a current I flows through the weak link the cosine potential is tilted to form a washboard and the band breaks up into a ladder of levels—the so-called Wannier ladder (Wannier 1960, 1962). Since the translation from well to well in the cosine potential is Φ_0 (i.e. in this translation a Φ_0 flux bundle cuts across the weak link for the limit where only nearest-neighbour contact ($n \to n \pm 1$) is made), the energy difference between adjacent rungs on the ladder is just

$$\hbar\omega_I = I\Phi_0 \tag{8.54}$$

or, by analogy with Josephson's second equation (8.19),

$$\nu_I = \frac{I}{q}. \tag{8.55}$$

The lifetime of the weak link, acting as a single quantum particle, will be finite and the stochastic decay from level to level down the tilt created by the applied current I leads to an average rate of change of flux across the link. The lifetimes in the levels (i.e. the level widths) will, of course, be a function of the tilt on the potential. Various possibilities for lifetime expressions have been reviewed by Avron (1982), see also Callaway (1974). The simplest choice is due originally to Zener (1943), see also Ziman (1964), and is the one we adopt here. We assume a barrier factor, $\exp(-I_0/I)$, for flux, in Φ_0 bundles, to cut across the weak link quantum mechanically, where I_0 can be termed a Zener breakdown current for these flux transitions. By definition, I_0 is not a critical current (I_c in equation (8.20)) beyond which the weak link can be driven into the normal resistive state. Indeed, if the link is driven normal the potential in equation (8.53) automatically disappears. Also, it is difficult to accept the traditional, quasi-classical view (Solymar 1972, Barone and Paterno 1982) that the voltage-carrying state of superconducting weak links occurs when the tilt in the potential is sufficient to remove completely the barriers between adjacent flux states in Φ space. In this situation it is difficult to argue that the quantum mechanical potential of equation (8.53) has any validity at all. If we place a voltmeter directly across the weak link, between one superconducting electrode and the other, and assume an attempt frequency f for the flux transitions, the Faraday law voltage developed across the link can be written in the Zener breakdown form as

$$V = K_\Phi \Phi_0 f \exp(-I_0/I) \tag{8.56}$$

where K_Φ is a fitting parameter.

In this view, the well known microwave-induced constant voltage-current steps in the DC current against voltage characteristics of a current-fed weak link arise through the synchronisation of the motion of Φ_0 flux bundles across the link by an external microwave field. This was first pointed out by Anderson (1964). The approximate Zener form, equation (8.56), for the ground state Φ-space band has recently been justified in terms of a quantised Sine–Gordon model of a weak link (Widom and Srivastava 1986).

Provided we restrict ourselves to nearest-neighbour ($n\Phi_0 \to (n \pm 1)\Phi_0$) flux changes, the matrix equation (8.41) for the superconducting flux mode, weak link ring can be modified to provide a reasonably accurate description of the singly connected weak link with bias-current-dependent lifetimes, $\Gamma(I)$. This takes the form

$$\left(n\hbar\omega_l + m\hbar\omega_c - \frac{i\hbar}{2}\gamma_m(I)\right)D_{nm} - \frac{\hbar\Omega}{2}\sum_{m'} C_{mm'}(D_{n+1,m'} + D_{n-1,m'}) = \left(E(I) - \frac{i\hbar}{2}\Gamma(I)\right)D_{nm} \tag{8.57}$$

where D_{nm} is the amplitude for the weak link to be in the mth excited level and the nth flux state.

In equation (8.41) we assumed that ω_c is set by the geometric inductance, Λ, and neglected the effect of the kinetic inductance, Λ_k, of the superconducting weak link. For the case of the singly connected weak link described by equation (8.57), we must identify ω_c with $1/(\Lambda_k C)^{1/2}$ where, conventionally, $\Lambda_k = \hbar/4e^2\nu$ (Barone and Paterno 1982, p 22). In the flux mode limit the ν_η values (equation (8.50)) will be large, so ω_c must also be large.

Equations (8.53), (8.55), (8.56) and (8.57) allow us to make the following predictions about the behaviour of current-fed singly connected weak link construction devices.

(1) When the cross section of the weak link is less than or equivalent to λ, v will be small and we may expect only one band (the ground state band of width $\hbar\Omega$) to be accommodated in Φ space (see equation (8.53)). In this apparent charge mode limit, flux will move across the weak link extremely easily but, to a first approximation, external charge space energy bands (figure 8.12(*a*)) will not exist unless I is much less than qv. The DC voltage against applied current characteristic will have a simple Zener lift-off shape, i.e. from equation (8.56) it will have an initial exponential growth in time-averaged voltage $\bar{V}$ at small currents, followed by a linearly increasing $\bar{V}$ at large currents. The $\bar{V}$ against I characteristic will show a current offset (an excess current) compared with an Ohmic line through the $\bar{V}=I=0$ origin. The high-current slope resistance will be large since, by definition, flux moves across the link much more easily than charge moves through the link. We would expect to be in this limit, roughly speaking, when flux, in units of Φ_0, cuts across the weak link at a rate an order of magnitude or more greater than charge, in units of q, transfers through the link. If we define a characteristic quantum resistance as (Gallop and Radcliffe 1985)

$$R_Q = \frac{\Phi_0}{q} = \frac{h}{4e^2} \tag{8.58}$$

then this single Φ-space band, Zener lift-off limit is reached when the high-current slope resistance $R \gtrsim 10R_Q$, i.e. roughly 100 kΩ or larger. If we identify the attempt frequency f in equation (8.56) with I/q, then the fitting parameter $K_\Phi \geqslant 10$. In this limit, where quantum mechanical flux transfer processes $n\Phi_0 \rightarrow (n\pm 1)\Phi_0$, $(n\pm 2)\Phi_0$, $(n\pm 3)\Phi_0$, ... etc are highly probable up to high order, the effect of an externally applied microwave source will be to synchronise with each of these processes, each producing microwave-induced structure in the DC $\bar{V}$ against I characteristic. The end result will be a smeared-out characteristic with no clearly defined step structures. However, all synchronisation processes will lead to a reduction in the lift-off current, I_0 (the critical current I_c in the traditional quasi-classical model of a weak link (equation (8.20)) since these all vary with the zero-order Bessel function whose argument contains the microwave current amplitude.

Figure 8.21 shows DC $\bar{V}$ against I data (triangles) for a high resistance ($R=$ 84.82 kΩ), superconducting niobium point contact weak link taken at $T=4.2$ K (Prance *et al* 1983, Mutton 1984, unpublished). The full line through these points is the single-band Zener lift-off characteristic calculated from equation (8.56) for a best fit with $K=13.15$ and $I_0=2.52$ nA. Extrapolation of this characteristic back on to the current axis yields a critical current I_c, as conventionally defined (Solymar 1972, Barone and Paterno 1982), of 0.75 nA.

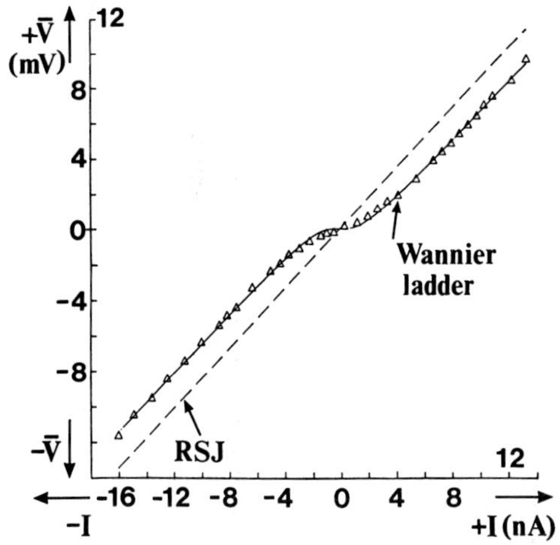

Figure 8.21 Best-fit Zener lift-off plot of $\bar{V}$ against I for a very high resistance current-fed niobium point contact weak link with $T=4.2$ K, $K=13.15$ and $I_0=2.52$ nA.

The fit between the data and the simple Zener lift-off curve is very good and the excess current at large values of applied current is quite apparent. The noise temperature of the measurement system used to plot this characteristic (essentially due to the very high imput impedance voltmeter measuring the voltage across the weak link) is less than or equal to 5×10^{-6} K and, consequently, noise-rounding of the characteristic can be regarded as negligible (Yoshida 1982). Microwaves have the effect of reducing the excess current and at large enough power levels the $\bar{V}$ against I characteristic becomes Ohmic. However, no microwave-induced constant voltage–current steps are observed.

(2) When the cross section of the weak link is greater than or equivalent to a few multiples of λ, Ω will be small and we may expect two or more bands (ground state plus excited state bands) to be accommodated in Φ space. In this flux mode limit macroscopic currents can be fed through the weak link from an external current source and flux will move across the link in Φ_0 bundles as transitions $n\rightarrow n\pm1$ take place. Since this is the only flux transfer process taking place, we expect good synchronisation with external microwaves to take place. This will lead to constant voltage–current steps in $\bar{V}$ against I accurately spaced in voltage by $\Phi_0 v_m$, where v_m is the microwave frequency. By contrast with (1) this flux mode limit requires that the charge, in units of q, transfers through the weak link at a rate an order of magnitude or more greater than flux, in units of Φ_0, can cut across the link. Thus we expect flux mode large-current slope resistances $R\lesssim0.1R_Q$, i.e. a few hundred Ω or less. In this limit we can compute a DC $\bar{V}$ against I characteristic for the onset of the voltage-carrying state of a weak link by considering a two band (ground state

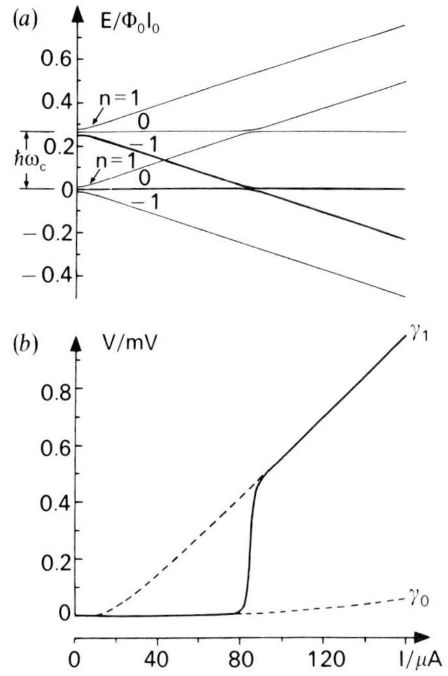

Figure 8.22 (*a*) Schematic Wannier energy level dependence on bias current I for a two-band Φ-space mode; $n=0$ to $n=-1$ transition, with splitting $\hbar\Omega$, shown in thick line. (*b*) Inverse lifetimes ($\gamma_0(I)$ and $\gamma_1(I)$) for the two bands of (*a*), showing crossing from γ_0 to γ_1 at the splitting point in (*a*). $\gamma_0(I)$ is the lower line.

plus first excited state) model in Φ space, each band developing into a Wannier ladder under the influence of the current I. Using the matrix equation (8.57) we can solve this system for the real part of the current-dependent energies in the two bands. This is shown schematically in figure 8.22(*a*) for just three representative levels ($n=-1, 0$ and $+1$) in each band. As can be seen, the two bands are spaced apart in energy by $\hbar\omega_c$, centre to centre (i.e. $n=0$ to $n=0$ level).

For all values of I we expect the lifetimes of the Wannier levels to be shorter (and the flux tunnelling rate to be higher) in the upper band than the lower band. To a good approximation each band can be described in its current dependence by a Zener lift-off function of the form of equation (8.56) but, for a given I, the inverse lifetime γ (which is proportional to $\bar{V}$) will be larger in the upper band than the lower band. The γ (or $\bar{V}$) against I curves for the two bands are shown schematically in figure 8.22(*b*) as dashed lines. In the flux mode limit, where we need only consider nearest-neighbout ($n\Phi_0 \to (n\pm 1)\Phi_0$) transitions, the weak link will cross from the lower to the upper γ against I curve in figure 8.22(*b*) when adjacent n levels (i.e. $n\Phi_0 \to (n\pm 1)\Phi_0$) in the real part of the energy $E(I)$ cross and split. As an example, the $n=0$ to $n=-1$ transition is shown as a thick line in figure 8.22(*a*). If the splitting ($\hbar\Omega$) is small, the crossing from lower to upper band $\gamma(I)$ in figure 8.22(*b*) can be very sharp, as depicted.

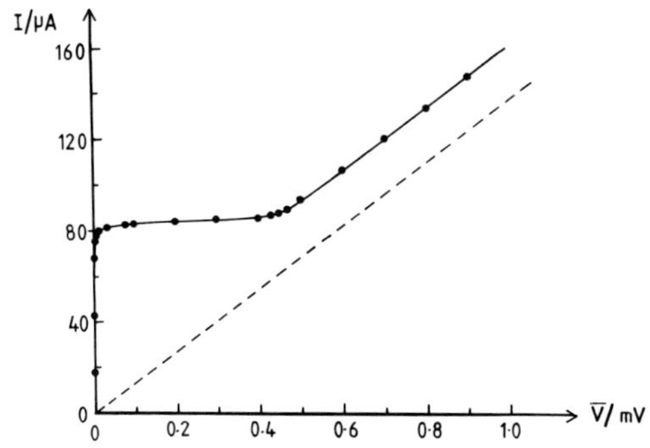

Figure 8.23 DC I against $\bar{V}$ data taken at $T=4.2$ K for a low resistance, superconducting niobium point contact; the full line is the theoretical best-fit for the two-band Zener lift-off model; the Ohmic line is shown dashed to emphasise the appearance of excess currents at finite voltages.

Figure 8.23 shows typical DC I against $\bar{V}$ data (as circles) for a low resistance, superconducting niobium point contact weak link. The solid line is the theoretical curve for nearest-neighbour ($n\Phi_0 \rightarrow (n \pm 1)\Phi_0$) transitions computed in the two-band model from equations (8.56) and (8.57) with $\hbar\omega_c = 0.27\Phi_0 I_0$, $\hbar\Omega = 0.016\Phi_0 I_0$, $I_0 = 319\ \mu$A, $I_1 = 29\ \mu$A, $K_{\Phi_0} = 3.9 \times 10^{-4}$, $K_{\Phi_1} = 1.2 \times 10^{-3}$ and $\lambda_c = 0.30$. The data clearly fit the theoretical curve extremely well and, by contrast, would not fit the standard quasi-classical RSJ hyperbolic characteristic (equation (8.20)). Nor would the RSJ model predict the excess current at large bias currents, which is quite apparent in figure 8.23. Application of microwaves generates good quality, constant voltage–current steps in the $\bar{V}$ against I characteristic of figure 8.23, as is to be expected for a low resistance, flux mode limit weak link. These steps are spaced accurately $\nu_m \Phi_0$ apart in voltage. It is worth noting with regard to this band model of $\bar{V}$ against I characteristics that, using the technique of microwave-enhanced tunnelling at very low temperatures, evidence for Φ-space energy levels has been provided recently by Clarke and his co-workers (Clarke *et al* 1985). A review of some of the theoretical problems associated with the biasing of weak links by external sources has been presented recently by Ben Jacob (1985).

We emphasise that $\bar{V}$ against I characteristics of the Zener lift-off type, as illustrated by the limiting form of figure 8.21 and the limiting form I against $\bar{V}$ characteristic of figure 8.23, are the general rule for superconducting weak link constriction devices such as point contacts and microbridges (Anderson and Dayem 1964, Kirschman *et al* 1970). Very rarely does the RSJ hyperbolic curve (equation (8.20)) approximate to the experimental $\bar{V}$ against I characteristics of these devices.

We have seen that it is possible to provide a realistic quantum mechanical description of singly connected weak links based on the Φ-space potential

(8.53) or its variants (see equation (8.50)). In all cases mentioned, charge is injected into the weak link from an external source. Experimentally, this appears sufficient to suppress what we will term singly connected charge mode behaviour. This should derive, under suitable bias conditions, from equation (8.39) which is conjugate to the quantum mechanical form of the Josephson pendulum equation (Josephson 1962, Mutton *et al* 1984, Widom 1984). By analogy with equation (8.49), we should be able to write down an equation describing a voltage-fed charge mode weak link, including a capacitive energy storage term, of the form

$$\left[\frac{-\hbar^2}{2\Lambda}\cdot\frac{\partial^2}{\partial Q^2}+\frac{(Q+Q_x)^2}{2C}-\hbar\Omega\cos\left(\frac{2\pi Q}{q}\right)\right]\psi(Q)=E(Q_x)\psi(Q) \qquad (8.59)$$

or its variant

$$\left[\frac{-\hbar^2}{2\Lambda}\cdot\frac{\partial^2}{\partial Q^2}+\frac{(Q+Q_x)^2}{2C}-\hbar\sum_\eta\Omega_\eta\cos\left(\frac{2\pi\eta Q}{q}\right)\right]\psi(Q)=E(Q_x)\psi(Q) \qquad (8.60)$$

which will generate energy bands in external charge space periodic in q (figure 8.12(*a*)), with reasonable energy separations, provided Λ is small. In order to remove the periodicity in Q_x space and form bands of width $\hbar\nu$ in Q space (if they exist) we need to remove the parabolic term in the potential in equation (8.59), i.e. we must somehow connect an external capacitance across the weak link. In principle this capacitance should be infinite, but it must be at least very large compared with C. If this can be achieved, then connecting a true voltage source V across the weak link yields the singly connected charge mode potential (cf equation (8.53))

$$U(Q,Q_x)=\hbar\Omega\cos(2\pi Q/q)-VQ \qquad (8.61)$$

or its variant through equation (8.60).

By analogy with current-fed weak links we can make the following predictions concerning singly connected, voltage-biased charge mode links.

(1) High resistance (a few tens of kΩ) weak links, where ν (or the ν_η values) is relatively large, will be restricted to a single band, of width $\hbar\nu$, in Q space. This will lead to a Zener-type lift-off characteristic in I against V of the form (cf equation (8.56))

$$I=K_Q q\frac{V}{\Phi_0}\exp\frac{-V_0}{V} \qquad (8.62)$$

where V/Φ_0 is the attempt frequency for pair charge transfer through the weak link, K_Q is a scaling parameter and V_0 is a Zener-type breakdown voltage. A current flows through the weak link as a consequence of the link, acting as a single quantum particle, falling stochastically down the rungs of the Wannier ladder in Q space, created when V is finite. The spacing between the rungs is

$$\hbar\omega_V=Vq \qquad (8.63)$$

or

$$\nu_V = \frac{2\,\mathrm{eV}}{\mathrm{h}} \tag{8.64}$$

which is just the original Josephson frequency–voltage relation (equation (8.19)).

In this view, synchronisation of these charge transitions down the Q-space Wannier ladder can generate constant current–voltage steps, conjugate to the well known microwave-induced steps from which the modern voltage standard has developed. These constant current–voltage steps were predicted in 1982 on the basis of the argued existence of energy bands in external charge space (Widom *et al* 1982a and b). As such, they could lead to a new quantum standard of current, since the steps would be spaced, as regards current, at exactly $\Delta I = \nu_m q$ intervals (Gallop and Radcliffe 1985, Widom *et al* 1982b). For weak links with resistances of the order of 10^4 ohms (i.e. a few multiples of R_Q), we would expect microwave synchronisation with a range of charge transfer processes ($Nq \rightarrow (N \pm 1)q$, $(N \pm 2)q$, $(N \pm 3)q$, ... etc). We would, therefore, not expect to see well defined, constant current–voltage steps. However, the excess voltage at large bias voltages in I against V plots would be suppressed if the microwave voltage levels were high enough.

(2) For extremely high resistance (of the order of 10^5 ohms) weak links, where only single pair transfer processes ($Nq \rightarrow (N \pm 1)q$) occur and ν is very small while Ω is correspondingly large, we would expect several energy bands to be accommodated in the Q space of the singly connected weak link. Taking a two-band (ground state and first excited state band) model, with small splitting $\hbar\nu$, we can generate the analogue in I against V terms of the flux mode $\bar{V}$ against I characteristic depicted in figure 8.22(*b*). When microwaves are applied, well defined, constant current–voltage steps should develop in this $\bar{I}$ against V characteristic. On a preliminary basis, we appear to have some evidence for rather smeared-out voltage steps, separated in current by $\nu_m q$, in the $\bar{V}$ against I characteristics of superconducting niobium point contacts with resistance of a few hundred kΩ. These point contacts are shunted by a high value (approximately equal to 10 pF) superconducting capacitor.

8.6 SUPERCONDUCTING WEAK LINK ELECTRONICS

In the wake of the run down of the superconducting electronics programme at IBM it seems rather certain that commercial superconducting digital computers, based on traditional, quasi-classical Josephson junction technology will never be realised. This does not seem to have deterred the Japanese, who are still very actively engaged in Josephson junction integrated circuit fabrication, particularly using refractory, high transition temperature materials such as NbN. It may be that their perseverance will be rewarded and that they will find solutions to the three major problems facing Josephson tunnel junction electronics. These are as follows.

(i) The problem of fabrication of very large scale integrated (VLSI) circuits with acceptable yields when the technology is based on tunnel barriers 10^{-9} m in thickness, i.e. just a few atomic layers across. Since any tunnel currents vary exponentially with barrier thickness, device quality control in VLSI Josephson tunnel junction circuits poses very great difficulties. These may be overcome by substituting thicker semiconducting layers, with lower barrier heights, for the conventional oxide tunnel barriers (Barone and Paterno 1982, p 222).

(ii) Electronic design problems when using active Josephson junction devices which have low input impedances and are current (i.e. by generating a magnetic field) controlled. This is to be compared with, for example, the semiconducting field effect transistor (FET), which is much more convenient for electronic design since it has a high input impedance and is voltage controlled.

(iii) The very rapid advances being made in high speed semiconductor devices combined with the huge in-built advantages that semiconducting technology has as the basis of the present day electronics industry.

On the positive side, it is possible to argue the general merits of low-temperature electronics, whether involving superconducting or semi-conducting (i.e. gallium arsenide based) devices. These include (Keyes 1975):

(i) Lower voltages, since in any switching circuitry the energy imparted to the charge carriers ($q_{\mathrm{eff}}=e$ for semiconductors, $2e$ for superconductors) must be large compared with $k_{\mathrm{B}}T$, i.e. voltage changes must be greater than $k_{\mathrm{B}}T/q_{\mathrm{eff}}$. Typically, semiconductor devices at room temperature run on switching voltage changes approximately equal to 1 V. For liquid helium compatible devices (semiconducting and superconducting) this could be scaled down to 10 mV or less. Lower voltages mean a much reduced likelihood of dielectric breakdown in the insulation layers in the integrated circuit and, of course, less power consumption. A considerable reduction in power dissipated leads directly to the higher packing density of components and, correspondingly, an overall increase in the operating speed of the circuit.

(ii) Elimination of the resistance of interconnections in low-temperature integrated circuits by the use of superconducting materials. Again, this reduces overall power consumption so that the packing density and speed of the circuit can be increased.

(iii) Virtual elimination of the long-term degradation in performance of electronic devices due to activated atomic motion and chemical degradation. This is of particular importance where extremely small (submicron) components are involved.

Thus, at least where VLSI is concerned, there may develop a convergence of superconducting and semiconducting technologies based on the advantages

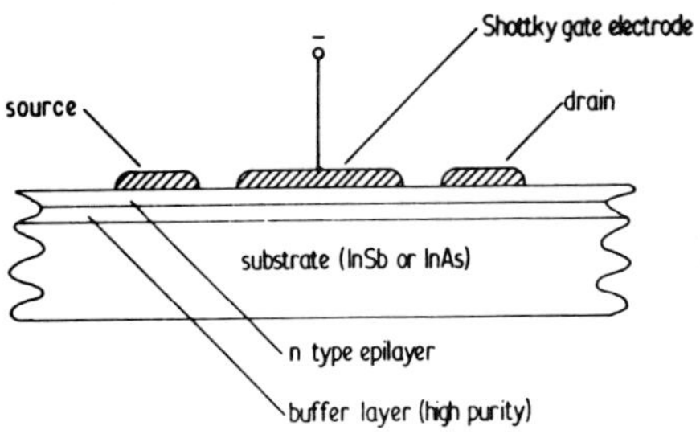

Figure 8.24 Proposed hybrid, weak link FET structure with superconducting source and drain electrodes and a semiconducting gate.

accruing from low-temperature operation. At the device level this may include hybrid structures incorporating both semiconducting and superconducting materials. One example, depicted in figure 8.24, is a proposed hybrid FET weak link (Clark *et al* 1980). In its original form, as shown, the source and drain electrodes are superconducting and the heavily doped channel acts as a weak link. Control of pair transfer through the channel is provided by the voltage applied to the Schottky barrier gate electrode. Although this concept was thought through originally with high mobility, small band-gap semiconductors such as InSb in mind, mobility, and hence device speed and gain, could be substantially improved by using structures of the HEMT GaAs–GaAlAs type for the channel region. Three terminal transistor devices of this type, and others based on quasi-particle injection into superconducting Josephson tunnel junctions, are being considered seriously (Gallagher, private communication, Buhrman 1985). It is quite conceivable that hybrid superconductor–semiconductor devices such as the weak link FET may, in the reasonably near future, make the best use of what now appear to be competing technologies. Even more exotic devices may arise from the interplay between superconductivity and the recently discovered quantum Hall effect in two-dimensional electron gas structures (Klitzing 1980). In this regard, the steady progress being made by Little and his co-workers in the development of miniaturised planar integrated refrigerators for cooling integrated circuits is of considerable significance (see, for example, Heiden 1985).

In the more distant future, the wealth of macroscopic quantum phenomena now being recognised in superconducting weak link circuits promises to generate a completely new form of electronic engineering, based on quantum mechanics rather than classical electrodynamics. At the moment, however, theoretical descriptions of quantum logic for a macroscopic quantum circuit computer are still at a rather early stage in their development (Feynman, priviate communication).

8.7 SUPERCONDUCTIVITY IN INSTRUMENTATION

Over the last fifteen years or so the remarkable coherence properties of superconductors have been applied in many ways to the design and construction of extremely-low-noise measurement systems. Two very good examples of such systems are provided by superconducting weak link ring (SQUID) magnetometers and superconducting cavities.

We have already discussed in considerable detail the dynamical quantum mechanical behaviour of one class of SQUID magnetometer, the so-called AC- or RF-biased type. The other form of SQUID magnetometer is referred to as DC-biased, since it consists of two superconducting weak links connected in parallel by a superconducting circuit, which are fed by an external direct current source. The current is large enough to put both weak links into the voltage-carrying state. In this state, at fixed bias current, the time-averaged voltage is a periodic function of the magnetic flux threading the area between the two weak links (Solymar 1972, Chapter 13, Barone and Paterno 1982, p 403).

In the last few years the sensitivity of both kinds of SQUID magnetometer has increased dramatically, with claims that the quantum limit in terms of sensitivity has been reached (see, for example, Ketchen and Jaycox (1982)). In discussing quantum limits in macroscopic electrical circuits it is essential to appreciate what such a limit implies. If a circuit actually operates in a quantum limit then its behaviour will be governed by canonically conjugate variables such as were introduced in equation (8.23) to describe a thick superconducting ring containing a weak link constriction. Clearly, if this quantum limit holds, then the circuit itself will act as a single macroscopic quantum object with quantum mechanical properties which are observable at the macroscopic level. For the AC-biased SQUID magnetometer (and its conjugate form, the charge mode AC-biased SQUID voltmeter) such properties have been established experimentally. Thus, it seems acceptable to invoke the uncertainty relation (8.24), between flux and charge, in any discussion of limits of sensitivity of AC-biased weak link ring systems. However, although the best DC- and AC-biased SQUID magnetometers now have comparable flux sensitivity, as yet only the AC-biased system has been described theoretically in a fully quantum mechanical way. We are, therefore, not able to discuss with any precision quantum limits of sensitivity in DC-biased SQUID systems.

For AC-biased SQUID magnetometers, operating as macroscopic quantum systems, the route to greater flux sensitivity appears straightforward, at least in principle. The relation (8.24) requires that improvements in the accuracy of measurement of the flux state of a weak link ring can only be achieved at the expense of increased delocalisation of charge at the weak link, i.e. by making thicker weak links which can carry larger macroscopic supercurrents. As we have shown, the AC-biased SQUID magnetometer actually functions by measuring the times at which flux (in Φ_0 bundles) enters or leaves the ring. The smaller the splitting energy $\hbar\Omega$, the more accurately can these times be measured. In the flux mode limit, Ω can be made smaller by increasing the cross section of the weak link. Thus it would appear that to improve the intrinsic flux sensitivity of an AC-biased SQUID system we are required to strike a balance between the flux transfer frequency and both the ambient (thermodynamic) temperature of the SQUID ring and the externally imposed electromagnetic noise temperature. With the electronic devices available at present, front-end (as seen by the SQUID tank circuit) receiver noise temperatures of a few kelvin can be achieved. This appears to be adequate where flux transfer frequencies of a few hundred gigahertz are involved (Prance *et al* 1981, 1983, Spiller *et al* 1985). For these state-of-the-art systems, measured flux sensitivities are found to be in the $10^{-7}\Phi_0/\sqrt{\mathrm{Hz}}$ range. Given the very rapid improvements being made in GaAs-based FET which can be operated at liquid helium temperatures, we might expect this flux sensitivity to improve by a factor of ten or more in the next few years. Sensitivities of this order will certainly be required if SQUID magnetomoeters, in Weber bar configurations, are to be used to detect gravitational radiation (Smarr 1979, Braginsky and Manukin 1977).

In their own way, superconducting microwave cavity resonators are just as remarkable as SQUID magnetometers (Silver 1976). It has already proved possible to construct these cavities with quality factors close to 10^{11}. These have been used to maintain an X-band (10 GHz) frequency stability approaching $1{:}10^{15}$. There seems no obvious reason why this stability should not be capable of improvemcnt, at least to $1{:}10^{16}$ or better. Obviously this degree of stability allows for the construction of extremely accurate clocks with all the possibilities this opens up for new physical experiments. Superconducting cavity resonators may also find application as gravitational wave detectors, since any incident gravitational radiation of sufficient intensity will cause an observable shift in resonant frequency.

ACKNOWLEDGMENTS

I would like to thank my colleagues Drs Helen Prance, Robert Prance, Tim Spiller and John Mutton at the University of Sussex for the outstanding contributions they have made to this work. I would also like to thank the BP Venture Research Unit for their generous support of our research programme at Sussex.

REFERENCES

Abrikosov A A 1957 *Zh. Eksp. Teor. Fiz.* **32** 1442

Anderson P W 1964 *Special Effects in Superconductivity—Lectures on the Many Body Problem 5th Int. School of Physics (Ravello, Italy) 1963* ed. E R Cainello (New York: Academic Press) vol. 2 p122

Anderson P W 1968 The Josephson effect and quantum coherence measurements in superconductors and superfluids *Prog. in Low Temp. Phys.* vol **V** ed. C. J. Gorter (Amsterdam: North-Holland)
Anderson P W and Dayem A H 1964 *Phys. Rev. Lett.* **13** 195
Avron J E 1982 *Ann. Phys., NY* **143** 83
Bardeen J 1961 *Phys. Rev. Lett.* **6** 57
Bardeen J, Cooper L N and Schrieffer J R 1957 *Phys. Rev.* **108** 1175
Barone A and Paterno G 1982 *Physics and Applications of the Josephson Effect* (Chichester: John Wiley)
Beasley M R and Geballe T H 1984 *Phys. Today* **37**(10) 60
Ben Jacob E 1985 Noise and dynamical behaviour of junctions and devices *Proc. IC SQUID Conf. (Berlin) 1985* ed. H D Hahlbohm and H Lübbig (Berlin: Gruyter)
Braginsky V B and Manukin A B 1977 *Measurement of Weak Forces in Physics Experiments* ed. D H Douglass (Chicago: University of Chicago Press)
Brandt N B, Kuvshinnikov S V, Rusakov A P and Semenov M V 1978 *Zh. Eksp. Teor. Fiz. Pis. Red.* **27** 37
Brovman E G, Kagen Y M and Kholas A 1971 *Zh. Eksp. Teor. Fiz.* **61** 2429
—— 1972 *Zh. Eksp. Teor. Fiz.* **62** 1492
—— 1977 *Zh. Eksp. Teor. Fiz.* **73** 1967
Buhrman R A 1985 Three terminal cryogenic devices *Proc. IC SQUID Conf. (Berlin) 1985* ed. H D Hahlbohm and H Lübbig (Berlin: Gruyter)
Callaway J 1974 *Quantum Theory of the Solid State Part B* (New York: Academic Press) pp 472, 566
Chapline G F 1972 *Phys. Rev.* B **6** 2067
Chu C W and Rusakov A P 1978 *Phys. Rev.* B **18** 2116
Clark T D 1973 *Phys. Rev.* B **8** 137
Clark T D Macroscopic quantum objects, Festschrift for David Bohm *Quantum Predictions* ed. B J Hiley and F D Pest (London: Routledge and Kegan Paul)
Clark T D, Prance R J and Grassie A D C 1980 *J. Appl. Phys.* **51**(5) 2736
Clarke J, Devoret M, Esteve D and Martinis J 1985 Quantum effects in the zero voltage state of the current biased Josephson junction *Proc. IC SQUID Conf. (Berlin) 1985* ed. H D Hahlbohm and H Lübbig (Berlin: Gruyter)
Cohen M L and Anderson P W 1972 *Superconductivity of d- and f-Band Metals: AIP Conference (Rochester, New York) 1971* p17
Cohen M H, Falicov L M and Phillips J C 1962 *Phys. Rev. Lett.* **8** 316
Coleman L B, Conen J J, Sandman D J, Yamagishi F G, Garito A F and Heeger A J 1973 *Solid State Commun.* **12** 1125
Cooper L N 1956 *Phys. Rev.* **104** 1189
Deaver B S and Fairbank W M 1961 *Phys. Rev. Lett.* **7** 43
Doll R and Näbauer M 1961 *Phys. Rev. Lett.* **7** 51
Dunne L J, Clark A D and Brändas E J 1983 *Chem. Phys. Lett.* **97** 573
Essmann U and Träuble H 1967 *Phys. Lett.* **24A** 526
Ferrell R A and Prange R E 1963 *Phys. Rev. Lett.* **10** 479
Feynman R P and Vernon F L 1963 *Ann. Phys., NY* **24** 118
Fiory A T 1971 *Phys. Rev. Lett.* **27** 501
Fröhlich H 1950 *Phys. Rev.* **79** 845
Gallagher W 1985 IBM Research Laboratories, IBM Yorktown Heights, New York, private communication

Gallop J C and Radcliffe W J 1985 A possible quantum current standard based on a superconducting weak link *Proc. IC SQUID Conf. (Berlin) 1985* ed. H D Hahlbohm and H Lübbig (Berlin: Gruyter)

de Gennes P G 1966 *Superconductivity of Metals and Alloys* (New York: Benjamin) p118

Giaever I 1960 *Phys. Rev. Lett.* **5** 147

Giaever I, Hart H R and Megerle K 1962 *Phys. Rev.* **126** 941

Ginzburg V L 1976 *Usp. Fiz. Nauk* **118** 315

—— 1985 *Physics and Astrophysics. A Selection of Key Problems* (Oxford: Pergamon) p15

Ginzburg V L and Kirzhnitz D A (ed.) 1982 *High Temperature Superconductivity* (New York: Plenum)

Ginzburg V L and Landau L D 1950 *Zh. Eksp. Teor. Fiz.* **20** 1064

Glover R E and Tinkham M 1956 *Phys. Rev.* **104** 844

—— 1957 *Phys. Rev.* **108** 243

Goodman W L, Willis W D, Vincent D A and Deaver B S 1971 *Phys. Rev.* B **4** 1530

Gorkov L P 1959 *Zh. Eksp. Teor. Fiz.* **36** 1918

Gorter C J and Casimir H B G 1934a *Phys. Z.* **35** 963

—— 1934b *Z. Tech. Phys.* **15** 539

Grassie A D C 1975 *Superconducting State* (Brighton: Sussex University Press) p70

Heiden C 1985 Progress in cooling facilities for SQUID applications *Proc. IC SQUID Conf. (Berlin) 1985* ed. H D Hahlbohm and H Lübbig (Berlin: Gruyter)

Hulm J K, Kunzler J E and Matthias B T 1981 The Road to Superconducting Materials *Phys. Today* **34**(1)

Jerome D and Schulz H J 1982 *Adv. Phys.* **31** 299

Josephson B D 1962 *Phys. Lett.* **1** 251

—— 1964 *Rev. Mod. Phys.* **36** 216

—— 1965 *Adv. Phys.* **14** 419

Ketchen M B and Jaycox J M 1982 *Appl. Phys. Lett.* **40** 737

Keyes R W 1975 Physical limits in digital electronics *Proc. IEEE* vol. 63 p740

Kirschman R K, Notays H A and Mercereau J E 1970 *Phys. Lett.* **34A** 209

Klein B M and Pickett W E 1982 *Superconductivity in d- and f-band Metals* ed. W Buckel and W Weber (Karlesruhe Kernforschungszentrum)

Klitzing K V, Dorda G and Pepper M 1980 *Phys. Rev. Lett.* **45** 494

Landau L D and Lifshitz E M 1970 *Course of Theoretical Physics, Volume 5, Statistical Physics* (Oxford: Pergamon Press) p430

Leggett A J 1985 Macroscopic quantum tunnelling and coherence: Their significance for the foundations of physics *Proc. IC SQUID Conf. (Berlin) 1985* ed. H D Hahlbohm and H Lübbig (Berlin: Gruyter)

Little W A 1964 *Phys. Rev.* A **134** 1416

—— 1967 *Proc. 10th Int. Conf. on Low Temp. Physics 1966 vol. IIA, Superconductivity* p115

—— 1967 *Phys. Rev.* **156** 396

Little W A 1971 *Superconductivity, Proc. Int. Conf. on the Science of Superconductivity* 1969 ed. F Chilton (Amsterdam: North Holland)

Liuken G, Smithey R and Meyer O 1984 *J. Phys. F: Met. Phys.* **14** L113

London F 1961 *Superfluids, Volume I, Macroscopic Theory of Superconductivity* (New York: Dover Publications) Section E

London F and London H 1935 *Proc. R. Soc.* A **149** 71
Long, A P, Clark T D and Prance R J 1980 *Rev. Sci. Instrum.* **51** 8
Martinoli P, Daldini O, Leeman C and Stocker E 1975 *Solid State Commun.* **17** 205
Matthias B T 1971 Anticorrelations in superconductivity in *Superconductivity, Proc. Int. Conf. on the Science of Superconductivity 1969* (Amsterdam: North-Holland) p69
Maxwell E 1950 *Phys. Rev.* **78** 477
McCumber D E 1968 *J. Appl. Phys.* **39** 3113
McMillan W L and Rowell J M 1965 *Phys. Rev. Lett.* **14** 108
Meissner W and Ochsenfeld R 1933 *Naturwissenschaften* **21** 787
Mutton J E 1984 *DPhil Thesis* University of Sussex, Brighton, Chapter VII
Mutton J E, Prance R J and Clark T D 1984 *Phys. Lett.* **104A** 375
Mutton J E, Spiller T P, Clark T D, Prance R J, Prance H, Croyden P and Nest R 1985 Schrödinger equation solutions for a Weak Link Constriction Ring *Proc. IC SQUID Conf. (Berlin) 1985* ed. H D Hahlbohm and H Lübbig (Berlin: Gruyter)
Onnes H K 1911 *Commun. Phys. Lab. Univ. Leiden* No 120b
—— 1913 *Commun. Phys. Lab. Univ. Leiden* Suppl. No 34
Patton B R and Sham L J 1973 *Phys. Rev. Lett.* **31** 631
Philips J C 1972 *Phys. Rev. Lett.* **29** 1551
Prance H, Prance R J, Mutton J E, Spiller T P, Clark T D and Nest R 1985a *Festkörperprobleme* **XXV** 337
Prance H, Prance R J, Spiller T P, Mutton J E, Clark T D and Nest R 1985b *Phys. Lett.* **111A** 199
Prance H, Prance R J, Mutton J E, Spiller T P and Clark T D 1985c Behaviour of an AC-biased SQUID ring in the quantum mechanical Josephson limit *Proc. IC SQUID Conf. (Berlin) 1985* ed. H D Hahlbohm and H Lübbig (Berlin: Gruyter)
Prance H, Spiller T P, Mutton J E, Prance R J, Clark T D and Nest R 1986 Quantum mechanical flux band dynamics of a weak link constriction ring Phys. Lett. **115A** 125
Prance R J *et al* 1981 *Nature* **289** 543
Prance R J, Mutton J E, Prance H, Clark T D, Widom A and Megaloudis G 1983 *Helv. Phys. Acta* **56** 789
Prance R J, Mutton J E, Prance H, Clark T D, Long A P, Widom A and Megaloudis G 1984 Quantum mechanical behaviour of large scale ($\sim$1 cm) superconducting circuits *Proc. Cryogenic Fundamentals Conf. (Wroclaw, Poland) 1983*
Prance R J, Clark T D, Mutton J E, Prance H, Spiller T P and Nest R 1985 *Phys. Lett.* **107A** 133
Prance R J, Prance H, Clark T D, Mutton J E, Spiller T P and Nest R 1985 Measurement of the electron pair transfer frequency in a superconducting weak link device in *Noise in Physical Systems and 1/f Noise* ed. A D Amico and P Mazetti (Amsterdam: Elsevier Scientific)
Reynolds C A, Serin B, Wright W H and Nesbitt L B 1950 *Phys. Rev.* **78** 487
Rowell J M and Kopf L 1965 *Phys. Rev.* **137** A 910
Saint-James D, Sarma G and Thomas E J 1969 *Type II Superconductivity* (Oxford: Pergamon)
Saltpeter E E 1972 *Phys. Rev. Lett.* **28** 560
Schön G 1985 Quantum noise in Josephson elements *Proc. IC SQUID Conf. (Berlin) 1985* ed. H D Hahlbohm and H Lübbig (Berlin: Gruyter)
Shirane S 1983 *Superconductivity in Ternary Compounds II: Superconductivity and Magnetism* ed. M B Maple and O Fischer (Berlin: Springer)

Shoenberg D 1965 *Superconductivity* (Cambridge: Cambridge University Press) p3

Silver A H ed. 1976 Navy study of superconductive electronics *Office of Naval Research* Report No NR 319–110 p76

Smarr L ed. 1979 Sources of gravitational radiation *Proc. of the Battelle Seattle Workshop 1978* (Cambridge: Cambridge University Press)

Solymar L 1972 *Superconductive Tunnelling and Applications* (London: Chapman and Hall)

Spiller T P, Mutton J E, Prance H, Prance R J, Clark T D and Nest R 1985 Macroscopic quantum mechanical behaviour of an AC-biased SQUID magnetometer *Proc. IC SQUID Conf. (Berlin) 1985* ed. H D Hahlbohm and H Lübbig (Berlin: Gruyter)

Spiller T P, Clark T D, Prance R J, Prance H and Mutton J E Susceptibility measurements of macroscopic quantum objects (in preparation)

Spiller T P, Clark T D, Prance R J, Prance H, Mutton J E and Nest R Quantum mechanical derivation of the V_{OUT}–I_{IN} characteristic for an AC-biased SQUID magnetometer (in preparation)

Steglich F, Aaarts J, Bredl C D, Lieke W, Meschede D and Franz W 1979 *Phys. Rev. Lett.* **43** 1982

Wannier G H 1960 *Phys. Rev.* **17** 432

—— 1962 *Rev. Mod. Phys.* **34** 645

Widom A, Megaloudis G, Clark T D, Prance H and Prance R J 1982a *J. Phys. A: Math. Gen.* **15** 3895

—— 1982b *J. Phys. A: Math. Gen.* **15** 3877

Widom A, Megaloudis G, Clark T D and Prance R J 1982c *J. Phys. A: Math. Gen.* **15** 1561

Widom A, Megaloudis G, Clark T D, Mutton J E, Prance R J and Prance H 1984 *J. Low Temp. Phys.* **57** 651

Widom A and Srivastava Y 1986 Electrical engineering implications of the quantised Sine–Gordon model *Phys. Lett.* **114A** 337

Yoshida K 1982 *J. Appl. Phys.* **53** 7471

Zener C 1943 *Proc. R. Soc.* A **145** 523

Ziman J M 1964 *Principles of the Theory of Solids* (Cambridge: Cambridge University Press) p 163

Zimmerman J E and Silver A H 1968 *J. Appl. Phys.* **39** 2679

9 Critical Phenomena

Julia Yeomans

9.1 INTRODUCTION

Theoretical physicists delight in seeing patterns in the data emerging from lots of different experiments and models. They sense that they are on the trail of an underlying principle that can unify a bewildering mass of numbers and results. The theory of critical phenomena and phase transitions was in this messy situation at the end of the 1960s. Sorting the tangle out led to the renormalisation group, a Nobel Prize for Ken Wilson of Cornell University and an understanding of how to cope with physics when all length scales are equally important. It led to an explanation of phase transitions as seemingly diverse as the superfluid transition in liquid ^{4}He at 2.17 K, the ordering of brass at 740 K and the ferromagnetic transition in Fe at 1043 K.

Moreover, it now seems that the ideas and formalism developed will apply not only to phase transitions but will also be useful in understanding turbulence and chaos.

9.2 CRITICAL EXPONENTS

The thermodynamic variable which most clearly characterises a phase transition is called its order parameter. For example, the order parameter of the gas–liquid phase transition is the density. This is small in the gas phase and larger in the liquid and generally jumps discontinuously at the transition itself. Such a jump in the order parameter is characteristic of a first-order phase transition in which extra heat, called the latent heat, is needed to change the state.

As the pressure is increased, boiling takes place at higher temperatures and there is a decrease in the latent heat and discontinuity in the density. They become zero at the critical point, beyond which it is possible to move continuously from a gas-like to a liquid-like fluid. At the critical point itself the order parameter is continuous but there is a discontinuity in its first derivative.

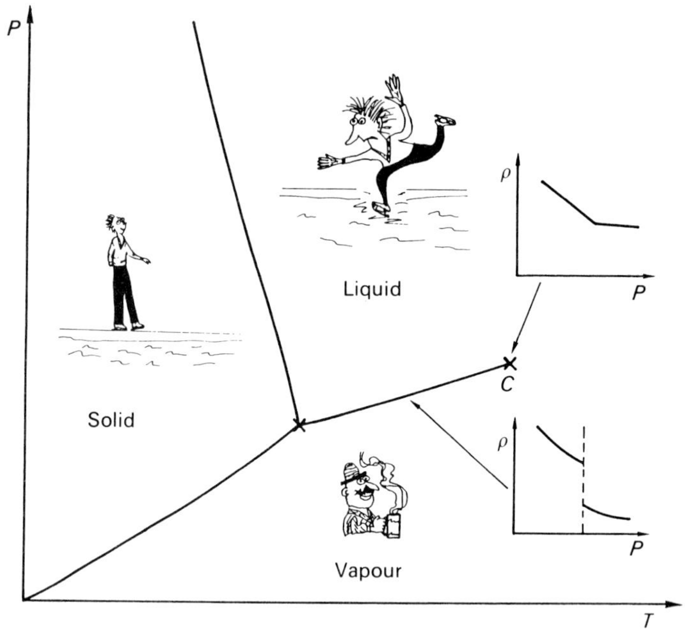

Figure 9.1 Phase diagram of water. All the phase transitions are first-order except at the point C, where the transition is second-order. Beyond C it is possible to move continuously from a liquid-like fluid to a vapour-like fluid. The boundary between the solid and the liquid phases is thought to be always first-order and not to terminate in a critical point.

This behaviour is characteristic of a second-order, or continuous, phase transition.

The critical point is one special point in the pressure–temperature diagram of a fluid (figure 9.1). Second-order transitions occur more generally in magnetic systems. The order parameter is the magnetisation, which is zero in the high-temperature phase, where the elementary magnetic dipoles point in random directions, but finite in the low-temperature, ferromagnetic, phase, where the dipoles are ordered.

Measuring the behaviour of thermodynamic variables, such as the order parameter, M, the specific heat, C, and the susceptibility, χ, near a second-order phase transition leads to a surprising result—singular behaviour like that shown in figure 9.2. Mathematically, the discontinuity in M and the divergences in C and χ can be written as simple power laws, as long as one is close enough to the critical temperature, T_c. Such laws are of the form

$$M \sim (T_c - T)^{\beta} \qquad C \sim (T - T_c)^{-\alpha} \qquad \chi \sim (T - T_c)^{-\gamma}$$

where β, α and γ are a measure of the sharpness of the cusp or the steepness of

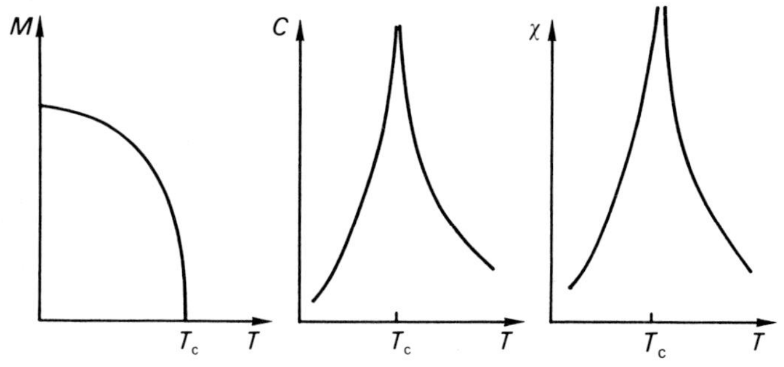

Figure 9.2 Behaviour of the thermodynamic functions at a phase transition. Note the discontinuity in the magnetisation, M, and the divergences in the specific heat, C, and the susceptibility, χ.

the divergence. They are called critical exponents and they play a central role in the theories of critical phenomena.

In 1907 Pierre Weiss proposed a theory of ferromagnetism which allowed him to predict values for the critical exponents. He assumed that each of the elementary magnetic moments in a solid was subject to a 'molecular field', λM, due to interaction with all the other moments, which was simply proportional to the magnetisation. Hence he hoped to account for the cooperative effect of the moments trying to align parallel to each other to minimise energy. He found that by working through the usual statistical mechanics, but with the external field, H, replaced or supplemented by an internal field, λM, he was able to predict ferromagnetism with a spontaneous total magnetic moment which varied with temperature like the plot shown in figure 9.2. Moreover, he found values $\beta = 1/2$, $\alpha = 0$ and $\gamma = 1$ for the critical exponents.

A connected theory, formulated by Landau in 1937, has been of great importance in the development of the subject and demonstrates mathematically how the magnetisation develops. Landau theory expands the free energy, F, as a power series in the order parameter,

$$F = aM^2 + bM^4.$$

The odd terms in the expansion, proportional to M and M^3, vanish because of the symmetry of the problem in zero magnetic field, since there is nothing to distinguish plus or minus values of M. Moreover, higher order terms in the expansion are unnecessary to obtain the essential physics.

Having written down an expression for the free energy the aim is to find its minimum value, which corresponds to the equilibrium state. F is plotted as a function of M for $a > 0, b > 0$ in figure 9.3(a) and for $a < 0, b > 0$ in figure 9.3(b). For $a > 0$, F has a single minimum at $M = 0$, showing that the stable state is paramagnetic and, hence, that positive values of a correspond to temperatures above the transition temperature, T_c. For $a < 0$, however, the minima of F

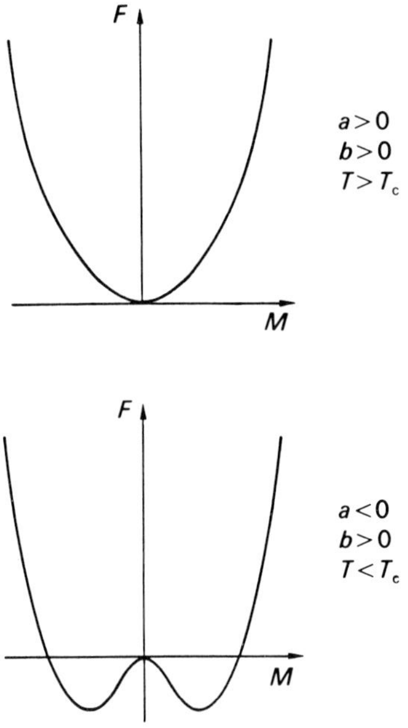

Figure 9.3 The form of the free energy assumed within Landau theory plotted against the order parameter for temperatures (*a*) greater than and (*b*) less than the critical temperature, T_c. In case (*a*) the free energy has its minimum value for a zero-order parameter ($M=0$). In case (*b*) the minimum value corresponds to a finite magnetisation of either sign.

occur at finite values of M, showing that there is a spontaneous magnetisation. Thus, $a<0$ corresponds to $T<T_c$ and, to describe this temperature dependence, we may write

$$a=a'(T-T_c).$$

The value of the critical exponent, β, can be found as follows. When the free energy takes its minimum value we can write

$$\frac{dF}{dM}=2aM+4bM^3=0$$

$$M^2=\frac{2a}{4b}=\frac{a'}{2b}(T_c-T)$$

thus

$$M \propto (T_c-T)^{1/2}$$

and this gives $\beta=1/2$. Similar reasoning leads to values of $\alpha=0$ and $\gamma=1$, the same values obtained using the Weiss molecular field theory. These are called 'mean-field' exponents and they emerge from many of the simple theories of phase transitions. Indeed, van der Waals, in his theory of critical points in fluids in 1873, had obtained the same results.

9.3 EXPERIMENTS

With so many theories pointing in the same direction, do the experiments agree? In the 1960s it first became possible to obtain accurate experimental values for the critical exponents. Figure 9.4 shows results from a paper by Heller and Benedek, which are used to obtain β for MnF_2. Note that to get close to a straight line the plot is of M^3, not M^2, as a function of temperature, giving $\beta = 0.335 \pm 0.005$, which disagrees with the mean-field value of 0.5.

Similar experiments have now been performed for many different ferromagnetic and antiferromagnetic materials and for other critical transitions, for example, at the λ-point where ^4He becomes superfluid, in liquid crystals and in quasi-two-dimensional solids. In general, the mean-field exponents are *not* obtained but a surprising pattern does emerge from the results.

The same exponents are obtained for water as for CH_4 and N_2, for the order–disorder transition in the alloy β-brass, for structural phase transitions and for some anisotropic magnets. This is the case even though the transition temperatures can differ by hundreds of degrees. But not all substances share these values for their critical exponents. Isotropic magnets, like Fe and Ni, fall

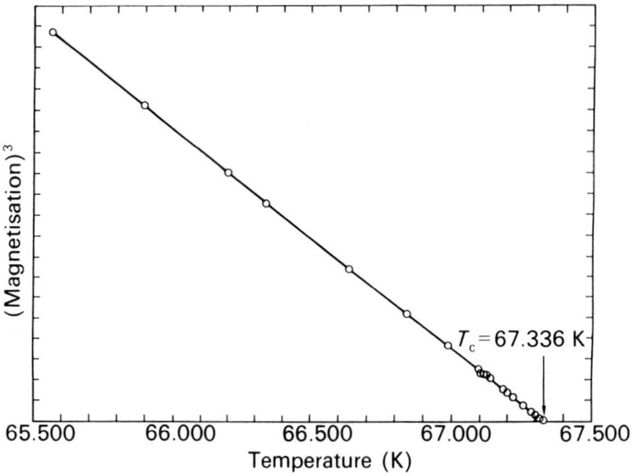

Figure 9.4 An experiment which measured the critical exponent, β, to a new degree of accuracy. The hyperfine splitting of the ions in MnF_2 monitors the magnetisation and can be measured very precisely from the nuclear magnetic resonance frequency. The cube of the (sublattice) magnetisation plotted against the absolute temperature gives a straight line suggesting $\beta \approx 1/3$, in contrast to the mean-field value, 1/2 (Heller and Benedek 1962).

into a different class and the superfluid transition in ^4He has yet another set of exponents.

As materials which appear to be totally unconnected can be grouped into classes which share the same critical exponents, we should ask what determines which class a given phase transition belongs to. It turns out that the two most important parameters are the dimensionality of space, d, and the number of components of the order parameter, n. Normally, in magnets like Fe, $n=3$ because the magnetisation is a vector which can point in any of the three space directions. However, magnets in which anisotropy constrains the order parameter to lie in a plane have $n=2$ and if the magnetisation can only point 'up' or 'down' along a specified axis, then we have a so-called uniaxial ferromagnet with $n=1$. Fluids have a scalar order parameter, the density, and hence have $n=1$. In liquid helium the order parameter is a complex number and, therefore, $n=2$. So ^4He will have the same critical exponents as planar magnets and water will have the same exponents as uniaxial ferromagnets. The descriptive jargon used is that these materials belong to the same 'universality class'. These results suggest that microscopic details are less important than symmetries and dimensionality in determining critical behaviour.

To understand what is happening on a microscopic scale let us look at a simple model of a magnet which undergoes a phase transition and which can easily be programmed on a microcomputer. This is the Ising model which was suggested to Ising by his supervisor, Lenz, in 1925 as a possible model which will have a phase transition. Ising solved it in one dimension where it only has a transition at $T=0$. He then left physics and it was not until 1945 that Onsager published his famous exact solution of the two-dimensional Ising model, which has served as a very important testing ground for the theories of phase transitions. In four or more dimensions the mean-field exponents are correct. Solving the three-dimensional model remains one of the outstanding problems in statistical mechanics, although a lot is now known about it from computer calculations. The three-dimensional Ising model, it is worth noting, has $n=1$ and so has the same critical exponents as water.

The Ising model is based on a lattice of sites. On each site there is a variable called a spin which can take two different values, $S_i=+1$ or -1, or thinking in terms of elementary magnetic dipoles, spin up $\uparrow$ or spin down $\downarrow$. Each spin interacts with its nearest neighbours on the lattice through an exchange interaction which favours any given pair of spins pointing in the same direction. At $T=0$ the spins are in their ground state, the state of minimum energy, with all the spins aligned. As the temperature is increased, entropy takes over, it becomes favourable for the spins to flip, and the system becomes increasingly disordered.

To see this happening it is easy to simulate the model on a computer with the fluctuations characteristic of finite temperatures being mimicked by a random number generator. The method for doing this was developed in 1953 and used on one of the first computers—the Los Alamos MANIAC. Now the programs can be fitted into even the smallest home computers†.

Figure 9.5 shows the results for a square lattice with 40 sites on each side for temperatures $T \ll T_c$, $T \gg T_c$ and T_c. The important question is 'what is so special about the critical temperature, T_c?' To answer this, imagine that you are looking at the pictures under a microscope and adjusting the controls on the instrument to decrease the magnification. In figure 9.5(*a*) the − areas will soon disappear (as the magnification is decreased) and the magnet will look exactly as it does at zero temperature with all the spins aligned. This is because at low temperatures interesting things only happen on short length scales and these soon disappear as we turn down the magnification. Now consider figure 9.5(*b*), where the temperature is greater than T_c. The small clusters of ordered spins which exist here will soon become indistinguishable from the totally disordered background characteristic of infinite temperature.

So as the magnification is decreased, low temperatures look increasingly like zero temperature, whereas high temperatures flow to infinite temperature. There must be a temperature in the middle that 'stays put', where the spin

† A booklet explaining the method, with programs, can be obtained from Dr C Windsor, Avebury, Millown Ave, Goring RG8 0AS, UK.

(*a*)

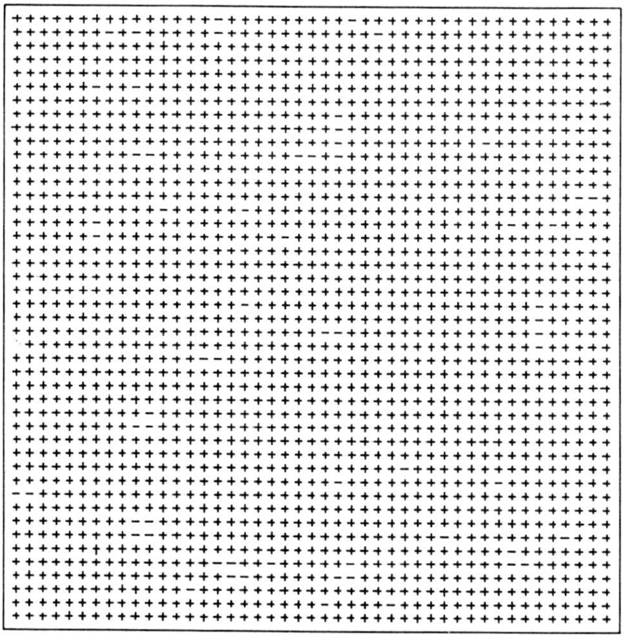

Figure 9.5 An Ising model at different temperatures. (*a*) $T \ll T_c$.

(b)

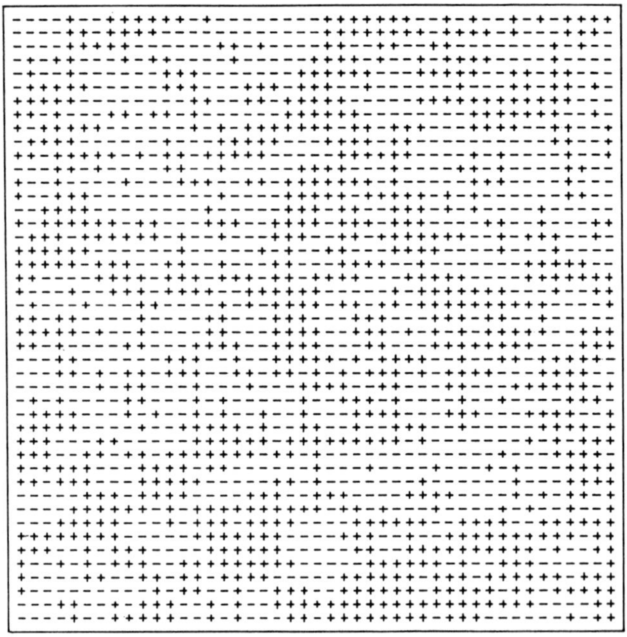

Figure 9.5(*b*) $T \gg T_c$.

(c)

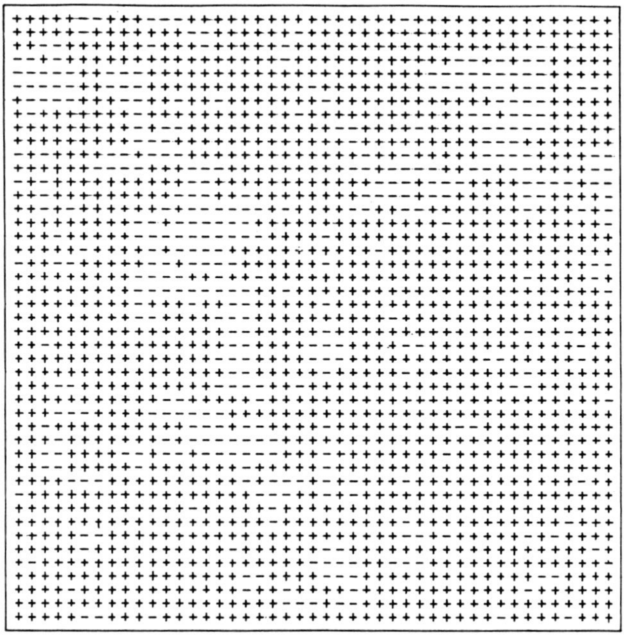

Figure 9.5(*c*) $T = T_c$. At T_c structures on all length scales are important.

pattern looks the same at all magnifications. That is to say, the spin system is the same on all length scales. The special temperature is the critical point, T_c (see figure 9.5(*c*)). At T_c all length scales are important and they are all going to have an effect on the physics of the system.

This is what makes the problem so difficult. We are quite good at doing physics when only large scales are important. For example, to describe the behaviour of waves in the sea the individual molecules can be ignored and only the bulk elastic constants of the medium considered. Similarly, if only small length scales are important that, too, is easy. It is quite justifiable to look at the properties of an atom without taking into account the effects of gravity. What are we to do when we cannot isolate a single significant length scale in this way?

9.4 THE RENORMALISATION GROUP

The resolution of the problem of how to treat systems in which all scales of length are important came in 1971 with the renormalisation group. As mentioned before, Ken Wilson of Cornell University was recently awarded the Nobel Prize for this work. The mathematics of the renormalisation group is based on the microscope idea described above—just keep on decreasing the magnification and see what happens. Ending up at zero temperature must have meant that the starting point was below T_c. A starting point above T_c flows to a totally disordered system at infinite temperature.

What about starting at T_c itself? Well, systems at T_c flow about on the critical surface, a special surface separating the flows to $T=\infty$ and the flows to $T=0$. They end up at one of a set of special points lying on the critical surface which are called fixed points. Their critical exponents are determined by which fixed point they flow to: there is one for each value of n and d, one for each universality class.

So, exponents are not determined by where you start from on the critical surface, which will depend on whether you are considering a magnet or a fluid and on all the details of the interactions in the system you are considering. Instead, they depend on which fixed point you flow to eventually, after looking at all possible magnifications. Because there is only one fixed point for each value of n and d the renormalisation group has divided systems into their different universality classes.

As a result of the many calculations, computations and experiments testing the scaling ideas which lie behind the renormalisation group we now think we understand many of the features of critical phenomena. We can work out the universality class to which a given material belongs and, at least numerically, predict the critical exponents. Some problems remain; for example, critical phenomena in disordered systems are still not well understood.

9.5 CHAOS

Cross fertilisation between different branches of physics is always vital, and it now looks like there is another problem where all length scales are important and where the ideas learned from critical phenomena can be applied. This is turbulence or chaos. One of the easiest visual examples of turbulent flow is water cascading down a waterfall. Calling a flow like this chaotic seems eminently sensible! Mathematicians have coined the term to describe any sort of motion where slightly different initial conditions can lead to very different outcomes. In waterfall terms this means the two drops of water which are neighbours at the top of the waterfall are likely to fall along entirely different paths and be far apart at the bottom. On their way, they get caught in eddies of all sizes and we again have to worry about all length scales to understand the physics of the waterfall.

The English weather is a rather splendid example of a chaotic system. Given a fine midsummer's day, it is impossible to predict whether it will be a white Christmas or, for that matter, whether it is a good idea to take a holiday in mid-July. All length scales are important, from tiny breezes to enormous depressions and the equations that govern the behaviour of the atmosphere are very complicated and highly non-linear. We have, at present, no hope of an exact solution and even the biggest computers can only produce a crude model of what is going on. Therefore it would be helpful if we could find some sort of universal behaviour which does not depend on the 'nasty' details of the equations.

We might be lucky! Recent work by Feigenbaum, looking at one particular aspect of turbulence, the transition from the linear, or non-turbulent, regime to the chaotic regime, has shown that the equations describing some chaotic systems possess universal exponents analogous to those of critical phenomena.

One of the beauties of his ideas is that many of them can be understood by looking at the simple recursion equation

$$x_{n+1} = ax_n(1 - x_n).$$

In this equation, a is a constant for which we shall choose various values between 0 and 4, x_n is a variable which can take values from 0 to 1, and x_{n+1} is the new value of x_n when the equation has been iterated. The value of x_{n+1} is then fed back into the right-hand side of the equation to obtain the next iterate x_{n+2}, and so on.

This equation is used by biologists to study the behaviour of animal populations. The number in the nth year is x_n and x_{n+1} the number in the subsequent year. The first term says that the larger the initial population the larger the new population whereas the second term cautions 'as long as they have enough to eat'!

Our aim is to see how successive iterates of this equation behave as the value of a is changed. This is most easily done using a pocket calculator—preferably a programmable one so that the x_{n+1} of one iteration can be fed straight back in as the x_n of the next—or a home computer. A BASIC program which can be used is

```
10    INPUT A
20    INPUT X0
30    LET X = X0
40    LET X=X*(1-X)*A
50    PRINT X
60    GOTO 40
```

Choose a value between 0 and 4 for the constant a and any value between 0 and 1 for x_0 (X0 in the computer program) and try a few iterations. Some typical results are shown in figure 9.6. You can easily reproduce these plots yourself on home computers with graphics by adding lines such as

```
35    LET N=0
45    LET N=N+1
55    PLOT N,X*100
```

Figure 9.6(a) shows that, for $a=2$, successive iterates flow towards a fixed value, $x_\infty=1/2$, and stay there. The situation for $a=4$, shown in figure 9.6(d), is much more complicated because now the system is chaotic. Although successive iterates are determined by the recursion equation, it is impossible to see a pattern emerge from them. Try two starting values of x_0, very close together. Although the iterates will be comparable for a few steps, they soon diverge and bear no resemblance to each other. One can see why the weathermen have so much trouble!

So how do things change from $a=2$ to $a=4$? Let us start with a close to 2 and see what happens as we gradually increase its value. For $2<a<3$ the flows still ends up at a fixed point. But as a moves above 3—try $a=3.3$—there is not one fixed point, but two, with every alternate iteration giving the same value as shown in figure 9.6(b). This is called a period 2 cycle and we say that the fixed point has bifurcated. Population-wise, it means that one year the greenfly will win, the next the roses will.

For a certain value of a, which turns out to be $a_2=3.4495$, the period 2 cycle bifurcates again into a period 4 cycle. Taking, for example, $a=3.5$, the solutions of our equations, once they have settled down, repeat every four iterations as shown in figure 9.6(c).

Now bifurcations come thick and fast. Given the patience to try enough

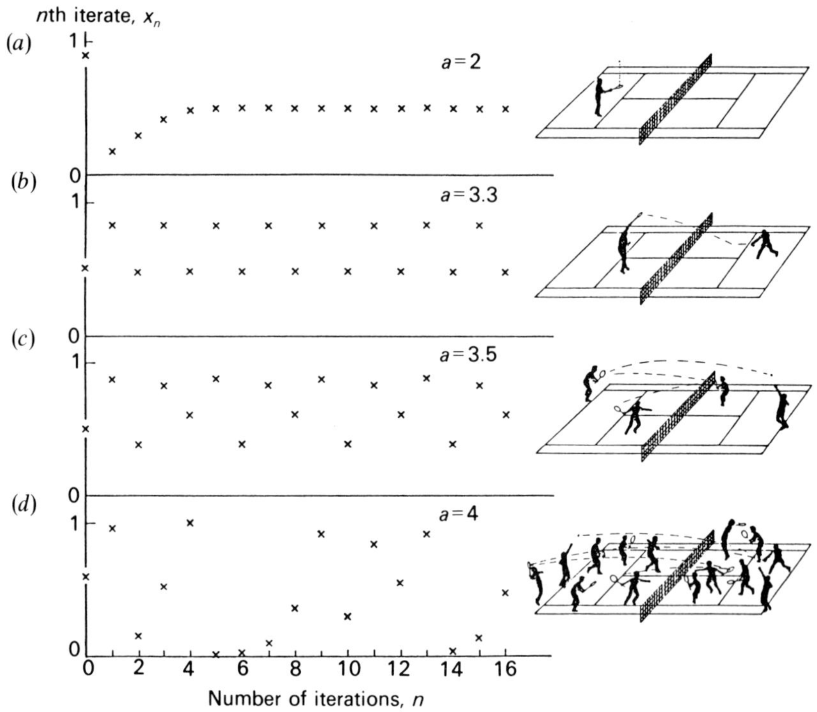

Figure 9.6 Successive iterations of the population equation for; (*a*) $a=2$, showing flow to a fixed point, (*b*) $a=3.3$, showing a period 2 cycle, (*c*) $a=3.5$, showing a period 4 cycle and (*d*) $a=4$, showing chaos.

values of a, one finds that the period 4 cycle is replaced by a period 8 cycle at $a_4=3.5441$ and the period 8 by a period 16 at $a_8=3.5685$, and so on. The points $a_2, a_4, a_8, \ldots, a_{2^n}, \ldots, a_\infty$ at which bifurcations occur become closer and closer together and accumulate at a value $a_\infty=3.5699$. At a_∞ there is a 2^∞ cycle with a period so long that one cannot pick out a pattern and we (effectively) have chaos.

For a large enough value of n, the values of a at which bifurcation occurs, a_{2^n}, approach the limiting value a_∞ in a way that can be described by the equation

$$a_\infty - a_{2^n} \sim \delta^{-n}$$

where δ is a number, rather like a critical exponent, equal to 4.669. This is called the period-doubling route to chaos.

So far, we have considered examples from mathematics, atmospheric physics and biology. It seems appropriate to look now at an experimental system somewhat closer to solid state physics. The phenomena of period-doubling can be observed in simple electrical circuits containing a non-linear

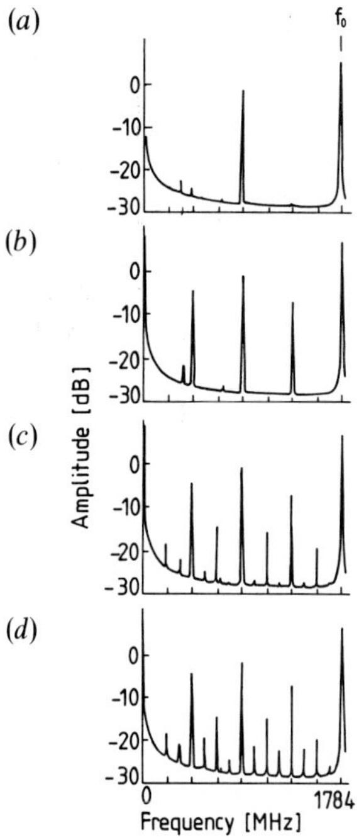

Figure 9.7 Period-doubling in a non-linear RLC circuit. The graphs show the power spectrum. A peak at frequency half the fundamental, f_0, shows the existence of a period twice the fundamental in the circuit oscillations. The driving voltages increase from (*a*) to (*d*). (Linsay 1981.)

element. Figure 9.7 shows results for an RLC circuit where the capacitor has a capacitance which varies as a function of the voltage across it. As the drive voltage, which is sinusoidal, is increased at a fixed frequency, harmonics corresponding to periods 2, 4, 8 and 18 times the fundamental successively appear in the power spectrum.

If a value for the critical exponent, δ, is extracted from these data, one obtains a value equal, to within experimental error, to that of the simple population equation described above. Similar behaviour with the same value of δ has now been observed in many different non-linear systems, such as Josephson junctions, a *Q*-switched CO_2 laser, chemical reactions, a driven semiconductor oscillator and convection in Rayleigh–Bernard cells.

So, it seems that if all length scales are important, we are able to find universal numbers which do not depend on the details of the Hamiltonian or differential equation or problem we are considering. Is there anything special about these numbers? Are they always rational fractions or can they take any value? Very recent work has shown that in two dimensions there is a connection between the values of the critical exponents and the invariance of statistical systems under a conformal transformation, a mapping which preserves local angles. Perhaps, ultimately, we shall find that the diversity of critical phenomena can be explained starting from just this basic symmetry, just as crystal properties rely so heavily on the invariance of the crystal lattice under translation by a lattice vector.

No aspect of solid state theory is hermetically sealed from the wider world of theoretical physics and applied mathematics. The study of critical phenomena, once a highly specialised preoccupation of solid state theorists, borrowed some of its seminal ideas from particle physics (in the work of Wilson) and now finds applications on a wider scale, touching many aspects of our lives. Beyond turbulence and chaos lie many other problems related to fractal geometry, computational science and much else, in which the theorists of critical phenomena can bring to bear their special insights.

REFERENCES

Heller P and Benedek G B 1962 *Phys. Rev. Lett.* **8** 428
Linsay P S 1981 *Phys. Rev. Lett.* **47** 1349

W R Hamilton (1805–65) and H Lloyd (1800–81)

10 Solid State Optics

S McMurry and
B Henderson

10.1 HISTORICAL REFLECTIONS

Until the very end of the last century, optical properties of materials were studied mainly to elucidate the nature of light. Consequently, attention was focused on describing the effects of a solid on light in terms of the most general macroscopic characteristics of that solid. Its external geometric shape determines the directions in which light is scattered from it, either through reflection (known to the Greeks around 400 BC) or, when the dimensions of the structure are small enough, through diffraction. Apart from its geometry, the property of a transparent solid most amenable to quantitative measurement is the degree to which it refracts light. This property, quantified by the refractive index of the material, was shown by Newton (1642–1727) to vary slightly with the colour of light, producing dispersion. Newton's interpretation of dispersion led to the explanation that the colours of solids were caused by the differential selective absorption of the components of white light. However, a method of measuring colour quantitatively was not available until the mid-nineteenth century, when an understanding of diffraction and the development of the spectrometer made it possible to measure optical wavelengths accurately. Of course, interest in the colour of solids, particularly of minerals and natural gemstones, dates from much earlier times. Optical absorption bands in diamond and other crystals have been known since the seventeenth century and by 1867 E Becquerel (1820–91) had measured the fluorescence lifetime of ruby with reasonable accuracy. However, in general the atomistic mechanisms of coloration of solids remained a mystery until this century.

Both Newton's corpuscular theory and the wave theory of Huygens (1629–95) could account for the laws of geometrical optics and describe the phenomenon of double refraction, if appropriate assumptions were made about the speed of light in solids. In particular, double refraction was explained by assuming that a beam of light is split by a crystal, into an ordinary ray—having a speed associated with the normal refractive index of the solid—and an extraordinary ray—which has a speed that varies with its direction in the crystal. Only along an optic axis of the crystal are the speeds of the two rays identical. The ordinary and extraordinary beams are plane-polarised at right angles to one another. Huygens noticed this when he observed that the relative intensity of the two rays produced in a second crystal, using either the ordinary or the extraordinary ray from the first, depended on the relative orientation of the crystals. But it was not until Fresnel (1788–1827) introduced transverse waves that polarisation could be incorporated into wave theory.

Newton's corpuscular theory had serious faults, such as the prediction that the speed of light in a material should increase with the refractive index. Its final downfall, however, was an inability to explain diffraction phenomena simply. On the contrary, wave theory accounted for all the observed effects in terms of no more than the geometry of the apertures or obstacles. Fresnel also used wave theory to predict the relative intensities of rays reflected and refracted at a surface, and gave an elegant explanation of double refraction in terms of wave and ray surfaces in a crystal. Since the energy flow in a crystal is not, in general, in the direction of wave propagation, the ray of light is not normal to the wave surface, so that two different surfaces, normal to the propagation vector and the ray respectively, can be defined. From the mathematical properties of these surfaces, Hamilton (1805–65) was able to predict the phenomenon of conical refraction in biaxial crystals: if the *propagation vector* of the refracted wave is parallel to an optic axis, the direction of the ray makes a fixed angle with that axis, so that instead of being split into two within the crystal, the beam of light is spread into a continuous distribution on the surface of a cone; conversely, if the *ray* direction in a crystal is along an optic axis the emerging beam is conical. H Lloyd (1801–81), at Hamilton's request, demonstrated both these effects using a crystal of aragonite and found remarkable agreement between the measured and predicted values for the angle of the cone in each case (Lloyd 1833).

In spite of such amazing successes, wave theory was plagued with serious problems. Fresnel and other proponents of the theory explained transverse waves in terms of the vibrations of an elastic aether. It was not clear whether these vibrations were perpendicular or parallel to the plane of polarisation of the light. Fresnel assumed them to be perpendicular to this plane, while McCullagh (1807–47) made the contrary assumption. To explain the behaviour of light in a solid they assumed that the properties of the aether were modified by the material. Alternatively, Helmholtz (1821–94), Thomson (1856–1940) and others suggested that the aether itself was unaffected, but that in some way the optical vibrations were transmitted to particles of matter in the solid, modifying the light waves by absorbing some of their energy (Glazebrook 1885). Since the properties of the aether were obscure, and even contradictory, the mechanism of the interaction between light and the solid was impossible to explain.

The dilemma was resolved around 1860–70 by Maxwell (1831–79) who treated light as an electromagnetic wave which interacts with charged constituents of matter through recognised forces. The direction of Fresnel's aethereal vibrations was shown to be the direction of the electric field in the wave, and consequently the direction of the vibrational motion induced by this field in electrically charged particles. In 1887 J J Thomson's discovery of the electron identified the constituent of matter which can respond to electric fields oscillating at optical frequencies. Maxwell's theory provided a relationship between the refractive index of a material and its dielectric constant, paved the way for simple resonance models which described the dependence of the refractive index on wavelength and explained the anomalous dispersion which Kundt (1829–96) had shown was always associated with the selective absorption of wavelengths in the visible range. Perhaps most important was the fact that Maxwell changed the emphasis of optical research to the study of the interaction of light with matter, rather than the nature of light. In due course this led to the development of quantum theory. Confirmation of the electromagnetic nature of the interaction between light and matter was provided by Zeeman (1865–1943), who observed that a strong magnetic field produced broadening of spectral lines. He confirmed experimentally the predictions, based on Lorentz's oscillator model (Lorentz 1909), concerning the polarisation of light at the edges and centre of the line. But measurements with better resolution disclosed the anomalous, as well as the normal, Zeeman effect, which was only explained fully using a further radical departure from classical physics—the introduction of electron spin. Slowly it had become evident that light could be used to probe the electronic structure of matter rather than the converse.

Interest in measuring the *optical constants* of matter has not waned during this century. However, the most dramatic developments have been in studies of the *optical spectra* of solids. Even with improved radiation sources and detectors, the major thrust for change was theoretical, rather than experimental, because spectroscopy was a major test flight for the fancies of quantum theorists. Although he was not one of the architects of quantum mechanics, Einstein (1879–1955) supplied the essential ideas of the quantum theory of the interaction of radiation and matter. In 1905 he introduced the photon in his explanation of the photoelectric effect and so paved the way for the great controversy of particle/wave duality. Quantum mechanics provided a detailed explanation of optical absorption and emission processes in solid matter, and Einstein's recognition of the importance of stimulated as well as spontaneous emission eventually led to the development of the laser. The colours of crystals were explained as the effects of impurity ions and defects in the regular crystalline lattice, and the dispute as to whether the chromium ion in ruby was the cause of the characteristic red colour or merely an activator for the regular components of the crystal was resolved.

Nevertheless, experimental innovation has been important. Detection systems and radiation sources have improved dramatically over the last 50 years and the laser has made possible the observation of non-linear effects never suspected in classical optics.

10.2 THE OPTICAL CONSTANTS OF SOLIDS

Classical ideas

For the present purpose, the discussion is restricted to observations made in the wavelength range 10^2–10^6 nm. The visible portion of this range is approximately 400–650 nm. Since the atomic spacing in a solid is only about 10^{-1} nm, light with these wavelengths is not diffracted and a perfect crystal can be treated as a continuous medium. Any defects or impurities with spacing much greater than the wavelength will scatter the radiation according to the laws of geometrical optics.

The energies of optical photons, ranging from $\hbar\omega \sim 10^{-3}$ eV in the infrared to ~ 10 eV in the ultraviolet, are very small compared with the rest mass of an electron (0.511 MeV c^{-2}). Consequently, the radiation may be treated classically. Since the solid behaves like a continuum at optical wavelengths its response may be described in terms of the macroscopic fields, which are averages of the rapidly varying atomic fields over a volume large on the atomic scale but small on the scale of optical wavelengths. For example, the macroscopic polarisation $\boldsymbol{P}$ can be related to the atomic polarisability, α_{at}, of the atoms in an insulator by

$$\boldsymbol{P} = N\langle \boldsymbol{p} \rangle = N\alpha_{\mathrm{at}}\langle \boldsymbol{E}_{\mathrm{loc}} \rangle \tag{10.1}$$

where $\langle \boldsymbol{p} \rangle$ is the average dipole moment induced in a single atom by the local (microscopic) field. ($\langle \boldsymbol{E}_{\mathrm{loc}} \rangle$ is not equal to the total macroscopic field $\boldsymbol{E}$.) This linear relationship is valid if the displacement of the atomic charge distribution is small enough for anharmonic effects to be negligible.

The optical constants—the electric and magnetic susceptibilities, χ_{e} and χ_{m}, the dielectric constant, ε, magnetic permeability, μ, and conductivity, σ—are defined with the assumption that the response of the material is linear. That is, the induced polarisation and magnetisation depend linearly on the electric and magnetic fields respectively:

$$\boldsymbol{P} = \varepsilon_0 \chi_{\mathrm{e}} \boldsymbol{E} = \varepsilon_0(\varepsilon - 1)\boldsymbol{E} \tag{10.2a}$$

$$\boldsymbol{M} = \chi_{\mathrm{m}} \boldsymbol{H} = (\mu - 1)\boldsymbol{H}. \tag{10.2b}$$

Similarly, the conduction current density is linearly related to the electric field through the conductivity

$$\boldsymbol{J} = \sigma \boldsymbol{E}. \tag{10.3}$$

This approximation is valid at normal radiation intensities but is not satisfied when the electric field in the radiation is comparable to that within atoms (approximately 10^6 V m^{-1}). Such large intensities are available with modern lasers and lead to non-linear effects.

One may conceive two main mechanisms through which an optical photon interacts with a solid; it can be absorbed in exciting an electron from one energy level to a higher (unoccupied) one, or it may polarise the material through which it passes in a predominantly elastic process. In fact, these two types of process are not independent—the absorption of a photon occurs through a polarised electronic state, and this polarisation will in turn affect the charge distribution over a relatively large region, so that absorption must be accompanied by a change in the polarisation of the material. Also, it is not physically reasonable to expect a photon to pass through a solid, polarising it, with no energy loss at all.

In the case of static fields the distinction between bound (polarisation) charges and the free charges which carry the conduction current is quite clear. In a constant electric field the bound charges reach a static equilibrium distribution which is maintained, without any net expenditure of energy, because of the balance between the force due to the applied field and the Coulomb interactions between the charges. On the other hand, the free charges must absorb energy from the field continuously in order to maintain a steady average linear momentum in spite of the dissipative effects of collisions. This distinction is, however, lost in the case of an oscillating field, such as that associated with the radiation, since both *bound* and *free* charges oscillate at the frequency of the applied field. Consequently, both contribute to the polarisation of the material and both can lose kinetic energy through collisions. Assuming a time dependence of the form $e^{-i\omega t}$ for the fields, Maxwell's equations, together with equations (10.2a) and (10.3), relate the magnetic and electric fields by

$$\nabla \times \boldsymbol{H} = (\sigma - i\varepsilon\varepsilon_0\omega)\boldsymbol{E} = -i\omega\varepsilon_0[\varepsilon + (i\sigma/\varepsilon_0\omega)]\boldsymbol{E}. \tag{10.4}$$

If ε and σ were real, as in the static case, the contribution to the total effective current, $\boldsymbol{J}_{\text{eff}} = \nabla \times \boldsymbol{H}$, of the bound and free charges would be $\pi/2$ out of phase. But, because of the inertia of the charges, neither the conduction current nor the oscillations of the bound charges can be in phase with the electric field, so that σ and ε must be complex. The physical distinction is now between the energy-conserving and the dissipative contributions to the effective current, rather than between bound and free charges, and it can be shown that the average rate at which energy is absorbed per unit volume of the solid is $(\text{Re}\,\sigma_{\text{eff}})\langle \boldsymbol{E}^2 \rangle$ where the effective conductivity is

$$\sigma_{\text{eff}} = \sigma - i\varepsilon\varepsilon_0\omega. \tag{10.5a}$$

With this notation all the charges are treated as free. Alternatively, all charges can be treated as bound and the effective dielectric constant is defined by

$$\varepsilon_{\text{eff}} = \varepsilon + (\mathrm{i}\sigma/\varepsilon_0\omega) = \varepsilon_1 + \mathrm{i}\varepsilon_2 \tag{10.5b}$$

where ε_1 and ε_2 are the real and imaginary parts of ε_{eff}, each of which depends on the (complex) values of both ε and σ.

Maxwell's theory determines the relationships between the dielectric constant and magnetic permeability and the more directly measurable reflectivity, absorption coefficient, and refractive index. The electric field of a plane monochromatic wave propagating through a solid along the z direction is

$$\boldsymbol{E}(\boldsymbol{r}, t) = \boldsymbol{E}_0 \exp(-\omega kz/c) \exp \mathrm{i}\omega(nz/c - t) \tag{10.6}$$

with

$$n^2 - k^2 = \mu\varepsilon_1 \tag{10.7a}$$

and

$$2nk = \mu\varepsilon_2. \tag{10.7b}$$

The phase velocity of this wave is obviously $v = c/n$, so that n must be interpreted as the (real) refractive index of the solid at frequency ω. The energy carried by the wave (which is proportional to $|\boldsymbol{E}|^2$) is damped exponentially as it penetrates the material and the absorption coefficient α is defined in terms of the rate of loss of intensity I of the beam by

$$\alpha = -(\mathrm{d}I/\mathrm{d}z)/I. \tag{10.8}$$

Equation (10.6) yields the expression

$$\alpha = 2(\omega/c)k = \mu\mu_0(c/n) \operatorname{Re} \sigma_{\text{eff}}. \tag{10.9}$$

The boundary conditions, which the fields must satisfy at the surface of the solid, determine the reflectivity, R, which is the proportion of the radiant energy that is reflected: at normal incidence

$$R = \frac{(1-n)^2 + k^2}{(1+n)^2 + k^2}. \tag{10.10}$$

It is possible to interpret the real part of the effective conductivity in terms of photon absorption. The average energy density in a beam of monochromatic radiation is $\xi = \varepsilon_0\langle|\boldsymbol{E}|^2\rangle$, which corresponds to $N\hbar\omega$, where N is the average photon density in the beam. Energy is lost from the beam by photon absorption at a rate $|\mathrm{d}\xi/\mathrm{d}t| = \hbar\omega|\mathrm{d}N/\mathrm{d}t|$. Therefore,

$$\frac{\mu}{\varepsilon_0 n^2} \operatorname{Re} \sigma_{\text{eff}} = \frac{1}{\varepsilon_0\langle \boldsymbol{E}^2\rangle} \frac{\mathrm{d}\xi}{\mathrm{d}t} = \frac{1}{N}\frac{\mathrm{d}N}{\mathrm{d}t} \tag{10.11}$$

which is the probability per second that a photon will be absorbed by the solid.

Although the optical constants have been defined classically, their frequency dependence reflects the quantum structure of the solid. The energy of an optical photon is of the same order of magnitude as the separation between energy levels associated both with the non-localised Bloch states of an electron in a crystal lattice and with localised atomic electron states. There is, therefore, resonant absorption of light by the electrons in a solid and this determines the optical constants. Consequently, measurements of these constants over a range of frequencies yield information on the band structure of the solid and also on any localised centres within it.

Fundamental optical constants

In the nineteenth, and early twentieth, century the optical spectra of many atoms and molecules were measured. The electronic structure of such atoms and molecules is deduced from the quantum mechanical interpretation of these spectra. The extension of spectroscopy to condensed matter coincided (more or less) with the first application of quantum ideas to crystalline solids. Hilsch and Pohl (1930) measured the fundamental absorption of evaporated thin films of alkali halides and found some similarities to the spectra of the corresponding diatomic molecules. However, there were also weak absorption bands in the visible region caused by *colour centres* in the thin films. For the next three decades research on fundamental optical spectra of solids was almost exclusively concerned with ionic crystals or with metals and their alloys. Not until the early 1950s did work on semiconductors begin to rival that carried out over the years in metallic and polar crystals.

To predict, even qualitatively, how the optical constants vary from one material to another, or how they depend on the frequency of the incident radiation, a model of the electronic structure of the solid is necessary. However, the general features of the frequency dependence near absorption peaks are not strongly model-dependent. As early as 1882 Lord Rayleigh (1842–1919) explained the anomalous dispersion close to an absorption band by analogy with the effects which occur near resonance in a pair of coupled mechanical oscillators. In the Lorentz (1853–1928) oscillator model of an insulator, harmonically bound charges are forced to oscillate at the frequency ω of the incident radiation, producing a polarisation proportional to the electric field. If a damping term, proportional to the velocity of the charge, is included in the equation of motion, the contribution to the electric susceptibility from N identical oscillators of charge e, mass m, natural frequency ω_0 and damping constant Γ is given by

$$\chi_e = \frac{Ne^2}{\varepsilon_0 m}\left(\frac{(\omega_0^2-\omega^2)+i\omega\Gamma}{(\omega_0^2-\omega^2)^2+\omega^2\Gamma^2}\right). \tag{10.12}$$

The frequency dependence of the dielectric constant, $\varepsilon = 1+\chi_e$, is shown in figure 10.1. The corresponding refractive index, extinction coefficient (proportional to the effective conductivity) and reflectivity are also shown.

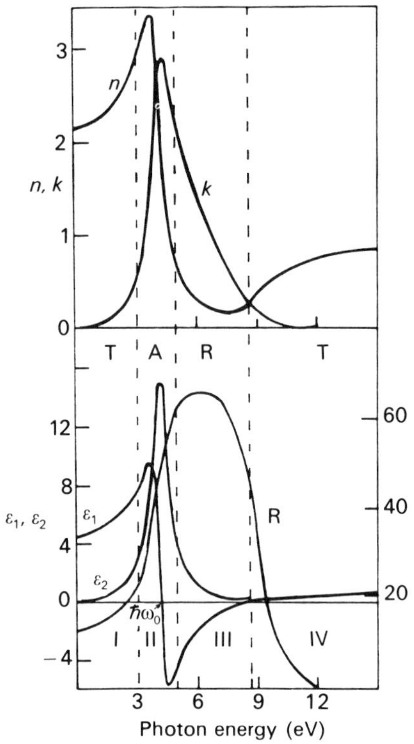

Figure 10.1 The spectral dependence of the real (ε_1) and imaginary (ε_2) dielectric constants, the refractive index (n), extinction coefficient (k) and reflectivity (R). The curves are calculated for an assumed damping constant, $h\Gamma = 1\,\text{eV}$, and elastic vibrational frequency $h\omega_0 = 4\,\text{eV}$. The regions I, II, III and IV are predominantly transmitting (T), absorbing (A), reflecting (R) and transmitting (T) respectively. (Adapted from Wooten 1972.)

The Lorentz model predicts four regimes for the frequency dependence of the optical constants, marked I to IV in figure 10.1. In the low-frequency region I, $\omega \ll \omega_0$, there is negligible absorption and low reflectivity. The solid is transparent, with a refractive index $n > 1$ which increases with frequency (normal dispersion). Region II, where $\omega \approx \omega_0$, is characterised by strong absorption, high reflectivity and a refractive index typical of anomalous dispersion which decreases with increasing frequency. For frequencies $\omega > \omega_0$, region III, dispersion is normal, although the reflectivity and effective conductivity are still large. In this region the insulator has the reflection characteristics usually associated with metals. The real part of the dielectric constant goes through zero at the plasma frequency, which for large frequencies is given by $\omega_p^2 = Ne^2/\varepsilon_0 m$. A vanishing refractive index implies an infinite phase velocity of the wave at this frequency, indicating an instantaneous correlation between all the oscillators. Physically, this means that there is a collective mode at the plasma frequency, in which all the charges oscillate in phase. The very small absorption at ω_p shows that this longitudinal collective mode is not excited by the transverse radiation in this simple model. However, longitudinal optical modes can occur in a crystal lattice in which there is strong coupling between the ionic oscillators. In region IV, $\omega \gg \omega_0$, the binding energy of the charges is negligible compared to the photon energy and

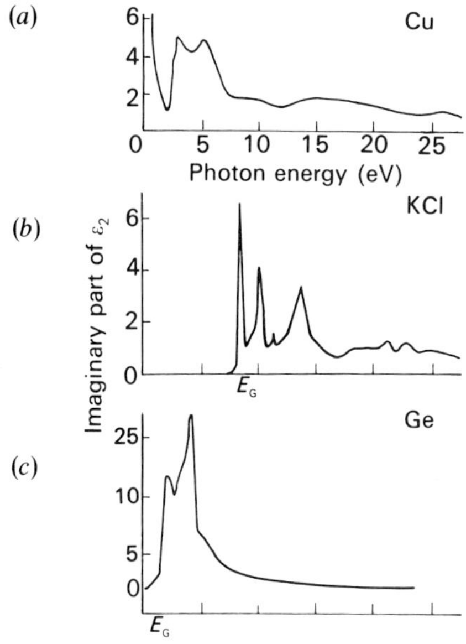

Figure 10.2 The spectral dependence of the imaginary part of the dielectric constant ($\varepsilon_2 = \sigma_{eff}/\varepsilon_0\omega$) as a function of photon energy for (*a*) Cu, (*b*) KCl and (*c*) Ge (Philip and Ehrenreich 1962, 1963a and b). E_G is the band gap energy.

they therefore respond as if they were free. The absorption is low in this region, and the solid again becomes transparent.

The conduction electrons in a metal are described semiclassically by the free electron model of Drude (1863–1906) (see Drude 1902). The dielectric constant is determined by the dissipative effects of collisions and the contribution to the electric susceptibility is represented by equation (10.12) with a zero-frequency resonance $\omega_0 = 0$. The general behaviour of the optical constants for a metal should, therefore, be similar to that for an insulator in regions III and IV of figure 10.1 where $\omega \geqslant \omega_0$. This predicts the large effective conductivity and metallic reflection properties at lower frequencies and the transparency of a metal to light at frequencies $\omega > \omega_p$.

The observed spectral dependence of the optical constants of metals, ionic solids and semiconductors are contrasted in figure 10.2, which plots $\mathrm{Re}\,\sigma_{eff}/\varepsilon_0\omega$ as a function of photon energy for Cu, KCl and Ge. These data (Philip and Ehrenreich, 1962, 1963) agree superficially with classical predictions: they also show a wealth of structure due to quantum mechanical effects. As expected, Cu (figure 10.2(*a*)) strongly reflects low-energy photons and transmits high-energy radiation. The broad bands at photon energies greater than 2.0 eV cannot be interpreted classically: they are associated with transitions between different energy bands. Cu and Au are unusual among metals since these band-to-band absorptions occur in the visible spectrum and produce the characteristic coloration of these metals. KCl is a typical ionic crystal and shows (figure 10.2(*b*)) the four regimes of behaviour expected on the Lorentz model. The region associated with perfect crystal transmission

($0 < h\nu < 10$ eV) is followed by fundamental electronic absorption at the band gap energy. The narrow absorption band reflects the tight binding in ionic crystals. The spectrum for Ge (figure 10.2(*c*)) is somewhat similar, but shows also the effects of a broad (0.7 eV) valence band. The spectrum of a solid should also show resonances associated with the vibrational motion of ionic dipoles. These would be expected in the far infrared (and microwave) regions, since ions are typically about 10^4 times more massive than electrons. Indeed, for ionic crystals such *reststrahlen* absorption (not shown in figure 10.2) is observed at the low energy end of the optical spectrum, superimposed on the classical electronic oscillator behaviour. *Reststrahlen* absorption is not observed in Ge because there are no permanent electric dipoles.

10.3 QUANTUM THEORETICAL INTERPRETATION OF THE OPTICAL CONSTANTS

Quantum mechanical models

Classically, energy exchange between a charged particle and electromagnetic radiation is a continuous process. The origin of quantum theory lies in Planck's (1858–1947) hypothesis that in a black body such energy exchange occurs in discrete units. Einstein, in introducing the photon, generalised this to a universal characteristic of the matter/radiation interaction. Quantum mechanics was developed to predict the energies of the emitted or absorbed photons and to calculate the strengths of such processes.

This was first applied to problems in atomic physics: the time-independent Schrödinger equation determines the discrete energies of atomic electrons and, hence, the energy $\hbar\omega$ of a photon absorbed or emitted in a transition from an initial occupied level of energy E_i to a final unoccupied level of energy E_f is given by $\hbar\omega = |E_i - E_f|$. Conservation laws determine whether a particular transition is observable. The strengths of optical transitions are calculated from Schrödinger's time-dependent equation, which also translates conservation laws into the selection rules of atomic spectroscopy (Condon and Shortley 1964). A transition which is forbidden through one mechanism may, however, occur through a different one. For example, the emission or absorption of a photon through an electric dipole transition can only occur if the initial and final states are of opposite parity, but transitions between states with the same parity may occur through magnetic dipole or electric quadrupole processes.

Quantum mechanical calculations related to the vast number of interacting atoms in a solid present a more daunting challenge. The simplest model (Kronig and Penney 1931) assumes that the (immovable) positive ions provide a periodic potential in which electrons move; the (mobile) electrons create an electric potential which is constant everywhere in the crystal. The periodicity of the ionic potential, with an appropriate boundary condition, predicts allowed bands of very closely spaced, discrete energy levels: energy values between these allowed bands are forbidden. This model leads to the familiar classification of the electronic properties of metals, insulators and semiconductors given in elementary texts.

The optical constants are defined for an equilibrium state of the system of atoms plus radiation, in which the energy levels of the solid are perturbed by the interaction with the radiation, and the polarisation is determined by the expectation value of the dipole moment operator, $\boldsymbol{D}=e\boldsymbol{r}$, in the perturbed state. The resulting quantum mechanical expression for the effective dielectric constant is

$$\varepsilon_{\text{eff}}=1+\frac{Ne^2}{\varepsilon_0 m}\sum\frac{f_{\text{fi}}}{\omega_{\text{fi}}^2-\omega^2-\text{i}\omega\Gamma_{\text{fi}}} \tag{10.13}$$

which has the same form as that resulting from equation (10.12), assuming that the system contains charges with different characteristic frequencies, ω_{fi}. The damping constant, Γ_{fi}, determines the width of the resonance peak, and is related to the lifetime of the excited state by the uncertainty principle. The proportion of charges of each type is determined by f_{fi}, the *oscillator strength*, which is proportional to $|\langle\psi_{\text{f}}|e\boldsymbol{r}|\psi_{\text{i}}\rangle|^2$. In the quantum mechanical case, $\hbar\omega_{\text{fi}}$ is interpreted as the energy difference between the final and initial (unperturbed) states $|\omega_{\text{f}}\rangle$ and $|\psi_{\text{i}}\rangle$, between which a transition occurs through emission or absorption of a photon of energy $\hbar\omega$. In a simple model in which there are only two energy levels, there can be no transition unless $\omega\approx\omega_{\text{fi}}$, and the classical picture of an oscillation in space of charges in the solid is replaced by an oscillation in time between the two energy levels at a frequency which depends on the electromagnetic interaction connecting them. The rate at which energy is absorbed in a transition is proportional to the oscillator strength and is just the product of f_{fi} and the semiclassically calculated rate of energy absorption. The oscillator strength can, therefore, be interpreted as the effective number of classical oscillators which absorb or emit energy at the same rate as the equivalent quantum mechanical transition. In accordance with this interpretation the oscillator strengths satisfy summation rules of the form

$$\sum_f f_{\text{fi}}=n \tag{10.14}$$

where $n=Z$ for a free atom with Z electrons. If the initial state is not the ground state the summation includes final states below, as well as above, the initial state; final states in the continuum must also be included. For one electron in an ionic lattice, $n=1-(m/m_{\text{i}}^*)$, where m_{i}^* is the effective mass of the electron in the initial state. In general, the summation rule, equation (10.14), does not apply to the transitions of a single electron but to the complete set of electrons in the initial state.

The energy band structure of a solid determines the number and positions of the electronic absorption resonances in ε_{eff} (equation (10.13)). In an insulator, the highest band of occupied levels—the valence band—is completely filled and separated from the lowest band of excited states—the conduction band—by a finite energy gap. The band gap (the minimum separation of these bands) is very much greater than kT, which is approximately 0.003 eV at normal temperatures. It is smaller in semiconductors than in insulators, so that at normal temperatures some electrons are thermally excited to states in the conduction band. A conductor is characterised by a partially filled conduction band in the ground state at all temperatures, so that electronic excitation can be produced even at zero frequency, and the electrons in the conduction band are, to a good approximation, free, as in Drude's semiclassical model. Equation (10.14) implies that there can be no transitions between the states of a free electron gas, where $m_i^* = m$. This is to be expected, since a single electron cannot, by itself, absorb a photon and conserve both energy and momentum. In the presence of a crystal potential, interband transitions between states of the same $\boldsymbol{k}$-vectors are possible. Lattice vibrations and/or defects are required to make intraband transitions possible in metals, the latter giving a Drude-like contribution to the optical constants, as discussed later.

Interband transitions in insulators and semiconductors

The distinction between insulators and semiconductors is essentially semantic, since at low temperatures semiconductors are good insulators. Most common semiconductors are predominantly covalent, with rather narrow band gaps (less than 3 eV). In these materials, impurities and defects may have energy levels within the band gap and optical transitions take place either between energy levels of the particular centre or between these levels and the valence and/or conduction bands. Significant experimental advances paralleling the development of solid state theory were crucial to the progressive understanding of optical phenomena in solids. Derivative spectroscopy, in particular, has provided great insight into band structure subtleties in many solids.

The threshold for the absorption of photons by electrons in an insulating solid is an indication of the minimum energy necessary to excite an electron from the valence band to the conduction band. The observed thresholds in Ge and KCl (figure 10.2), at 0.8 eV and approximately 6 eV respectively, reflect the much lower band gap to be expected in a semiconductor, compared with an insulator. Above the threshold there is a continuous distribution of possible transitions because of the large number of closely spaced levels within a band. The absorption rate at any frequency, ω, depends strongly on the density of states in the valence and conduction bands which are separated by $E \sim \hbar\omega$, through the oscillator strength, f_{fi}. 'Van Hove singularities' in the density of states produce edges in the absorption spectra, which may increase or decrease with an increase in frequency and may give peaks if an upward and a downward edge almost coincide. Such edges can be seen in the reflectivity of Ge, figure 10.2. In the late 1960s much effort was devoted to analysing structure in the optical spectrum to obtain information about the energy bands in a solid.

As already mentioned, the photon momentum is so small that interband absorption normally occurs through 'direct' transitions in which the crystal momentum $\boldsymbol{k}$, is unchanged. However, the coupling between the electronic states and lattice vibrations cannot always be neglected, even in this case. The absorption of a photon may excite an electron and simultaneously change the vibrational state of the lattice. Such 'indirect' transitions can be described as the absorption of a photon by an electron, accompanied by the emission or absorption of one or more phonons. The total (conserved) energy and momentum must now include that of the phonons, so that the crystal momentum in the initial and final electron states can be different.

A further over-simplification in the one-electron model is the neglect of the Coulomb interaction between an electron excited to the conduction band and the 'hole' it leaves in the valence band. The absorption edges already mentioned correspond to the production of free electrons and holes. However, the interaction between them may be strong, particularly in good insulators where the shielding effect of other electrons is small, and a bound state, or 'exciton', may be produced. The Coulomb interaction between electron and hole produces a bound Rydberg series of levels just below the band gap. These excitons are self-trapped because of the strong polarisation of the region in which they are generated. In ionic materials such as the alkali halides, the excitons spread over several ions. Such *Frenkel excitons* give rise to sharp peaks in the absorption spectrum below the band edge, as is shown in figure 10.2(*b*) for KCl. In general, the optical spectra of pure insulators are dominated by direct transitions producing sharp exciton peaks below the corresponding edges. The width of the peak depends on the lifetime of the exciton, which may decay radiatively to a lower exciton level or through interaction with the lattice. A well developed Rydberg series of discrete lines is shown in the absorption spectrum of excitons in Cu_2O, figure 10.3; these lines correspond to absorption transitions to 2p, 3p etc, exciton states.

Self-trapping is not effective in semiconductors and excitons are free to move with well defined $\boldsymbol{k}$-vectors, or to be trapped at defects. In semiconductors, the electron–hole interaction is better screened and the lowest frequency structure is broader and associated with edges. Exciton peaks can be observed at higher energies (greater than or equal to 10 eV) in materials such as Ge (figure 10.2(*c*)) and Si. These broader transitions are best studied using the techniques of derivative spectroscopy, in which (for example) the fractional change in reflectivity ($\mathrm{d}R/R$) is detected synchronously as the wavelength is modulated at some characteristic frequency. The resolution of such techniques can be better than 10^{-3} eV. Seraphin *et al* (1965) were the first to use derivative techniques to enhance structure in the

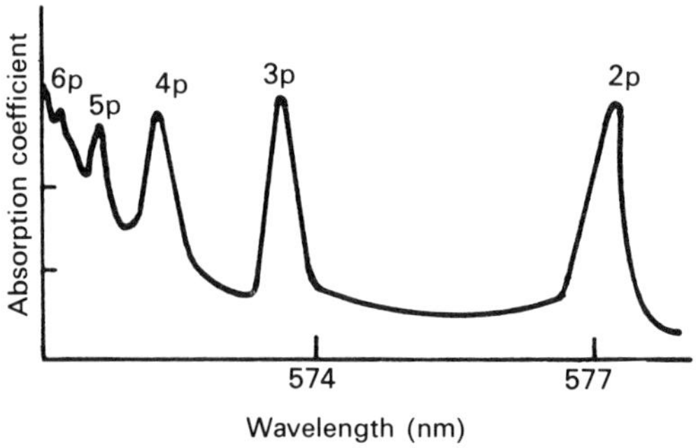

Figure 10.3 The absorption spectrum of np excitons in Cu_2O (adapted from Grosmann (1963)).

reflectivity spectrum. As figure 10.4 shows, the resulting spectra reveal in spectacular detail the structure associated with both excitons and band structure effects.

Optical properties of metals

In general, the optical reflectivity and absorption spectra of metals show much less structure than those of insulators or semiconductors. At low frequencies the absorption is high and decreases smoothly and rapidly with increasing frequency in the way suggested by the Drude free-electron model. The energy dissipation occurs through the scattering of conduction electrons by one another and by phonons but, as calculations of these intraband transitions are very complex, it is usually introduced through a phenomenological relaxation time. The free-electron model can be improved by using an effective mass to take account of binding of the electrons.

Above the interband threshold rather broad peaks appear, associated with transitions from the conduction band to higher excited bands and from core

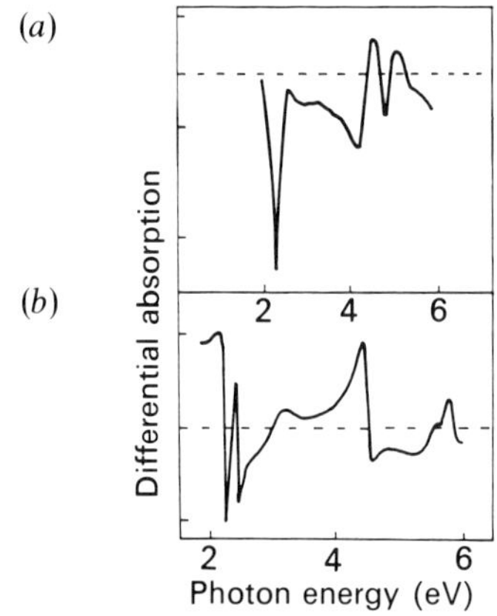

Figure 10.4 Modulation spectra at low photon energies for (*a*) Cu (Fong *et al* 1970) and (*b*) Ge (Zucca and Shen 1971) measured at 5 K.

states to empty states in the conduction band. Where this threshold is in the infrared, the reflectivity is comparatively small in the visible region, producing a dull metallic appearance, as for instance in tungsten and zinc where the thresholds are around 0.3 eV and 0.2 eV respectively. In contrast, the appearance of the noble metals is bright because the threshold for excitation from the d-band to the Fermi surface is in the visible region (approximately 2 eV for Cu, 2.5 eV for Au) or the ultraviolet region (approximately 4 eV for Ag). The reflectivity decreases above the absorption threshold so that it is largest at the red end of the spectrum in Cu and Au, giving rise to their characteristic colouring. Peaks in the visible and ultraviolet have been interpreted in terms of the band structure of Au but for most metals many-body interactions between the conduction electrons make accurate energy level calculations difficult.

Ultraviolet radiation can provide enough energy for electrons in the conduction band (or the d-bands and lower levels in some cases, and with hard ultraviolet light) to escape. Although analysis of the energy distribution of these photoelectrons is complicated by the need to take into account energy losses as the freed electron diffuses to the surface, this is the most important method for investigating the band structure of metals, giving much more accurate and detailed information than reflectivity or absorption measurements. For example, photoemission studies indicate that the transitions from the d-bands to the conduction band in the noble metals are mainly indirect.

Band gap states in insulators and semiconductors

In addition to information about the electronic structure of perfect crystals, optical spectra reveal defects and impurities in insulators and semiconductors. Even though present in comparatively low concentration, such localised optical centres can be observed in thick samples (approximately 1 mm) by measurement of the absorption coefficient or the photo-luminescence intensity. These centres are responsible for the coloration of crystals. Prior to 1940 spectra were usually recorded point-by-point. Accurate wavelength calibration, continuous wavelength scanning and synchronous detection only gradually became part of the spectroscopist's armoury. More recently, perturbation spectroscopy has been widely applied to localised centres in semiconductors and insulators. Furthermore, optically detected magnetic resonance has made seminal contributions to the understanding of radiative recombination processes in non-metals. In addition, the emergence of crystal growth as a science has given scientists access to samples of carefully controlled purity and/or composition.

The situation in ruby, so long a matter for controversy, is a graphic illustration of the continuing development. Although E Becquerel (1886) measured the fluorescent lifetime of the R-line he was unable to distinguish whether the bright red fluorescence was a property of 'pure' Al_2O_3 or the

Cr^{3+} ion. Not until the early 1900s was the essential role of the Cr^{3+} ion confirmed. By 1908 high resolution measurements of the Zeeman effect in the R-lines were published. Although the time was ripe for a major breakthrough, this was delayed until quantum mechanics demonstrated the importance of symmetry in the selection rules of optical spectra.

In 1929 Hans Bethe published his classic paper on crystal field theory, in which he treated the electrostatic field of the host lattice as a perturbation on the energy levels of the free ion and classified the new levels as representations of the symmetry group of the crystal field (Bethe 1929). Although Deutschbein (1932) reported studies of Cr^{3+} spectra in different crystals, he appeared to be unaware of the importance of Bethe's theoretical work. However, he did distinguish between the various zero-phonon lines and their vibrational side-bands. The extensions of Bethe's crystal field theory by Sugano *et al* (1970) provided the theoretical basis for modern transition metal ion spectroscopy. Nevertheless, it is remarkable that, 120 years after Becquerel's papers, publications on ruby still abound, albeit of ever increasing sophistication. In the meantime, optical absorption and luminescence studies have been extended to all transition metals and the rare earth, actinide and transuranic ions.

Bethe's crystal field theory and its developments are concerned with electronic energy levels in the *static* crystal field, determined by ions in their average position in the electronic ground state. In real crystals the ions are in a state of continual vibrational motion and the coupling between the electronic states of the optical centre and lattice vibrations has a profound influence on optical band shapes. In a seminal paper, Born and Oppenheimer (1927) showed that electronic and nuclear motions were separable and they were able to explain the band shape of electronic spectra of simple molecules. In its application to solid state systems, the Born–Oppenheimer approximation finds its simplest expression in the configurational coordinate model: the electronic states are coupled to a single vibrational mode of frequency ω. The energy of the electronic system is then given by

$$E_{en} = E_{e0} + (n + 1/2)\hbar\omega \tag{10.15}$$

where E_{e0} represents the purely electronic state and n is an harmonic oscillator quantum number ($n = 0, 1 \ldots$). This theory gives a simple qualitative interpretation of many of the observable features in solid state spectra. However, it has been much modified in the modern era, requiring large computers to solve the complexities of coupling of the electronic states to the $3N$ vibrational modes of a real crystal (Fowler 1968, Stoneham 1975). A variety of experimental band shapes, which can be interpreted using such models, is displayed in figure 10.5.

Colour centre spectroscopy developed alongside that of the transition and rare earth ions. The initiators were Hilsch and Pohl whose interest was the visible coloration of alkali halides containing intrinsic defects. The ubiquitous

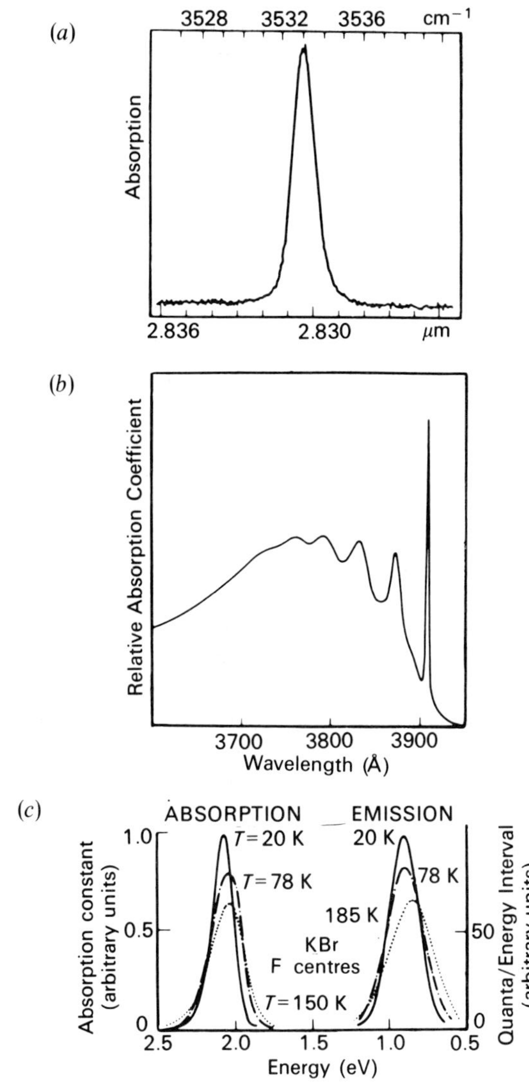

Figure 10.5 Typical optical spectra of electronic centres in ionic crystals (Hayes and Stoneham 1985). (*a*) Emission of the Ni^+ ion in cubic ZnS at 5 K. (*b*) Absorption of R centres in LiF at helium temperatures. (*c*) Absorption and emission of F centres in KBr as a function of temperature.

F centre was named from the German word *farbzentrum* (colour centre). This centre, consisting of a single electron trapped in an anion vacancy, is recognised by broad absorption and emission bands in the visible spectrum. Aggregates of F centres, e.g. F_2, F_3 centres, have a multiplicity of absorption bands because several electrons are involved in the optical transitions. The antimorph of the F centre, a positive hole trapped in the field of a positive ion vacancy, has only been observed in LiF. However, related trapped-hole centres are recognised in other alkali halides by their broad absorption bands in the ultraviolet region. After almost six decades the structure and properties of many colour centres are now well understood, largely as a result of the interaction between theory and experiment. Generally speaking, the spectra of defects and impurities in ionic solids cover most of the ultraviolet, visible and near infrared parts of the spectrum. A wide range of band shapes, transition strengths and radiative lifetimes have been observed. It remains one of the more remarkable successes of quantum mechanics that such spectra can be interpreted, often to the finest detail and experimental nuance.

Radiation with band gap energy, incident upon a semiconductor, produces electrons in the conduction band and holes in the valence band, which are subsequently trapped at donor and acceptor levels respectively in the band gap. Spectroscopic studies of the complex variety of donor and acceptor processes began in earnest in the early 1950s. Interpretation of the various optical phenomena again relied heavily on developments in theory. Shallow donors and acceptors may be treated in the so-called effective mass approximation (Kohn 1953): a single donor electron or acceptor hole is assumed to move with effective mass m^* in a Coulombic potential screened by the dielectric constant, ε, of the crystal. The orbital radii, a_{H}^*, and energy eigenvalues, E, are then obtained by scaling the well known values for atomic hydrogen: hence,

$$a_{\mathrm{H}}^* = \frac{a_{\mathrm{H}}}{\varepsilon} \times \frac{m}{m^*} \quad \text{and} \quad E = \frac{E_{\mathrm{H}}}{\varepsilon^2} \times \frac{m^*}{m} \tag{10.16}$$

in which E_{H} and a_{H} refer to atomic hydrogen, m is the free electron mass and ε is the dielectric constant. In consequence, this model predicts large orbital radii (approximately 5–10 nm) and a Rydberg series of transitions for both donors and acceptors. The observed, sharp-line spectra with photon energies of the order of a few millielectron-volts are in keeping with these predictions. Donor–acceptor (D–A) recombination is an important process in practical semiconductor devices. Recombination occurs when a donor electron annihilates an acceptor hole from which it is separated by a distance R_{ij}. The energy released, either as photons (radiative recombination) or phonons (non-radiative recombination), is then given by

$$E = E_{\mathrm{g}} - (E_{\mathrm{d}} + E_{\mathrm{a}}) + \frac{e^2}{4\pi\varepsilon_0\varepsilon R_{ij}} \tag{10.17}$$

where E_{d} and E_{a} are the donor and acceptor ionisation energies respectively, E_{g} is the band gap energy and the final term represents the Coulomb interaction between the ionised donor/acceptor pair. In consequence, radiative recombination spectra should display as many transitions as there are discrete values of R_{ij}. Furthermore, because donor and acceptor wavefunctions are spatially diffuse and insensitive to lattice vibrations, their spectra are dominated by sharp zero-phonon lines and phonon-assisted replicas. Such D–A recombination luminescence was first observed in GaP (Hopfield *et al* 1960) as several series of sharp lines, the positions of which were determined by discrete values of the D–A pair and the energy of the longitudinal optic phonons coupled to the trapped carriers. Figure 10.6 is an example of donor–acceptor recombinations in GaP. The strikingly rich spectrum of lines is due to recombination at Mg acceptors and S donors, separated by discrete values of R_{ij} determined by the GaP crystal structure.

Well resolved spectra are also observed for excitons recombining at traps in III–V, III–VI and II–VI crystals. However, D–A recombination at deep traps is often revealed as rather broad bands. Effective mass theory (Kohn 1953), so successful in interpreting the spectra of excitons and shallow traps, breaks down for deep levels (e.g. vacancy centres in Si, ZnSe and GaAs) because the localised carriers are strongly coupled to lattice vibrations. The optical spectra of deep levels, being rather broad and structureless, are amenable to interpretation using the configurational coordinate model, developed originally for application to the absorption/luminescence cycle of centres in ionic crystals. The advantages of studying recombination using optically detected magnetic resonance, whether narrow lines or broad bands and shallow or deep traps are involved, have been reviewed by Cavenett (1981).

10.4 OPTICAL COMPONENTS AND DEVICES

Many optical components (e.g. mirrors, prisms, lenses, birefringent and polarising crystals) in a modern optics laboratory have been familiar from the days of Newton. However, modern components are fabricated with far greater precision: even simple components are given antireflection coatings. Interference filters are increasingly used to replace traditional glass filters. Multilayer elements consist of alternate layers of high (ZnS) and low (Na_3AlF_6) refractive index material, the numbers and thicknesses of which are varied to modify the band width and centre wavelength of the filter. Reflectors are made with different metallic coatings appropriate to particular wavelength ranges. The important changes in optical detectors since the first RCA931 phototube (Zworykin and Roychman 1938) show that research into photoemission has paid handsome dividends. The early photocathode materials (Ag–O–Cs, Cs_3Sb) have been replaced for many applications in the range 200–950 nm by bi-alkali antimonides (e.g. Na_2KSb:Cs) and GaAs:Cs–O. For the infrared region, p–i–n semiconductor photocells have been developed. However, the greatest change has been in replacing traditional radiation sources by lasers.

In a prophetic paper, Townes and Schawlow (1958) described ways of establishing the population inversion necessary for stimulated emission to

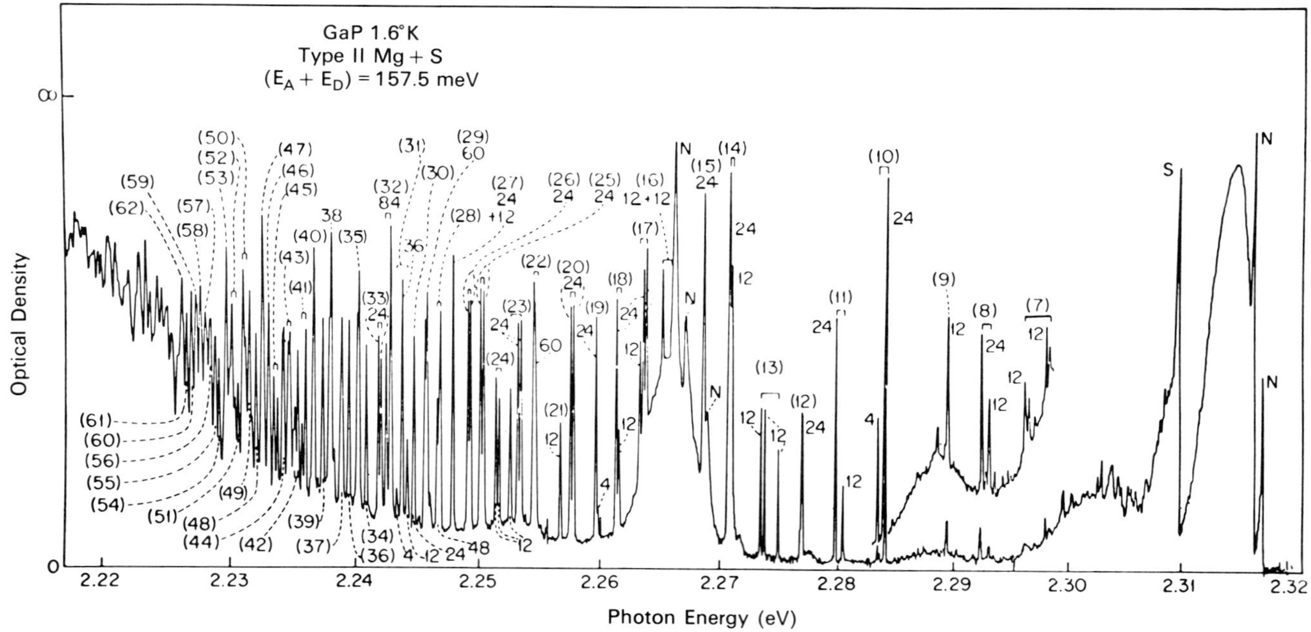

Figure 10.6 Line spectra due to recombination luminescence at Mg acceptors and S donors in GaP. (After Dean *et al* 1970.)

dominate over spontaneous emission. Maiman (1960) operated the first laser based on the R-lines in ruby. From the vast catalogue of spectroscopic data it was clear that numerous materials doped with transition metal (V^{2+}, Co^{2+}, Ni^{2+}) or rare earth (Nd^{3+}, Co^{3+}, Ho^{3+}) ions were capable of sustaining laser action. Many lasers based on rare earth ions were produced in research laboratories. However, until about five years ago ruby, Nd^3:YAG and Nd^{3+} in glass were the only commercially available solid state lasers. Recent development of broad-band tunable lasers, based on colour centres and Cr^{3+} in alexandrite ($BeAl_2O_4$), has stimulated research on several other promising systems (e.g. Ti^{3+} in Al_2O_3, Ni^{2+} in MgO and MgF_2 and Cr^{3+} in garnets).

Semiconductor diode lasers capable of up to 10 mW of power began to appear in 1962 (see Parish and Hayashi 1971). The materials used are the same as those in light emitting diodes (LEDS) and the same process of injection electroluminescence, in which a current drives electrons and holes into the junction region of a p–i–n diode where they recombine radiatively, is operative. The wavelengths spanned by III–V alloy lasers based on (In)Ga(Al)As(P) range from 400 nm to 1300 nm. The elecricity-to-light conversion efficiency can be as high as 20%.

10.5 TOWARDS THE FUTURE

Scientists sometimes participate in star-gazing: happily few are hell-bent on star wars! Nevertheless, the future of solid state optics revolves around the laser. The twin delights of exceedingly narrow lines (less than 500 kHz) and ultra-short pulses make all kinds of new experiments possible.

Intense laser beams generate non-linear optical effects in solids. Because the required fields are available in very narrow lines or in ultra-short pulses, non-linear spectroscopy is much in vogue (fluorescence line narrowing, optical hole burning (figure 10.7), transient grating spectroscopy etc). Time-resolved spectroscopy may uncover details of the mechanisms of energy relaxations in solids. The recent development of the soliton laser with pulses down to 25 fs takes spectroscopists into the realm of vibrational lifetimes in molecules and solids. Now that femtosecond pulses are available, it seems that non-radiative decay processes will be the focus of renewed activity. Effects which can be detected when the non-linear terms in the series expansion of the polarisation, as a function of the electric field, become measurable relative to the linear term include both passive effects (e.g. harmonic generation, optical rectification, parametric amplification) and active effects (stimulated Brillouin and Raman scattering).

Great excitement has been generated by the observation that picosecond pulses, at λ of approximately 1500 nm, propagate as solitons in optical fibres; i.e. pulses which either do not change shape or which change shape periodically during propagation. Data transmission rates of tens of gigabits

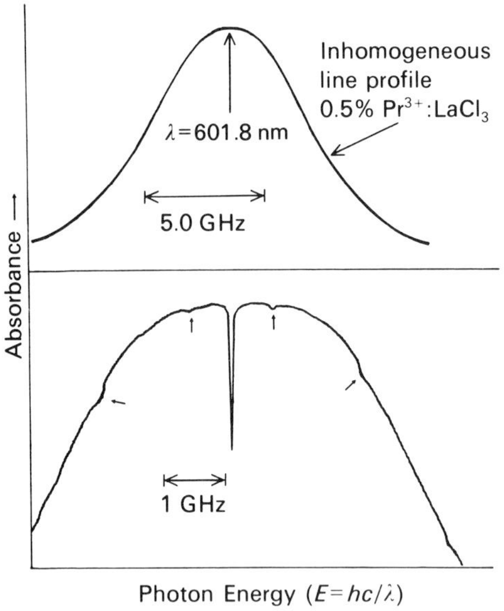

Figure 10.7 An example of non-linear optical spectroscopy. This figure shows an optical zero-phonon line at 601.28 nm due to Pr^{3+} ions in $LaCl_3$ in which a narrow hole is 'burned' by selective bleaching of a small set of crystal field ions using a single-mode ring dye laser. The presence of side holes due to hyperfine structure effects are also shown. (Harley and Macfarlane, unpublished.)

per second, with soliton pulses of about 50 fs duration, offer dramatic potential in fibre optic telecommunications and signal processing systems (Mollenauer and Stolen 1984). In semiconductor materials, resonant effects due to the formation of electron–hole plasmas make for very effective non-linear properties. Figure 10.8 shows optical bistability in InSb, i.e. for a critical range of CO laser input power there are two stable output levels under identical input conditions. This type of non-linearity can be used for optical switching and all-optical signal amplifiers (so-called transphasors). The recent development of interacting optical logic elements adds to the speculation that the optical computer is just around the corner.

Optical materials research will thrust ahead on a relatively broad front. The search for suitable solids for vibronically tuned lasers is growing rapidly. There is enormous potential in multilayer semiconductor materials, so-called multiquantum well structures, for precise tailoring of the electronic and optical properties of devices. High-quality junction diodes, field effect transistors, photopumped and heterojunction lasers have already been produced by MBE and MOCVD techniques. The possibility of multijunction solar cells, to take advantage of most of the solar spectrum, has been proposed

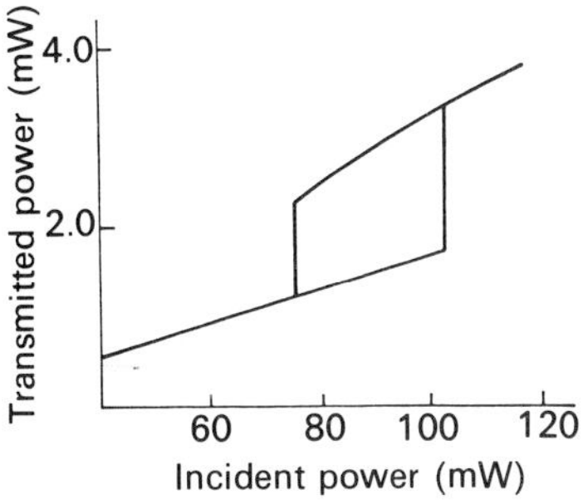

Figure 10.8 Optical bistability observed in the InSb/CO system by changing the detuning of the InSb etalon (adapted from Seaton (1982)).

recently (Vook 1984). Other low-dimensional structures include (organic) Langmuir–Blodgett films on semiconductors, which have very large, non-linear optical coefficients, so offering exciting possibilities for new optoelectronic devices.

REFERENCES

Becquerel E 1886 *La Lumière, ses Causes et ses Effets*

Bethe H A 1929 *Ann. Phys., Lpz* **3** 133

Born M and Oppenheimer R A 1927 *Ann. Phys., Lpz* **84** 857

Cavenett B 1981 *Adv. Phys.* **30** 485

Condon E and Shortley C P 1964 *Theory of Atomic Spectra* (Cambridge: Cambridge University Press)

Dean P J, Schönherr E G and Zetterstrom R B 1970 *J. Appl. Phys.* **41** 3434

Deutschbein O 1932 *Ann. Phys., Lpz* **7** 712

Drude P 1902 *Ann. Phys., Lpz* **7** 687

Fong C Y, Cohen M L, Zucca R R L, Stokes J and Shen Y R 1970 *Phys. Rev. Lett.* **25** 1486

Fowler W B 1968 in *Physics of Colour Centres* (New York: Academic Press)

Glazebrook R T 1885 *Report on Optical Theories*, British Association Meeting, London

Grosmann M 1963 Polarons and excitons *Proc. SUSSP Edinburgh* ed. C G Kuper and G D Whitfield (London: Oliver and Boyd)

Harley R T and Macfarlane R M unpublished data 1986

Hayes W and Stoneham M 1985 *Defects and Defect Processes in Solids* (New York: Wiley)

Hilsch R and Pohl R W 1930 *Z. Phys.* **59** 812

Hopfield *et al* 1960 *Phys. Rev. Lett.* **10** 163

Kohn W 1953 *Sol. State Phys. Vol. 5* ed. E Seitz and Turnbull (New York: Academic Press)

Kronig R de L and Penney W G 1931 *Proc. R. Soc.* A **130** 499

Lloyd H 1833 *Trans. R. Irish Acad.* **XVII** 145

Lorentz H A 1909 *Theory of Electrons* (Leipzing: Teubner)

Maiman T M 1960 *Nature* **187** 493

Mollenauer L F and Stolen R H 1984 *Opt. Lett.* **29** 13 and references therein

Parish M B and Hayashi I 1971 *Sci. Am.* **225** 32
Philip H R and Ehrenreich H 1962 *Phys. Rev.* **128** 1622
—— 1963a *Phys. Rev.* **129** 1550
—— 1963b *Phys. Rev.* **131** 2016
Seaton C 1982 *PhD Thesis* Heriot-Watt University, Edinburgh
Seraphin B *et al* 1965 *Phys. Rev. Lett.* **14** 138, **15** 103
Stoneham A M 1975 *Theory of Defects in Solids* (Oxford: Oxford University Press)
Sugano S, Tanabe Y and Kanimura H 1970 *Multiplets of Transition Metal Ions in Crystals* (New York: Academic Press)
Townes C H and Schawlow A L 1958 *Phys. Rev.* **112** 1940
Vook F 1984 *Phys. Bull.* **35** 474
Wooten F 1972 *Optical Properties of Solids* (New York: Academic Press)
Zucca R R L and Shen Y R 1971 *Phys. Rev.* B **1** 2668
Zworykin V K and Roychman J A 1938 *Proc. IRE* **27** 558

11 Nanometre Physics and Microelectronics

M J Kelly

11.1 INTRODUCTION

In the following paragraphs a situation without precedent is outlined. During the next decade, condensed matter physics, as we know it, is going to be substantially revised following the fabrication and study of solids that have been designed in all three dimensions with nanometre precision. Concurrently, the future of microelectronics into the next century is to be determined after a critical revision of the fundamental assumptions that have driven the industry for the last two decades. These two activities are really one, and that is what makes the situation unprecedented.

During the period 1962–82 the linear feature size of elements in commercially available integrated circuits has halved over each two-year period, and it stands at 1 μm today. With no let-up in this rate of miniaturisation we would reach angstrom-scale devices soon after the year 2010, but of course by that stage the conventional modes of device operation will have long ceased to be operative. During the last two decades many non-solid-state electronic devices (e.g. cathode ray tubes, lasers and now photomultipliers) have been replaced by their semiconductor analogues and whether this trend can continue is a part of the same critical assessment of microelectronics.

Elementary concepts of solid state physics, for example the resistivity or magnetic susceptibility of a metal or the optical absorption of any solid, are built on theories that date back a century to Ludwig Boltzmann. The theory for the response of a medium to an external perturbation is extremely subtle in its full generality but admits great simplications if one can separate by several orders of magnitude the three length scales represented by, firstly, the interatomic spacing, secondly, the electron mean free path, or wavelength of light, and thirdly, the size of the sample. Prototype test structures are, just now, becoming available in which no such clear separation exists and novel phenomena have been reported.

In this chapter I shall summarise developments with reference to layered semiconductor structures and fine-line lithography, with which one controls the feature size in the depth (z-axis) and lateral (x, y axes) directions respectively. It is anticipated that the independent technologies will be married in the next decade, with the consequent production of structures that are atomically engineered in all three dimensions to nanometre design rules. Just some of the novel phenomena that can be anticipated in these tailored structures are described here.

11.2 LAYERED SEMICONDUCTOR STRUCTURES

The advent of molecular beam epitaxy as a method for growing layered structures of semiconductors of ultra-high quality has produced a new generation of transistor and laser devices as well as the prototype for a solid state photomultiplier. Approximately one quarter of the entire semiconductor physics community is studying phenomena in the layered systems. The independent spatial control of the semiconductor band gap and level of doping is the most important advance that molecular beam epitaxy (MBE) allows. The significance of this new control is easily understood.

To date, bipolar and FET transistors based on a single semiconductor (typically Si and GaAs) have relied on n- and p-doping to achieve the desirable device functions. While doping allows the introduction of free carriers (electrons or holes), the parent impurity remains fixed in the host lattice as a charged site capable of strongly scattering the carriers through Coulomb interactions. Within semiconductor laser structures, the wavelength of the emitted light is tied to the host semiconductor band gap, which is in the near infrared rather than in the visible.

It is fortunate that the mismatch in lattice constant between GaAs and $Al_xGa_{1-x}As$ is small for all x and becomes very small as $x \to 0$. The technique of molecular beam epitaxy allows one to grow layers of $Al_xGa_{1-x}As$ of very high (device-grade) crystalline integrity at the rate of approximately one atomic layer per second. Atomic beams of Ga and Al and molecular beams of As impinge on a substrate kept hot enough so that the As molecules crack and the Ga and Al atoms can migrate reasonable distances to take up lattice positions. The Ga/Al ratio can be shuttered on the one-second time scale so that single crystals of layered semiconductors are produced with monolayer precision (see figure 11.1). The physics of a single interface between GaAs and $Al_{0.3}Ga_{0.7}As$ and series of layers of the two semiconductors have produced the new generation of device structures. The schematic diagram in figure 11.2 shows the reason why. As a function of depth normal to the layers, the semiconductor band gap varies as shown. For x expressed as a fraction, the discontinuity of the conduction band edge is approximately x eV and between the valence bands the discontinuity is approximately $0.15x$ eV. These

Figure 11.1 A high resolution image of a single quantum well. (Courtesy of C J Humphreys, Liverpool.)

basic quantities are not well established, as recent research shows, but the discontinuities are important in several ways. It is important to note that ΔE and the individual layer thicknesses can be predetermined for any structure. In figure 11.2 we show schematically the presence of electron states above the conduction band of GaAs but below that of $Al_xGa_{1-x}As$: these and equivalent hole states can be understood qualitatively as 'electrons or holes in a box', although the precise determinations of the energies requires a more detailed incorporation of the finiteness of the band gap discontinuities and the real energy band structures of GaAs and $Al_xGa_{1-x}As$.

Given the crucial role of interband excitation, conventional semiconductor laser structures operate at wavelengths closely tied to the host semiconductor band gap, as in the early but inefficient p–n junction laser. Even within the double heterojunction laser, where a thick GaAs layer is sandwiched between $Al_xGa_{1-x}As$ layers, the latter serve only to confine the photons via the increased refractive index. In a single or multiquantum well structure, the transitions between the electron and hole states defined by the quantum well parameters can be used. To a limited extent one can move the lasing frequency towards and into the visible region from the infrared. The precise wavelength can be designed into the quantum well parameters for MBE growth. The confinement of the carriers in the quantum wells, in addition to the confinement of the photons in cladding layers of high x-value $Al_xGa_{1-x}As$, results in lasers with threshold currents down by a factor of five and the low operating powers translate to longer life lasers.

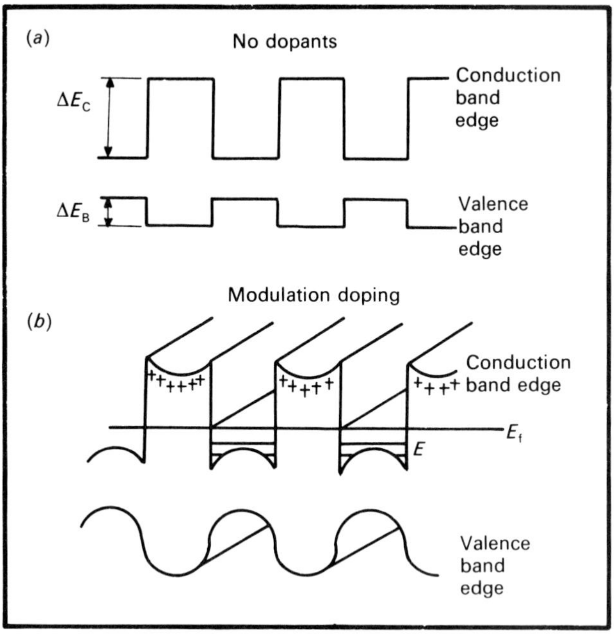

Figure 11.2 The energy band structure of a superlattice as a function of position; (*a*) idealised and undoped and (*b*) with modulation doping.

For x values less than 0.3, Si acts as a shallow donor in $Al_xGa_{1-x}As$, as it does in GaAs. Because one grows the MBE structures layer by layer, one can incorporate the dopant at selective positions during growth. The principle of modulation doping, which has led to the high electron mobility transistor (HEMT), is quite simple—the dopant is only placed in the middle of the $Al_xGa_{1-x}As$ layers, with the GaAs and the first 10 nm from the interface of $Al_xGa_{1-x}As$ being grown as intrinsic as the thermodynamic and kinetic conditions of growth allow. Since 1978, when 'modulation doping' was first introduced, the low temperature mobility of GaAs in a thin layer has increased from about 5×10^4 cm^2 V s^{-1} to more than 2×10^6 cm^2 V s^{-1} in leaps of a factor of approximately three each year for the first four years. Even at room temperature the mobility is enhanced by 50%. The reason is that the electrons from the Si in $Al_xGa_{1-x}As$ fall into the GaAs quantum wells and can be controlled in these layers by source-drain fields as in conventional field effect transistors. The fact that the carriers need never approach their parent charge sites results in the reduced scattering and higher mobility. Indeed, for all temperatures about 10 K and for all electric fields the limitation on the mobility of HEMT structures is limited by piezoelectric and, more importantly, polar optic phonon scattering. Any further improvements in intrinsic performance will need a manipulation of the interaction between carriers and the vibrations of the host lattice.

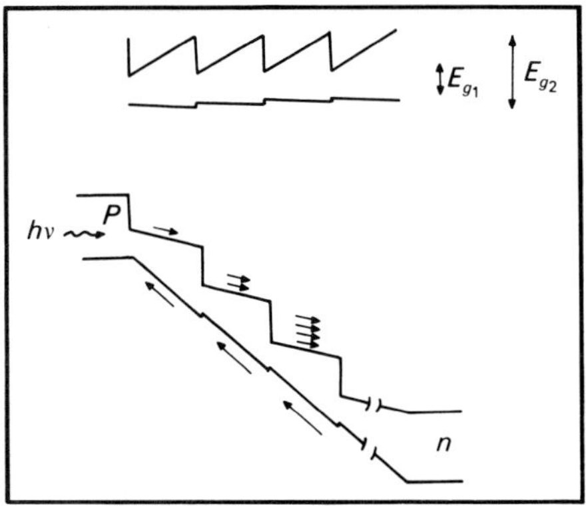

Figure 11.3 The saw-tooth superlattice profile that becomes a solid state photomultiplier under a laterally applied electric field. A single electron can cascade over subsequent band gap discontinuities after being excited by an incident photon.

At the very lowest temperatures, the extremely high mobilities translate to distances of the order of 10 μm, over which an electron can retain memory of its quantum phase. The addition of a fine-geometry gate structure on top of the $Al_xGa_{1-x}As$ capping layer offers the prospect of reproducing the physics of vacuum valve tubes at the micron scale in a solid. This topic will be discussed further below.

In addition to improvements in laser and transistor performances that MBE-grown heterojunctions and multilayers have afforded, new device concepts have emerged. One is the solid state photomultiplier whose operation can be understood with reference to figure 11.3. One can vary the x fraction of $Al_xGa_{1-x}As$ linearly as a function of growth and produce a saw-tooth profile for the band gap. A bias applied laterally turns the conduction band edge into a staircase. A single electron, excited by an incident photon, can cascade over subsequent band gap discontinuities. Provided the discontinuity is greater than the band gap, the electron becomes hot enough to excite another electron across the band gap as shown schematically. Since the valence band discontinuity is much smaller, the equivalent hole multiplication is not possible energetically. The avalanching occurs within 10 nm of the interfaces with a high probability, so that this structure, due to Frederico Capasso of A T and T Bell Laboratories, is a low-noise avalanche photomultiplier; yet another vacuum tube function is replaced by a solid state counterpart!

The physics of confined carriers is of great interest in its own right, as the Conference Series published in volumes 58, 73, 98, 113, 142 and 170 of *Surface Science* testify. The quantum confinement in the z direction produces a number of effects. Since the diameter of an exciton in bulk GaAs is approximately 20 nm, the equivalent exciton in a thin layer of GaAs bounded by $Al_xGa_{1-x}As$ must 'feel' the confinement. The enforced proximity of the electron and hole pair results in an enhanced binding energy. Excitonic effects are still seen in quantum wells at temperatures above those where thermal ionisation destroys them in bulk GaAs. Again, a free-electron plasma oscillation in three dimensions has a finite frequency at long wavelengths, while $\omega_p \sim q^{1/2}$ in two dimensions. At long wavelengths, novel interactions between two-dimensional plasmons and other bosons are observed. The magneto-transport properties in two dimensions differ dramatically from those in three dimensions, just because of the restrictions in phase space for the various scattering events. The quantum Hall effect and the fractional quantum Hall effect are the most notable examples. In the presence of a strong, normal magnetic field the free motion of electrons in the x–y direction, characterised by an energy–momentum relation $E = \hbar^2(k_x^2 + k_y^2)/2m^*$, with respect to the bottom of the band, is altered into discrete Landau levels for which the energy is $h\omega_c(n + 1/2)$. As the magnetic field is swept, successive Landau levels (of degeneracy eB/h) are emptied and, under the condition that the Fermi energy is within the localised states at the extrema of the Landau level, a fixed (quantised) Hall resistance, $\rho = e^2/hi$ (where i is an integer), results. To date, this is known to a precision of one part in approximately 10^7 and seems set to become a new resistance standard. The observations of quantised Hall resistances for fractional values of i ($i = 1/3$, $2/3$, $2/5$, $3/5$, $2/7$ . . .) demand a deeper explanation and the theory of a novel quantum fluid due to R B Laughlin currently predominates.

All this work has been possible because of the ability to grow simple crystal heterojunctions in the system GaAs–$Al_xGa_{1-x}As$. Work in heterojunctions where Sb replaces some, or all, of the As, or In replaces some, or all, of the Ga, is in progress. The requirement for good lattice matching seems not to be necessary if the layers are sufficiently thin and the range of potential semiconductor heterojunctions seems very wide. An alternative method of growth, metal–organic–chemical vapour deposition (MOCVD), in which trimethylgallium (or trimethylaluminium) and arsine (or phosphine) is cracked over a hot substrate, is also capable of growing sharp interfaces. Some of the highest performance laser structures are grown by this technique. However, the slightly inferior interface quality has always resulted in a decrease in low-temperature electron mobilities by a factor of 5–10 compared with those structures grown by MBE. Embryonic research has reported semiconductor/metal/semiconductor epitaxial structures in Si/silicide/Si systems and semiconductor/dielectric/semiconductor structures in the GaAs/(Ca, Ba, Sr)F_2/GaAs systems. Over the next decade one can expect the advent of heteroepitaxial layers of metals, semiconductors and dielectrics, with a great widening of the phenomena observed in the precursor example of the GaAs/$Al_xGa_{1-x}As$ system.

11.3 FINE-LINE LITHOGRAPHY

Quite independently of the advances in control of the semiconductor properties of layered structures, the miniaturisation of silicon devices and the increased integration on a chip have been made possible through the ability to define doped regions and metal interconnects between these regions with increasingly fine geometry and high precision. The basic resist/mask/exposure/development procedure for the definition of vias (holes in the resist), for the ion-implanation of dopants or the deposition of metals, is still in wide use. The exposure to light that cracks the polymeric resist material is now replaced by x-ray exposure for the finer resolution in the research fabrication of nanometre-scale structures, which are inaccessible via visible or UV light techniques because of diffraction effects.

The main alternative to this technique is the use of direct writing. A focused beam of high-energy electrons or ions is deflected by electric and magnetic fields to trace out patterns that selectively develop the resist in the case of electrons, or directly implant in the case of ions. This technique has been developed as a high precision method of pattern definition but suffers from the sequential nature of the beam writing. This technique is slow in comparison with the 'single shot' exposure of a mask pattern. The ion-beam writing has the distinct advantage of not requiring resists and this makes it directly compatible with an MBE high-vacuum chamber.

Wet chemistry for the development of fine patterns has been replaced largely by reactive ion and plasma etching. The etching takes place in a low-pressure atmosphere (typically Ar) and is still not directly compatible with the layered semiconductor growth technologies. The typical organic materials used as resists leave a carbon debris that is inimical to the subsequent growth of device-grade epitaxial semiconductors. Inorganic resists, based on the photosensitivity of the atomic-scale structure of chalcogenide glasses and a subsequent differential etch rate, have been suggested. Here, again, the debris represents dopant elements which would be inimical to the overgrowth of III–V semiconductor layers.

The best ion-implanted channels available in 1986 were of dimensions 0.1 μm wide, 20 nm deep and approximately 200 μm long. It is anticipated over the next decade that the width could decrease by a factor of ten. The metal wires on semiconductor or insulator substrates have reached the size of approximately 6 nm × 10 nm × (several microns), i.e. 20 × 30 atoms in cross section. Few physics experiments have been reported on these wires. Studies of wires of cross section greater than or equal to 30 nm have shown quantum corrections to Ohm's law at low temperatures. Theoretical explanations of these corrections are based on, firstly, localisation effects (brought on by the disorder that results from the mass fluctuations along the wire) and, secondly,

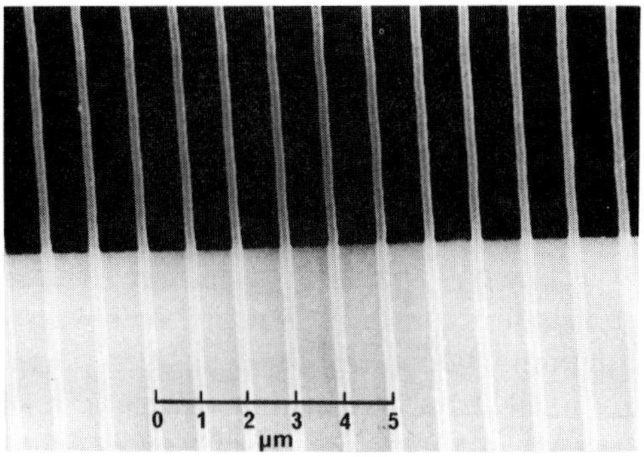

Figure 11.4 Free-standing wires of silicon nitride, of approximate cross section 0.14 μm, as they leave the silicon support (shaded white) to span a hole 120 μm across. (See Lee K L *et al* 1984 *Electron. Lett.* **20** 289–91.)

the effects of quantum interference between forward and back-scattering electron–electron interactions in a restricted momentum phase space.

The quantum corrections to Ohm's law suggest a search for similar corrections to Joule's law of heat conduction—even if the effects were not fundamental but merely a manifestation of quantum size effects in the thermal diffusivity. Experiments could only be performed on free-standing wires of an insulator if the phonon effects are of primary interest, as any substrate would be a thermal short-circuit. In figure 11.4 we show a picture of what we believe are the finest free-standing insulating wires; thin engineered filaments of silicon nitride of dimensions 0.14 μm × 0.10 μm × 300 μm, of which the central 125 μm is free-standing. Below a temperature of approximately 0.5 K, vibration modes that represent standing waves across such a wire are too energetic to be excited. One expects one-dimensional phonon transport. The electron micrograph was taken after the wires had been coated with gold. Free-standing gold–palladium wires of square cross section of side 0.03 μm and length 20 μm have been fabricated. Here, too, at low temperatures the phonons become one-dimensional and studies of the corresponding quantum corrections to Ohm's law are in progress. (All wires on substrates couple to a three-dimensional phonon bath, even at low temperatures.) The free-standing wires are produced by adding a further plasma etching step to remove the (Si_3N_4) substrate after the fabrication of the fine wires. The wires act as nanoampere fuses, so that electric transport measurements need to be extremely sensitive!

These wires are well into the region where the lattice constant is no longer much less than sample size. In a case where the wire, in turn, was used as the plasma etch shadow to produce a silicon MOSFET with a gate of 0.1 μm × 1 μm, the effects of single electrons entering and leaving trap sites under such fine-geometry gate structures have already been observed, as has the signature of one-dimensional electronic transport in sufficiently narrow gate devices.

Highly stochastic and irregular wires of aluminium, achieved by a selective reduction of β-alumina, reach dimensions of 1–2 nm. It seems probable that such structures are capable of regular engineering on the time-scale of a decade. The elastic properties of the free-standing wires show no peculiar anomalies that would suggest the approach of fundamental limitations. Over the next decade the linear dimensions should be capable of further reduction by a factor of approximately five.

11.4 THE MARRIAGE OF MBE AND FINE LITHOGRAPHY

The study of fine-geometry structures in solid state physics is not new. In the past, however, the systems included metal particulates in dielectrics or emulsions etc, where any experimental data could be interpreted only with assumptions about the statistical distribution of the size and position of the particulates. The advent of structures tailored in all three dimensions, e.g. metal inclusions in the shape of a cube of side 2 nm placed at each point of a cubic lattice of side 4 nm and embedded in a semiconductor host, is anticipated to exhibit novel optical and electrical properties. Similarly, a semiconductor doped to its metal–insulator transition, embedded in a dielectric, would also be quite unusual. The marriage of fine-line lithographic techniques with MBE growth methods should provide such structures within a decade.

The forerunner of these structures was fabricated and studied recently. A high-mobility, two-dimensional electron gas (2DEG) was formed at the GaAs–$Al_xGa_{1-x}As$ interface which etched down well into the GaAs to leave an array with 5 μm from the centre of one dot to the next, each dot being 3 μm in diameter. The submillimetre radiation in the presence of a magnetic field provided evidence of electromagnetic interactions between the adjacent dots of the 2DEG!

The next sequence of structures, configured gates on the capping layer above a high mobility 2DEG, are already under production. A fine-geometry interdigitated finger structure, for example, could be used to induce a one-dimensional periodic potential along the direction between the source and drain of a FET. The resulting nearly-free-electron band structure is controlled by the voltage of opposite polarity on the fingers, while an average voltage on the fingers will alter the electron density over a narrow region. One should thus be able to study band structure effects as a continuous function of the size of the periodic potential! With a coarser geometry, 'polaronic' instabilities are anticipated with electrons being trapped under the fingers of one gate. The range of optical and electrical properties that could be studied in this system, and especially in the presence of a magnetic field along the source–drain axis, is enormous.

Very recent studies have succeeded in growing epitaxial (Cu, Ba)F_2 on GaAs and defining patterns in the fluoride by first electron-beam damaging selected areas and then removing the damaged area by wet chemistry. If a differential plasma etch rate could be established, the system might be clean enough to attempt epitaxial GaAs overgrowth of the defined layer of fluoride. Inclusions of dielectric in the semiconductor will alter the optical properties by inducing quantum-size effects into the electron states of the semiconductor. A system with regular inclusions of semiconductor in a dielectric should exhibit strong excitonic effects when the inclusions are of a size smaller than the exciton diameter.

The electronic transport in grids of very fine parallel or crossed wires, embedded in dielectrics but sufficiently close that appreciable tunnelling can occur, should be unusually sensitive to fluctuations in the definition of the wires and non-Ohmic over a wide range of temperatures and electric fields.

These are just a few of the systems in which one expects novel phenomena because of the ability to tailor the electronic structure in a way that betters nature.

11.5 DEVICE IMPLICATIONS

The scaling down of the linear feature size of a Si MOSFET or GaAs FET, while maintaining the power dissipated per unit area in an integrated circuit, presents formidable problems for gate areas below one square micrometre. Not only do the tolerances on the fabrication become increasingly tight, but the inherent statistical fluctuations in small devices (less than or equal to 500 electrons per gate-volt injected into the channel) manifest themselves in the inability of some adjacent elements to drive each other. Small structures also have a small inertia and so all probes, electrical, magnetic or optical, drive them into non-linear regimes of response. The carriers in adjacent devices can leak via quantum mechanical tunnelling. All these, and many other phenomena, point to radically different device phenomena and operational methods on the small scale. The lifetime of these small devices, especially when driven hard, will be short, as electromigration or interdiffusion processes need occur only on a small scale before becoming deleterious.

The study of specialised test structures resulting from MBE and fine lithography should help in quantifying some of the above effects that spell doom for conventional devices. At the same time, these studies should show up new and exploitable phenomena. Just as methods of optical data storage in k-space, rather than discrete storage in real space, are being investigated, the systematic intercommunication of small structures, albeit complicated by the fluctuations, should provide novel methods of data storage and, for example, optically induced transfer of charge between structures as a form of contactless information processing. Indeed, the definition of (say) GaAs wires in

$Al_xGa_{1-x}As$ will allow the use of light, rather than electrons, for transmitting information on a chip.

The constraint of the two-dimensional topology of a conventional chip (with the slight freedom allowed by multilayer metallisations) should be freed in the next decade. Three-dimensional chips will be able to exploit the extra dimension to perform parallel processing in ways not possible at present. Indeed, some of the recent advances in neuroscience that shed light on the brain as an information processing machine may suggest novel interconnections in three dimensions.

ACKNOWLEDGMENTS

The above ideas represent a collage derived from many discussions with colleagues at GEC and many other laboratories.

FURTHER READING

Applied Physics Letters, *Electronics Letters* and *Physical Review Letters* contain the lastest announcements in this area. The *Surface Science* volumes cited in the text (p267) give a good overview of physics and structures in two dimensions. The most recent advances are discussed in *The Physics and Fabrication of Microstructures and Microdevices*, edited by M J Kelly and C Weisbuch (Berlin: Springer, 1986).

12 Solid State Instruments

Colin Windsor

12.1 INTRODUCTION

Several transformations in the instruments for studying solids, and in their use, have been seen over the last hundred years (see figure 12.1). In the period before 1914 it was the technique which was being investigated, as much as the sample. The interpretation of experiments on lodestone had to await a theory of magnetism. Maxwell's work on crystal optics was directed more to understanding light than to understanding solids. The excitement caused by von Laue's x-ray diffraction pattern came from the interpretation of the nature of x-rays which it gave.

Between the wars came that golden period of physics when the fundamental ideas were sorted out. The attentions of the experimental physicist could be directed away from the technique and towards the sample. Experimental solid state physics could begin. Other chapters of this book detail the progress made during this period, mostly made with simple and inexpensive instruments.

A change in the character of experimental apparatus throughout physics took place in the years following 1932 and throughout World War II. The year 1932 marked the discovery of the neutron—albeit with the simplest of physical apparatus. It also marked the beginning of accelerator physics with the Cockcroft–Walton experiments. Within 10 years, the first of many solid state physics techniques to depend on nuclear technology emerged. The electron microscope was one of the first fruits of accelerator technology. Neutron scattering emerged with the first reactor—later came the Mössbauer effect, positron annihilation, muon spin resonance and, most recently, the synchrotron radiation source. The radiofrequency and microwave research conducted on radar during World War II was to prove equally fruitful. The resonance techniques—nuclear, spin, ferromagnetic and cyclotron resonance—were all developed within a few years of sources being available. Microwave spectroscopy was to have an unexpected child. The detailed information it gave on energy levels and their

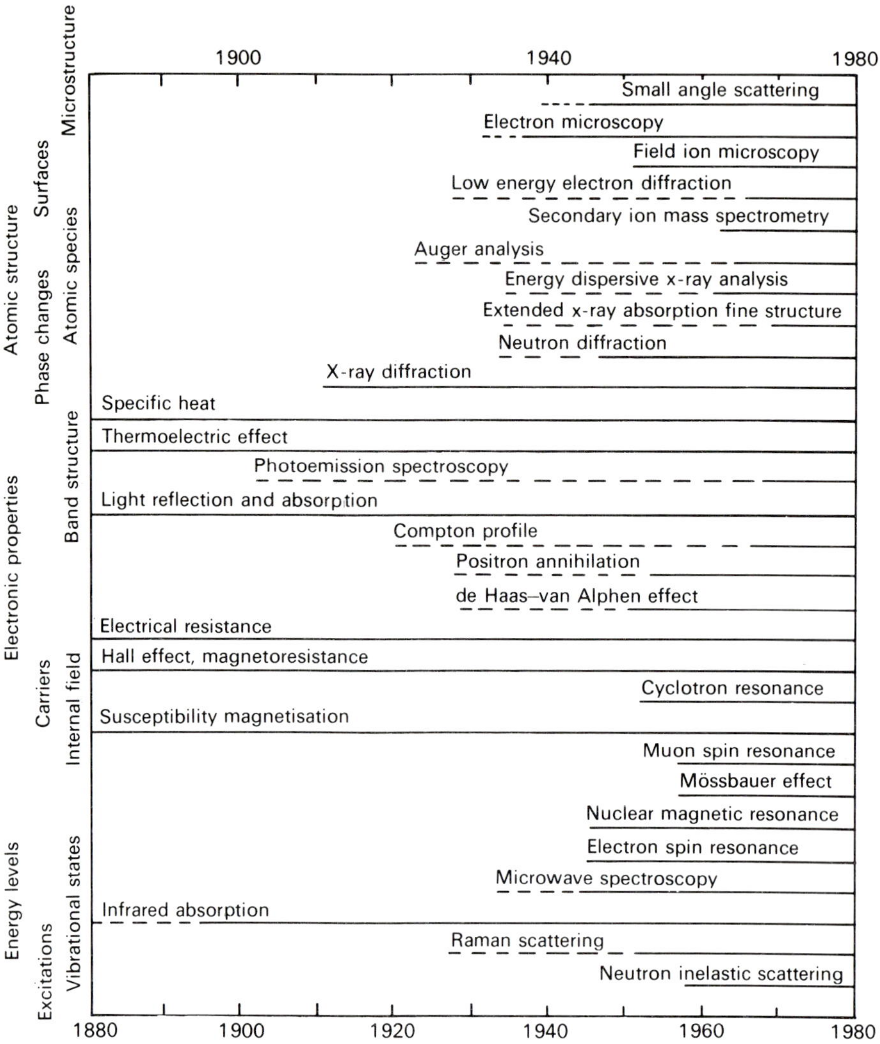

Figure 12.1 The principal techniques of solid state physics and the information they give. Dashed lines indicate the period between the discovery of a technique and its application.

relaxation led to the MASER—microwave amplification by stimulated emission of radiation. The laser soon followed and solid state optics entered a resurgent period, which it still enjoys. Previously difficult techniques, such as Raman scattering, became routine. As laser powers grew, the present growth field of non-linear optics was able to flourish. Semiconductor optics with its promise in communications was to be a growth point.

During the 1960s a new revolution occurred as ultra-high vacuum techniques revealed the surface as an object of study and experimentation in its own right. Again, old techniques, such as low-energy electron diffraction, emerged with new importance, even if disguised by the strange acronyms which characterise surface analysis. It is these surface techniques which have allowed the microchip revolution to continue.

12.2 SOLID STATE INSTRUMENTS BUILD THEIR OWN FOUNDATIONS

The apparatus for solid state physics has always been characterised by a pleasantly human scale. Most of the early instruments, and indeed most of the modern ones, fit into a corner of a laboratory. The foundations of solid state physics were built largely by single scientists working with simple apparatus.

Optical dispersion, absorption and reflectance

Optical experiments on solids had been performed as long as science had existed. Euclid discovered diffraction, and Snell its basic law in 1621. However, the early textbooks could make little sense in the list of refractive powers. Newton noticed that inflammable bodies such as sulphur and phosphorous possessed the greatest refractive power and because of the similar behaviour of diamond he guessed that diamond itself was composed of some inflammable substance. Even if the understanding was deficient, data on the optical properties of solids were steadily collected.

Much of the early apparatus was made with exquisite care. Figure 12.2 shows an optical spectrometer constructed around the beginning of this century by Hilger and Watts (of whom Adam Hilger publishers were once a subsidiary). The spectrometer, measuring as a function of wavelength, was the crucial tool. Its data were to prove invaluable in testing the theory of metals. Optical spectroscopy has continued to this day to be one of the foremost techniques of solid state physics.

Figure 12.2 An optical spectrometer constructed by Hilger and Watts around the turn of the century. It was used for many years by the Physics Department of Birmingham University.

Figure 12.3 A large-scale central facility from the 1850s. A static electricity machine used at the Panopticon, Leicester Square, London. (From *Circle of the Sciences*, The London Printing and Publishing Co Ltd.)

Conductivity

The beauty of the old instruments was matched only by the work of the engravers who recorded them. Figure 12.3 shows a rare exception to the usual size scale. It is a static electricity machine with a rotating glass plate 10 feet in diameter. It was used daily in the Panopticon, Leicester Square, London, around 1850. Driven by a steam engine, it produced sparks two feet long. However, in those days, electrical instruments, like most others, were not directed primarily towards the solid state. Electrical science was highly developed in Faraday's time (the 1800s), yet even that basic parameter—conductivity—was very poorly understood. To quote from a textbook of the mid-nineteenth century, outlining the effects of dampness, heat, spatial continuity and shape on conduction,

> From these facts our readers will at once see that the idea of conduction and non-conduction can only be sustained in a relative or comparative point of view, and that special circumstances can entirely modify, and occasionally change, the character of any substance.

However, electrical conductivity was, and is, one of the easiest properties of solids to measure with precision. Its measurement, particularly its temperature dependence, remains among the most important properties of any new material. Thermal conductivity measurement was active in the nineteenth century, but is less active today. Wiedemann and Franz realised the rough equivalence between thermal and electrical conductivity in 1853

(Wiedemann and Franz 1853). The temperature dependence of thermal conductivity was to prove useful to the theory of metals.

Crystallography from crystal shapes

Of the old sciences, crystallography was perhaps the only one directed always at the solid state. In the pre-x-ray days of the nineteenth century it required the considerable skills of crystal growing, followed by careful measurements of the angles between crystal faces made on a goniometer. Figure 12.4(*a*) shows the Wollaston goniometer, developed around 1790, which could measure the angles of faces less than a millionth of a square inch in area to an accuracy of less than a minute of a degree. By the 1860s, crystallography was a very highly developed science and diagrams such as figure 12.4(*b*) enabled the formal classification of crystal symmetries to proceed simply from the external form of the crystals. The explanation of the structures in terms of the packing of

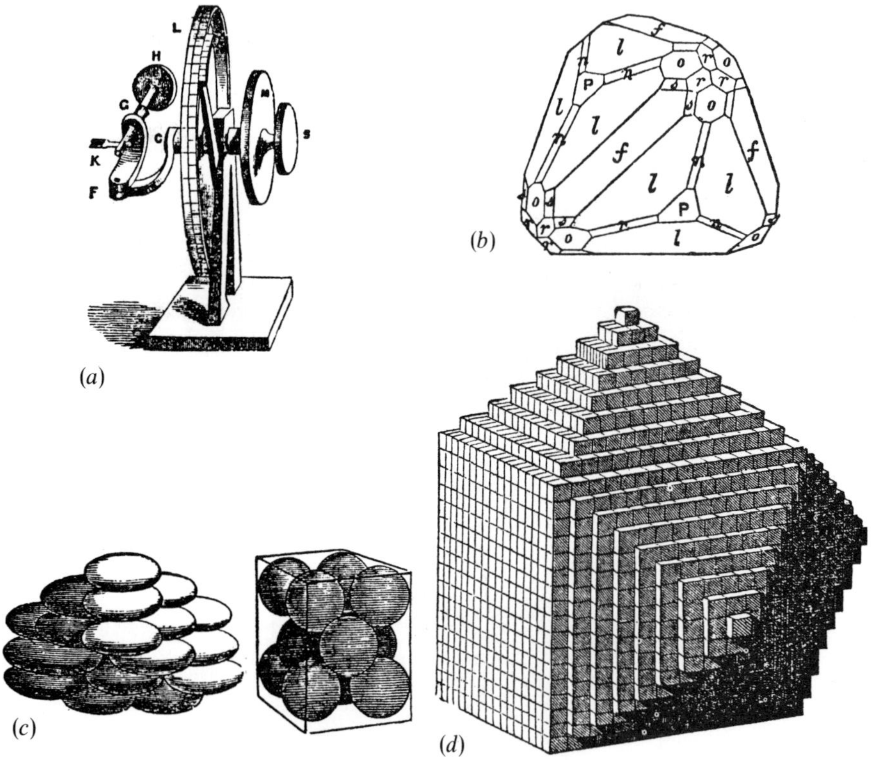

Figure 12.4 (*a*) The crystal goniometer constructed by Wollaston (1809) used to produce the crystal shape diagrams illustrated in (*b*). The crystal units made by Wollaston, (*c*), were understood as leading to the crystal structures of (*d*). (From *Circle of the Sciences*, The London Printing and Publishing Co Ltd.)

molecules was realised even in Wollaston's time. Figure 12.4(*c*) shows some of Wollaston's packing units. They gave us a view of the structure of crystalline solids which was basically correct.

X-ray diffraction

X-rays were to lead crystallography into the pre-eminent position it now holds for giving the detailed structure of solids. X-rays were discovered quite accidentally—probably many years after they were first produced in quantity. Röntgen in 1895 noticed the fluorescence of certain heavy metal salts during his experiments on electrical discharges in gases. It was some seventeen years later that von Laue conceived the idea of trying crystal diffraction with x-rays, and the first experiments by Friedrich, Knipping and Laue (1912) were successful. Figure 12.5 shows Bragg's early x-ray diffractometer (for the first x-ray diffraction pattern see figure 3.1 in Chapter 3). The crystallographic community embraced x-ray diffraction. The subject developed very rapidly and since the 1930s few solid state studies have been hindered by a lack of knowledge of the basic crystal structure of their samples.

Figure 12.5 Bragg's x-ray diffractometer. (Courtesy of the Science Museum Library, London.)

Magnetisation and susceptibility

Our understanding of the origins of magnetism awaited careful measurements of susceptibility and magnetisation. Although Faraday gave us the method, the measurements of Curie (1895) of the static susceptibility as a function of temperature for several materials provided the key data. Figure 12.6 shows a diagram of his high-temperature susceptibility apparatus. The tiny pull of the specimen in the field gradient produced by the angled magnet had to be measured outside the furnace with great precision. Weiss and Forrer (1929) made even more detailed measurements on the magnetisation of iron and nickel near the 'Curie' point. This led to the Curie–Weiss law and our modern understanding of the ferromagnetic transition. It is remarkable that even in the 1970s the old data of Weiss and Forrer were being re-analysed in the search for the theory of phase transitions.

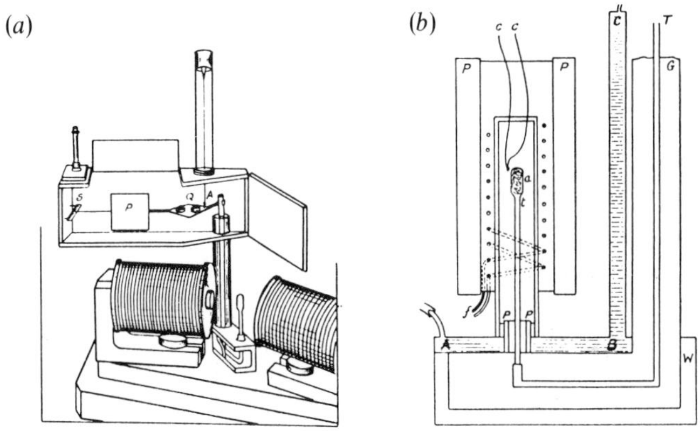

Figure 12.6 (*a*) Curie's high-temperature susceptibility apparatus used to deduce Curie's law. The angled magnet produced the field gradient. The force was measured within the furnace (*b*).

Specific and latent heat measurement

These simple measurements were all relatively routine by the nineteenth century. Figure 12.7 shows Regnault's apparatus from 1840. Its simple method, of mixing hot and cold matter and observing the resulting temperature, sufficed at room temperature. The electrical calorimeters following Nernst's calorimeter (1911) enabled the important temperature dependence of the specific heat to be studied. It proved to be the most useful clue to our understanding of the nature of heat. The specific heat had a temperature-cubed dependence, and Debye (1912) explained this in terms of atomic vibrations. Born and von Karmann (1912) had put forward the full theory of lattice vibrations 40 years before the predicted phonon dispersion curves were measured by inelastic neutron scattering. Specific and latent

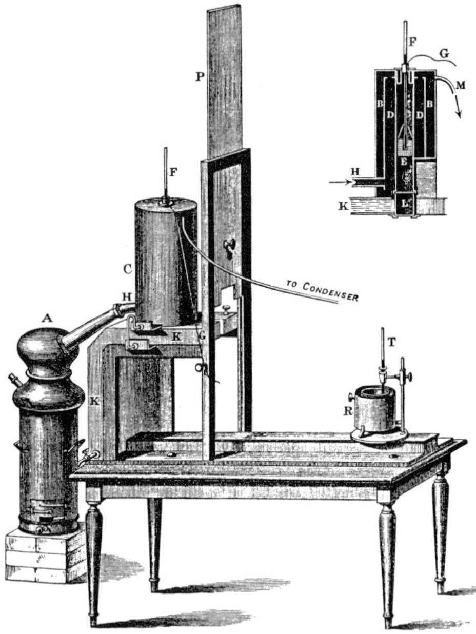

Figure 12.7 Regnault's 'method of mixtures' calorimeter for specific heat measurements. The sample was heated in the steam jacket C and dropped quickly into the calorimeter R. (From Preston 1894.)

heats, measured with the automation of a modern differential thermal calorimeter (DTC) instrument, provide one of the most important clues to the phase transitions in solids.

Heat and electricity: The Seebeck, Peltier and Thomson effects

The thermo-electric effect was discovered by Seebeck (1826) and by the middle of the century was in widespread use for thermo-electric batteries and for temperature difference measurement. Peltier (1834) showed the reverse effect—the emission of Peltier heat, or cooling, as current passed through a thermo-electric junction. Not long afterwards, Thomson (1855) found the related, but more subtle, 'Thomson heat' emitted by a homogeneous wire in which a temperature gradient is maintained. It is reversible and so is easy to determine by making measurements with the current flow in both directions relative to the temperature gradient. The Joule heating remains constant but the Thomson heat is reversed. All three effects are tied together and the 'Kelvin equations' relating them were of considerable importance to thermodynamics, as well as to solid state physics. The rather indirect information the effects gave on the metallic state was only appreciated in the 1930s after the development of the electronic theory of metals.

Magnetism and electricity: The Hall effect and magnetoresistance

Hall (1880) applied a magnetic field across a current-carrying metal strip and noted a voltage across the strip. The voltage is proportional to the current and the 'Hall effect' is the coefficient. It is a special case of a general magnetoresistance effect when a field is applied at an arbitrary angle to the current. The Hall effect has had a lasting usefulness in solid state physics. It gives the charge carrier sign and density—crucial parameters for semiconductors. It has remained important to this day.

Infrared and Raman scattering

The heating effects of the region beyond the red end of the spectrum were well known in the nineteenth century. By the end of the century black body sources, crystal prisms and windows, thermopile detectors, and other necessities for accurate infrared absorption and reflection measurements, had been developed. The data were not always understood at the time but have continued ever since to be important to solid state physics. Infrared energies lie in the 10 to 1000 meV range of molecular and crystal vibrations and are important to chemistry as well as to physics.

The inelastic scattering of light by solids was discovered by Raman (1928). Monochromatic light is incident on the sample. The scattered light wavelength is scanned stepwise away from the intense elastic Rayleigh scattering. The energy transferred from light to crystal can be positive (anti-Stokes) or negative (Stokes) and has the same range as infrared spectroscopy. However, each technique has its own intensity factor and transition probabilities, making the two techniques complementary. Raman scattering was to come into its own as a routine technique only after the development of the laser in the 1950s. The heat emitted by hot solids was to become a vital clue to physics. The spectral distribution of the radiation was remarkable for being independent of the solid. The careful measurements of the limiting 'black body radiation' from Lummer and Pringsheim (1897) were to provide key evidence for the breakdown of the classical theory of Rayleigh and Jeans. It was replaced by Planck's oscillator theory which led to quantum theory.

The photoelectric effect: the evolution of quantum mechanics

Discovered by Hertz in 1887, the photoelectric effect was to prove seminal to the whole of physics. While experimenting with electrical discharges, Hertz found the sparks would jump most easily if the negative terminal was illuminated with ultraviolet light. The effect was soon traced to electrons liberated by the light. In the hands of Einstein (1905) it was to prove one of the key experimental facts behind quantum mechanics. In the years 1924 to 1926 the theory of physics was placed on its present firm foundation.

Electron diffraction: the development of the theory of metals

The experiment by Davison and Germer (1927) showing that electrons had the wave properties expected from quantum mechanics, opened up a new experimental field and gave theorists the confidence to seek a solution to the understanding of electrons in metals. Bloch showed in 1928 that they behaved as lattice waves, while Sommerfeld (1928) applied the new Fermi–Dirac statistics to the electron gas defining the Fermi surface. Brillouin (1931) gave us his zones in reciprocal space for picturing the band structure effect. By 1936 Mott and Jones were able to review 'The Theory of Metals and Alloys' (Mott and Jones 1936). Old experimental data on optical absorption, conductivity and specific heat were readily explained by the new theories.

The electron microscope

Our understanding of the electron was to bear fruit with the new subject of electron optics and in 1931 the electron microscope was born in Berlin. The first compound magnetic lens transmission electron microscope was built by Ruska in 1934. Development was exceedingly rapid and by 1936 the Metropolitan Vickers Company in Britain, and soon after the Siemens Company in Berlin, were producing the first commercial instruments. Microstructure was there for all to see—in real space!

The neutron, the reactor, and neutron diffraction

The neutron's wave powers were demonstrated by Mitchell and Powers (1936) using a simple laboratory radium–beryllium source and diffracting crystals. Within the next hectic decade the slow neutron beams from reactors had given solid state physics one of its most productive new techniques. Figure 12.8 shows Shull's first neutron diffractometer on the Oak Ridge

Figure 12.8 The first neutron diffractometer, built by Shull and Wollan in 1948. (Photograph from Oak Ridge National Laboratory.)

reactor in 1948. The apparatus itself—a converted x-ray diffractometer—continued the personal scale of other solid state instruments. However, there was a real difference as its operation depended on the vast organisation needed to build, fuel and run the reactor. In those days, during and just after the war, nuclear physics and solid state physics became entwined as never before, or since. The neutron slowing-down measurements in moderating materials, so essential to the neutron chain reactor, were shown to be full of information on the solid state. Spurred on by Fermi's genius and the practical necessity of getting the right answers to reactor physics problems, neutron scattering theory made enormous leaps in a few years which laid the groundwork of the field for decades to come.

12.3 POST-WAR: SOLID STATE MEASUREMENTS DELIVER THE GOODS

As figure 12.1 shows, the period after the war marked a dramatic increase in the number and power of solid state experimental methods. Physicists had a new confidence, given by the quantum mechanics textbooks on their shelves and by the successes of the scientists (if not of the politicians) during the operations in the war. Governments supported science in all directions as never before and the number of professional physicists grew rapidly. Solid state physics made up an increasing fraction of the total.

Microwave resonance

Throughout the war years Griffiths and his colleagues at Oxford had worked on the development of the klystron microwave oscillators for airborne radar. In 1945, experienced in the new microwave plumbing of waveguides and resonant cavities, the group turned to microwave spectroscopy. The early work was done by scanning the wavelength point by point and measuring absorption along a waveguide. The technique was changed by the addition of a scanned magnetic field. The splitting in energy levels, discovered by Zeeman (1897), produced the scan. The technique was compact, easy to use and the field and sample anisotropy scans were full of information. Only simple apparatus was used by Griffiths (1946) and his colleagues in the discovery of ferromagnetic resonance and its paramagnetic analogue—electron spin resonance (Zavoisky 1945). Penrose (1951) was soon to discover hyperfine structure from the interaction between the electron spin and the nuclear spin. Owen and Stevens (1953) discovered the superhyperfine structure caused by bonding between the paramagnetic ion and its neighbouring ligands. As the complexity of samples increased, electron spin resonance, like nuclear magnetic resonance, moved towards chemistry. Today it is closer to chemistry than to physics.

Nuclear magnetic resonance

During the war years from 1939 to 1945, countless physicists were employed on radio research and operations and emerged afterwards with Terman's *Handbook* (1943) under their arm and plenty of war-surplus equipment. The Zeeman splittings of the nuclear spin in a field of a few kilogauss correspond to quanta of the order of 10 MHz. In absolute terms the absorption is very small and, once again, it is the damping of highly resonant circuits which made nuclear magnetic resonance (NMR) observable. The first measurements were made by Purcell *et al* (1946) and the subject grew rapidly as hyperfine structure and transferred hyperfine structure gave it great capacity for solid state measurements as well as for detailed chemical finger-printing. Electron nuclear double resonance, ENDOR, in which the strong spin resonance signals are switched by nuclear frequencies, gave enormous accuracy to our knowledge of the low-lying energy states of atoms. The ability of pulsed NMR to measure spin–spin and spin–lattice relaxation extended the field still further.

Cyclotron resonance

The circular paths made by any charge carrier in a magnetic field can be made to resonate with an electric field, as in the cyclotron accelerator. Dresselhaus *et al* (1953) first applied the technique to carriers in a solid using microwave frequencies. The method gives the effective mass tensor, m^*, of any charge carrier in a solid and has become a standard technique for semiconductor physics.

de Haas–van Alphen effect

Like any quantum mechanical oscillator, the cyclotron frequency is quantised into discrete frequencies, corresponding to discrete orbits, under an applied magnetic field. As the field is swept, these orbits pass one by one through the Fermi surface. This gives an oscillatory effect depending on the maximal or minimal Fermi surface cross sectional area normal to the field. The oscillations occur in a host of properties but the most quantitative technique arises from the oscillations in the susceptibility—the de Haas–van Alphen effect (1930). After the war the technique was brought to a fine art by Shoenberg and his colleagues. It has given us the detailed Fermi surface diagrams metal physicists now take for granted.

Microwave spectroscopy—the maser

Microwave spectroscopy had been studied before the war in high pressure gases by Cleeton and Williams (1934). Penrose and Bleaney (1946) used the new microwave sources to map out the detailed narrow-line spectrum of low-pressure ammonia gas. It was to yield a strange and unexpected dividend. Armed with this knowledge of the energy levels, Gordon *et al* (1954) in Columbia and Basov and Petrox in 1955 in Moscow, set about upsetting the usual Boltzmann distribution which defines the relative population of each level. By using electric discharges, many gas molecules could be left in excited energy levels which decay only slowly to their ground states. The conditions then exist for a 'population inversion' with high occupation of an excited energy state. The maser (acronym for microwave amplification by the stimulated emission of radiation) was born. The first masers were based on ammonia gas. The solid state was soon to enter the field with the ruby laser (light amplification by the stimulated emission of radiation) based on the energy levels of the Cr^{3+} ions in ruby (Maiman 1960). The laser was to dominate solid state optics indefinitely.

Semiconductor optics: the solid state laser

During the 1960s, optics and solids were to become more interlinked than ever before. In 1962, several American groups almost simultaneously invented the semiconductor laser. Born from the semiconductor diode, part of the electron–hole recombination energy is emitted directly as light. The $GaAs_{1-x}P_x$ direct band gap semiconductors gave a tunable wavelength of good intensity within a tiny volume. These devices were soon to play a large role in communications.

Non-linear optics

Non-linear devices are a familiar part of electronics. Quite soon after the discovery of the laser it was realised that its high intensity would enable the development of similar devices in optics. Bloembergen and his colleagues at Harvard soon developed the variety of techniques known collectively as non-linear optics. Harmonic generation and the tunable dye laser led to still more possibilities for optics experiments. Stimulated Raman emission is important in giving another set of transition probabilities to conventional Raman scattering.

Optical bistability—the optical computer

The very high instantaneous power levels of a pulsed laser have led to a new field of optical switching. Szoke *et al* (1969) proposed a method for producing optical bistability by using a saturatable absorber. A Fabry–Perot interferometer contains a light-absorbing gas with the property that its absorption decreases as the incident light intensity increases. A bistable device using sodium vapour was eventually demonstrated by Gibbs *et al* (1976). The excitement lies in the extremely fast switching times between states. Using the GaAs free excitation resonance, Gibbs later made a device with a switching time below one nanosecond and a switching energy of only 0.2 pJ per metre. This corresponds to switching frequencies hundreds of times faster than conventional semiconductor computers. The fact that they are driven by light, and that they have intrinsically a two-dimensional extended structure, gives them enormous potential for pattern recognition applications.

The new light and x-ray source—synchrotron radiation

The theory of synchrotron radiation appears in all the standard textbooks on electromagnetic theory. Any accelerating electron, such as those in a circular accelerator undergoing centripedal acceleration, radiates light and x-rays, which are polarised, and have an energy distribution depending on the electron energy and curvature of its orbit. Seen by the stationary absorber, the spatial distribution of light suffers a Lorentz transformation. For a multi-GeV electron beam the transformation results in almost all the radiation being emitted in the forward direction of the beam, to within minutes of radius of curvature. Every bending magnet in an electron synchrotron therefore emits an intense electromagnetic beam along the tangents to the beam line. The first light beams were seen by means of a quick glance through a mirror around the shielding of the cyclotron at the General Electric Laboratories in Schenectady (1947). Figure 12.9 shows a photograph of the light beam taken shortly

Figure 12.9 A photograph of the light emitted from the General Electric Company's synchrotron. The electrons are accelerated in a circular orbit between the magnetic pole pieces and the light is seen at the tangential position on the left. (Courtesy of General Electric Laboratories.)

afterwards. Until about 1970, interest in the radiation was slight. Access was dictated by the whims of particle physicists. Within a very few years several existing synchrotrons were converted into x-ray beam facilities. The first synchrotron dedicated to solid state uses was commissioned in 1981 at Daresbury in Britain.

Ultraviolet spectroscopy—electron photoemission

The lower energy ultraviolet light quanta, below 20 eV, are produced copiously by synchrotron sources. The analysis of the electrons emitted by the photoelectric effect have opened an important window into metallic solids, semiconductors and even their surfaces. Monochromatic radiation is incident on the sample. The emitted photoelectrons can be analysed in energy, angle and even spin, as a function of the incident photon energy to give very detailed information on the energy bands within a solid. The problem, as always with electrons, is the strong interaction they have with the solid under investigation. The photoemission depends on both the initial and final states of the electron in the solid. Extensive computation can only partially unravel these problems. Nevertheless, photoemission gives important information on electronic densities of states away from the Fermi surface.

EXAFS: electron spectroscopy around a site deep in a solid

In the x-ray energy range, the transmission of solids shows the sharp edges in the spectrum where the x-ray absorption jumps, as the possibility of photoemission from one of the inner shells of the solid becomes possible. A careful study of the edges revealed a fine structure extending upwards by some hundreds of electron-volts. The effect was successfully explained back in the 1930s by Kossel and Kronig. The extended x-ray absorption fine structure (EXAFS) is caused by interference between the electron wave extending from a given atom and the reflected electron wave from its immediate neighbours. It therefore monitors local structure around an atom in a solid. Because the edge energy is atom-dependent, the local structures around selected atom types can be individually monitored. EXAFS has been particularly useful for understanding the local structure in glasses and amorphous phases. Surface studies (SEXAFS) have become a major feature of the technique because it is very sensitive.

X-ray crystallography

Already highly developed by the end of the war, this subject has gone from strength to strength, solving the structure of ever more complex molecules. The double-helix structure of DNA, with its obvious visual explanation of one of the outstanding riddles of life, has inspired a generation. Nowhere have powerful computers and control systems found a more fruitful home. A modern crystallographic laboratory exists in unattended silence. Automated x-ray sets are controlled by one set of computers, whilst a second set solves the structures from previous measurements. The synchrotron sources have not taken over the field. One application is for protein crystallography, where the thousands of weak Bragg spots from a single crystal orientation can be recorded on a single photographic plate.

X-ray topography: observing deviations from perfection

The finely collimated x-ray beam from a synchrotron can give diffraction spots a spatial extent by using an extended incident beam and sample. The magnified sample image seen via diffraction is very sensitive to local crystal orientation. Even the tiny distortions from a single dislocation or stacking fault are clearly visible. No wonder the technique has assumed great importance in the semiconductor industry! With magnetic samples, domain walls and their movements can be studied. Synchrotron sources reduce exposures from days to seconds and allow quasi-real-time studies.

Small angle scattering: the diffraction of defects (SAXS, SANS)

As the scattering angle in diffraction is reduced, so the size of the features seen is increased. When the defects are hundreds of atoms across or more, the scattering is concentrated into a narrow distribution close to the forward beam. This is the small angle scattering which, in principle, gives the size distribution and amount of any defects present in a solid. The experimental method was developed using laboratory x-ray sources by Guinier and Fournet (1955). Its great advantage is to provide a good statistical average over the defect population, rather than the myopic view given by the complementary electron microscope. This advantage is even more apparent using neutrons, whose penetration allows a true bulk average from an actual component to be obtained.

Small-angle neutron scattering (SANS) has become a very wide field, spanning technology, chemistry and biology. Further information on the 'contrast', or scattering power, of the defects is given by isotopic substitution (Kostortz 1982). A famous instrument is the D11 SANS spectrometer at the Institut Laue Langevin (ILL) in Grenoble. Its 128×128 element counter gives many spectra in a few minutes, allowing thousands of experiments per year over a very wide field. Even faster exposures are possible using small-angle x-ray scattering (SAXS) on a synchrotron source. Solid state reactions can be followed in real time.

The flowering of neutron scattering: excitations in solids

The energies of thermal neutrons are, by definition, just the energies of the phonons, the normal modes of vibration of a crystal. Magnons, their analogues in magnetic solids, cover a similar, if broader, range of energies. If both the incident and the scattered neutron direction and energy are measured during scattering, the momentum transfer, $\hbar Q$, and energy transfer, $\hbar\omega$, needed to excite a particular phonon or magnon in the specimen are defined. The triple axis spectrometer was invented by Brockhouse (1961) for this purpose. Its elegant use—you simply dial a Q value and energy scan—meant that the excitations in solids could be mapped out in the same routine way that crystallography could map their structure. The four-dimensional nature of the inelastic scattering cross section, $S(Q_x, Q_y, Q_z, \omega)$, made the technique very fruitful. By the 1960s neutron beams had become the driving force towards a new generation of high flux reactors. The ILL, opened in 1970 as a French–German–British partnership, was to give Europe a lead in this field.

Inelastic x-ray scattering using synchrotron radiation

The triple axis spectrometer, with its careful definition of incident and scattered energies, is being explored with x-rays. With incident energies 10 000 times higher than with neutron scattering, the resolution has to be better by the same factor. The vast intensities from synchrotron sources make this a possibility. X-rays have an important advantage over neutrons for mapping out excitations with energy transfers above about one electron-volt. The momentum transfer remains low, so that low scattering vectors are easy to achieve. This field shows great promise for the future.

Inelastic x-ray scattering: the Compton effect

The inelastic scattering of x-rays by the electrons in a solid give rise to the effect discovered by Compton (1923). The momentum distribution of the electrons can be analysed from the Doppler shifts revealed by the 'Compton profile' of scattered x-ray energies. The subject was transformed in the 1970s by the solid state detectors which could give the profile using gamma-ray sources. The higher energy given by a gamma-ray source allowed the impulse approximation, that the electron binding energy was negligible, to become valid. The momentum distribution is a vital clue to understanding the bonding in a solid.

The Mössbauer effect

The absence of any recoil effects by gamma-ray resonant absorption when the emitting and absorbing atoms are firmly held in a solid was discovered by Mössbauer (1958). The resulting absorption line is so narrow (10^{-5} eV) that hyperfine splitting is easily observed. The apparatus is very simple (given the gamma-ray source) and a Doppler shift, using loudspeaker coils, is all that is needed to give the energy scan. The effect has come to be dominated by the ^{57}Fe gamma-ray and iron-based samples. In particular, the chemical coordination of the atom and its neighbours can often be given.

Positron annihilation

When a positron enters a solid, it soon (within 10^{-12} s) loses its energy, and then for a further 10^{-10} s behaves as a sensitive probe within the solid before being annihilated with some nearby electron and producing the two opposed 0.51 MeV gamma-rays by which it is detected.

The positron was first used for solid state physics by de Benedetti *et al* (1949) to investigate the momentum distribution of electrons in a solid, including the anisotropy characteristic of its Fermi surface. The positrons are obtained from some artificial isotope, such as ^{22}Na. The measurement depends on the small angular or energy shift in the emitted gamma-rays. The positively charged positrons tend to seek out and be trapped at any centre deficient in positive ions. This causes a slight decrease in the width of the gamma-ray energy distribution. This ability to seek out defect centres has given the technique considerable importance in technology. Plastic deformation and fatigue in metals causes an increase in defects detectable, non-destructively, by this technique.

Muon spin resonance (μSR)

Muons introduced into a solid quickly come into thermal equilibrium within the solid, where their magnetic moment causes them to precess in any internal or external magnetic field. The method began only in the late 1950s with the production of the polarised muon beams from the new 'meson factories' with high current proton beams. Formed in a thin target placed in the proton beam, they can be separated from other particles and channelled into the sample as a short pulse. They decay, with a 2.2 μs decay time, into positrons, which are easily detected and separated from the background. The precession frequency appears as an oscillation in the decay rate as a function of time and gives the internal magnetic field. The precession damping rate indicates the diffusion of muons and gives extra information.

The electron microscope reveals all

Throughout its life, the electron microscope has continued to improve year by year in resolution, ease of use, and in the detail of the information given by its ancillary equipment. Solid state physics was to benefit greatly. In 1956, the dislocation was revealed exactly as had been predicted back in 1936. The traditional transmission electron microscope (TEM) imaged electrons transmitted by the sample, which had, therefore, to be of submicron thickness. A dramatic improvement in visual quality came with the development of the scanning electron microscope. A focused spot is scanned over the sample surface and the electron current emitted by the surface (SEM), or transmitted through the specimen (STEM), was collected and displayed as the brightness of the image on a TV raster following the electron beam. In 1964, Cambridge Scientific Instruments sold its first 'Stereoscan' model. Objects, even replicas, appeared as truly three-dimensional.

A feature of modern electron microscopes is the ability to change rapidly from imaging to diffraction mode. The electron diffraction pattern reveals the structure and orientation of the grains illuminated by the beam. A recent development has been the use of the convergent beam technique. A conventional diffraction pattern is composed of sharp spots. By increasing the angular divergence at the image plane, each spot turns into a zone axis pattern (ZAP), revealing both the point and the space group of the crystallite.

Energy dispersive x-ray analysis (EDX)

The scanning electron beam ejects electrons, leaving the sample atoms in excited states which decay with the emission of x-rays. The x-ray energy is atom-dependent—in fact just the characteristic radiation of the ejected electron's atomic shell. The effect was recognised long ago by O'Bryan and Skinner (1934) but lay dormant. The x-ray analysis is now very conveniently performed by energy-dispersive semiconductor detectors. With the electron beam focused on a particular part of the scanning image, it enables the atomic composition of any desired region to be measured quantitatively. By scanning the beam, the concentration of any chosen atomic species can be displayed to the full resolution of the instrument.

Field ion microscopy

Individual atoms on a surfaces can be seen in the field ion microscope developed by Muller in 1951. The specimen must be in the form of a pointed wire with a tip radius of curvature of only a few tens of nanometres. An electric field of tens of keV is applied between the tip and a fluorescent screen. Some 10^{-3} Torr of helium gas present is ionised wherever the electric field distribution is strongest. The ions are accelerated down the lines of force to give a modified image of the tip. It gives unique information of the location of individual atoms on a surface to a few angstroms resolution. However, the detailed structures are often difficult to interpret.

Low-energy electron diffraction (LEED): surface crystallography

Although dating from the 1920s, low-energy electron diffraction, Auger electron spectroscopy and other techniques sensitive to only the first few layers of surface atoms, were unusable for many years. They had to await the development, during the 1960s, of the ultra-high-vacuum techniques (less than 10^{-10} Torr) which enable the true surface structure and composition to be separated from the impurities which otherwise cover all surfaces. Low-energy electron diffraction (LEED) came into its own as the technique for surface crystallography. Even some pure element, single-crystal faces have 'reconstructed' surfaces, in which the first layer or two has a different crystal lattice from the bulk. Almost all metals have different interlayer spacings for the first few layers. Since the electron is a strongly interacting probe, quantitative analysis of LEED intensities is not easy and relies on extensive computation.

Auger electron spectroscopy (AES): atomic decay by electron emission

Atoms excited by an ion, electron or x-ray beam have quite a high probability of returning to the ground state by simultaneously dropping an electron from an outer to an inner shell and emitting an electron. The excess energy of the transition is carried by the Auger electron (Auger 1925). While LEED is the tool of the surface crystallographer, Auger electron spectroscopy (AES) is the tool of the surface chemist. The emitted electron energies are characteristic of the atomic species and the low electron range ensures that the technique is sensitive to the surface, rather than the underlying layers. Most Auger systems use beams which can be focused onto a spot on the sample with spatial resolutions of a micrometre or so.

Secondary ion mass spectrometry (SIMS): probing into the surface

A beam of low-mass ions, He^+ or Li^+ atoms, bores into a surface much like a sand blasting tool. The splattered ions can be collected and their masses analysed in a technique called secondary ion mass spectrometry (SIMS). The method is extraordinarily sensitive, factors of 10^{-8} in concentration can be detected. As the ion beam removes surface atoms, it drills a depth profile of the order of some micrometres depth. It is finding use in the semiconductor industry to probe the complicated surface depth profiles now possible with molecular beam epitaxy techniques.

Rutherford back-scattering: ion beam analysis

Several supersensitive techniques for surface analysis open up when use is made of megavolt ion accelerators. Ions, such as protons or alpha-particles, can be focused to give micrometre spatial resolution on the sample. The Rutherford method, as in his original experiments, relies on detecting those few ions which are scattered by large angles because they have undergone a near collision with a heavier ion. The scattered ion energy, measured with a surface barrier detector, is dependent on both the mass of the scattering ion and its depth in the sample. Thus, either the mass number, or its depth profile (but not both), can be found. The method is sensitive to 10^{-3} of a monolayer on the surface.

Rutherford back-scattering is not sensitive to light elements in a heavy element matrix. A complementary technique is nuclear reactions analysis (NRA), in which the ion beam excites particular nuclear reactions, especially those with a large energy output, giving easily detected products. For example, ^7Li in silicon is an important reaction, giving 8 MeV particles. Particle-induced x-ray emission (PIXE) is the ion analogue of the electron microscope compositional x-ray analysis. However, the high ion energy makes the technique some 100 times more sensitive at levels down to 0.1 PPM. All three ion beam techniques can be applied simultaneously.

12.4 FROM 1985: THE CHANGING FACE OF SOLID STATE INSTRUMENTATION

Returning to figure 12.1, we can see several lessons to be learnt from the history of solid state instrumentation. Many techniques took several decades to mature—or to await the right conditions for their exploitation. New developments outside solid state physics—particularly the reactor, the accelerator, the laser and ultra-high-vacuum technique—have all been responsible for opening up new solid state fields. Some old and simple techniques remain important—electrical resistance for electronic properties, susceptibility for magnetism. Others may change their role. Specific heat measurement was once important for its information on vibrations in solids. This role dropped after 1945, but low-temperature specific heat became important as a measure of the density of states at the Fermi surface. Today this role has dropped, but the specific heat remains a valuable source of information on phase changes. Other old techniques, apparently important and easy to measure effects like thermal expansion, turned out to depend on anharmonic effects not important to solid state fundamentals.

A general point, illustrated by figure 12.10, is that the most successful techniques, whether new or old, are those giving us the information as a function of the largest number of variables. Most of the nineteenth century techniques measured only a single function of the sample state—for example, specific heat, electrical resistance or susceptibility. The exceptions in those days were 'spectrometers', which measured as a function of optical wavelength. Now, there are all kinds of spectrometers measuring as a function of magnetic field, neutron wavelength, electron energy, atomic mass, scattering angle etc. The list is almost endless. The most productive techniques, like photoemission spectrometry and electron microscopy, measure as a function of several variables. Thus, neutron scattering measures a function equivalent to a generalised susceptibility over three dimensions of momentum transfer, as well as energy transfer.

A growing number of techniques seem to depend on large, expensive, central facilities. Once this was confined to reactors for neutrons. Now accelerators are needed for neutrons, for synchrotron radiation and for muon spectroscopy. Figure 12.11 shows the new Spallation Neutron Source (ISIS) at the Rutherford Appleton Laboratory near Oxford, which has just started experimental work. It is a pulsed neutron source using a high-current proton synchrotron system which uses the buildings and staff from an earlier high-energy physics proton accelerator. The pulsed neutron sources, with their broad spectrum of neutron energies, widen still further the spectrum of neutron beam experiments. Unlike reactors, accelerator-based sources are not yet close to their ultimate performance.

This trend to central facilities is a little worrying, both for conventional laboratory techniques and for the flowering of other, as yet unexploited,

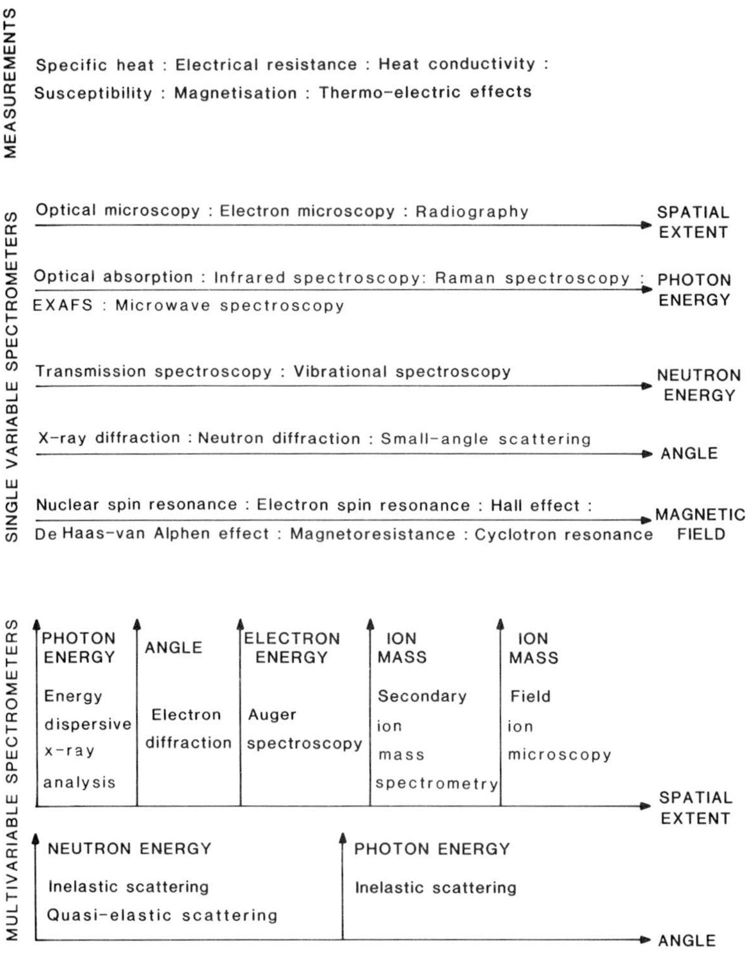

Figure 12.10 Spectrometers traditionally measured as a function of optical wavelength but modern instruments can measure as a function of one or more other variables. Here, the techniques are shown as a function of the variables being changed.

techniques which require appreciable funds to bring to fruition. The worry is that a decision made at one time is a major financial and personal commitment for perhaps several decades. The expenditure does not stop after the opening ceremony. Once built, the investment is wasted if it is not properly funded with instruments and scientists. Once funded, all is lost if the scientists are not given time to use the instruments. A 'silver lining' is the common occurrence in the later stages of the life of a facility—that its instruments

Figure 12.11 The experimental hall of the new Spallation Neutron Source, commissioned at the Rutherford Appleton Laboratory in 1985. Many beam lines radiate from the neutron generating target and moderator assembly at the centre of the circular shield.

become as useful for technology as for physics. Both neutron and synchrotron sources now earn appreciable sums from industrial research.

In fact, most solid state instruments remain very much part of 'small physics', even when they have become big business. The flowering of the microchip, of magnetic storage and optical communications have all taken place in conventional laboratories. It remains a characteristic of solid state physics that individual scientists can switch on their apparatus in a corner of a laboratory in the morning and make fundamental measurements before going home.

FURTHER READING

Arecchi F T and Salieri P 1982 Optical bistability *Phys. Bull.* **33** 20
Barfoot K M 1984 Ion beam analysis *Phys. Bull.* **35** 511

Berry F J 1983 Mössbauer spectroscopy *Phys. Bull.* **34** 517
Bleaney B 1984 Electron paramagnetic resonance: The early days *Phys. Bull.* **35** 466
Eaton G H 1979 The muon: A new tool *Phys. Bull.* **30** 152
Farge Y 1977 Synchrotron studies *Phys. Bull.* **28** 114
Glaser A M and Hart M 1980 A new era in X-ray crystallography *Phys. Bull.* **31** 278
Holt R S 1983 Whatever happened to the Compton effect? *Phys. Bull.* **34** 513
King D A, Richardson N V and Grimley T B 1981 Recent advances in surface science *Phys. Bull.* **32** 288
Mitchell I V 1979 Rutherford backscattering *Phys. Bull.* **30** 23
Mulvey T 1983 Fifty years of high resolution electron microscopy *Phys. Bull.* **34** 274
Standley K J 1983 James Griffiths and ferromagnetic resonance *Phys. Bull.* **34** 115
West R N 1977 Positron annihilation *Phys. Bull.* **28** 214
Willis R F and Feuerbacher B 1977 Probing solid surfaces *Phys. Bull.* **28** 117
Woolfson M M 1983 Structural crystallography in the 1980s *Phys. Bull.* **34** 330

REFERENCES

Auger 1925 *C. R. Acad. Sci., Paris* **180** 65
de Benedetti S, Cowen C E and Konneker W R 1949 *Phys. Rev.* **76** 440
Born M and von Karmann T 1912 *Phys. Z.* **13** 297
Brillouin L 1931 *Die Quanten Statistik* (Berlin)
Brockhouse B N 1961 *Inelastic Scattering of Neutrons* (Vienna: International Atomic Energy Agency) p113
Cleeton C E and Williams N H 1934 *Phys. Rev.* **45** 234
Compton A H 1923 *Phys. Rev.* **21** 207
Curie P 1895 *Ann. Chem. Phys.* **5** 289
Debye P 1912 *Ann. Phys., Lpz* **39** 789
Dresselhaus G, Kip A F and Kittel C 1953 *Phys. Rev.* **92** 827
Davison C and Germer L H 1927 *Phys. Rev.* **30** 705
Einstein A 1905 *Ann. Phys.* **17** 132
Friedrich, Knipping and Laue 1922 *Nature* **16** 365
Gibbs H M, McCall S L and Venkatesan T N C 1976 *Phys. Rev. Lett.* **36** 1135
Gordon J P, Zeiger H J and Townes C H 1954 *Phys. Rev.* **95** 282
Griffiths J H E 1946 *Nature* **158** 670
Guinier A and Fournet G 1955 *Small Angle Scattering of X-rays* (Chichester: Wiley)
Hall E J 1880 *Phil. Mag.* **12** 157
de Haas W J and van Alphen P M 1930 *Proc. R. Soc. K. Ned. Akad. Wet.* **33** 680
Joliot F 1933 *C. R. Acad. Sci., Paris* **197** 1622
Kostortz G 1982 *Treatise on Materials Science and Technology* vol. 15 *Neutron Scattering* (New York: Academic Press)
Lummer O and Pringsheim E 1897 Annalen der Physik **63** 395
Maiman T 1960 *Nature* **187** 493
Mitchell D P and Powers P N 1936 *Phys. Rev.* **50** 486
Mössbauer R 1958 *Z. Phys.* **151** 124
Mott N F and Jones H 1936 *The Theory of the Properties of Metals and Alloys* (Oxford: Oxford University Press)

Nernst W 1911 *Ann. Phys., Lpz* **36** 395
O'Bryan H M and Skinner H W B 1934 *Phys. Rev.* **45** 370
Owen J and Stevens K W H 1953 *Nature* **171** 836
Penrose R P 1951 *Nature* **163** 992
Penrose R P and Bleaney B 1946 *Nature* **157** 339
Peltier 1834 *Ann. Chem. Phys.* 2nd Series, **56**
Preston T 1894 *Theory of Heat* (London: Macmillan)
Purcell E M, Torrey H C and Pound R V 1946 *Phys. Rev.* **69** 37
Raman C V 1928 *Indian J. Phys.* **2** 387
Seebeck T J 1826 *Pogg. Ann.* **VI**
Sommerfeld A 1928 *Z. Phys.* **47** 1
Szoke A, Daneu V, Goldhar J and Kurnit N A 1969 *Appl. Phys. Lett.* **15** 376
Terman E 1943 *Radio Engineers' Handbook* (New York: McGraw-Hill)
Weiss P and Forrer R 1929 *Ann. Phys., Paris* **12** 297
Wiedemann G and Franz R 1853 *Pogg. Ann.* **89** 497
—— 1854 *Ann. Chim.* 3c **41** 107
Zavoisky E 1945 *J. Phys. USSR* **9** 211
Zeeman P 1897 *Phil. Mag.* **43** 226

Index